Communication Systems Engineering

Communication Systems Engineering

Prof. Sanjay Kumar Choudhary

ARCLER PRESS

www.arclerpress.com

Communication Systems Engineering

Prof. Sanjay Kumar Choudhary

Arcler Press
4164 Lakeshore Road
Burlington, ON L7L 1A4
Canada
www.arclerpress.com
Email: orders@arclereducation.com

ISBN: 978-1-77956-968-4 (Hardcover)

ABOUT THE AUTHOR

Prof. Sanjay Kumar Choudhary is a senior academician and experienced university administrator with a long and distinguished career in higher education. Born on 4 February 1964, he currently serves as the Vice Chancellor of Lalit Narayan Mithila University (LNMU), a position he has held since 24 January 2024. Over the years, he has played an important role in strengthening academic systems, governance, and research culture in Indian universities.

Prof. Choudhary has a strong academic background in Physics and Material Science. He holds an M.Sc. in Physics and Material Science, an M.Tech. in Surface Science and Engineering, and two Ph.D. degrees—one in Physics and Material Science and another in Electronics. He has also qualified the GATE examination in Material Science, reflecting his solid technical foundation. His main area of specialization is Material Science and Engineering.

Before joining LNMU, Prof. Choudhary served as Vice Chancellor at Munger University and T.M. Bhagalpur University, where he contributed significantly to academic planning and institutional development. He has also worked as Principal and Professor at institutions such as J.P. University, Sanjay Institute of Engineering and Management, and Bengal College of Engineering and Technology. Earlier in his career, he served as a Lecturer and Research Fellow at NIT Jamshedpur and B.I.T. Mesra, including work on an ISRO-sponsored project.

Prof. Choudhary has published 26 research papers, guided two Ph.D. scholars, and presented his work at international conferences, including events held in Germany. His academic contributions have been recognized with honors such as the Shiksha Ratna Puraskar (2010) and the Rastriya Gaurav Award (2011).

TABLE OF CONTENTS

LIST OF FIGURES

PREFACE

Communication has been central to human progress, governing the way in which societies communicate, exchange information, and conduct commerce. From the first forms of signaling to the global networks of today, with every technological innovation there is greater extent, pace, and effectiveness. Communications no longer comprise voice or text messages exclusively today but encompass multimedia services, data transmission, satellite systems, and wireless connections that enable the global economy and lifestyle. Their design requires a precise mixture of theory, mathematical modeling, and practical design to ensure reliability, efficiency, and responsiveness in diverse applications.

The communication systems engineering improves at an exponential pace, driven by the geometrical explosion in digital technologies, demand for increased data rates, secure transmissions, and energy efficiency. Analog methods began it, but digital techniques prevail today, enabling complex modulation, error control, and multi-user access in heterogenous networks. Information theory, coding methods, and advanced transmission media are reasons for the innovations of the day such as 5G, optical communications, and satellite constellations. As the world's network coverage continues to grow, engineers must not only understand the fundamentals of system design but also anticipate the challenges of noise, interference, and bandwidth limitations.

Communication Systems Engineering provides a comprehensive treatment of classic concepts and current trends in the field. It establishes the connection between elementary theory and contemporary practice, enabling readers to have a rational approach to valuing communication technologies. It covers analog and digital approaches, introduces statistical and probabilistic models to analyze noise and random processes, and identifies major technologies that characterize future systems. The book has been designed to serve as both an intellectual guide and professional reference, enabling students, scholars, and practitioners to build a solid foundation and also reach the frontiers of communication engineering.

Chapter 1 provides a background of communication systems and outlines the broad system structure—transmitter, channel, and receiver—and their mutual functions. It establishes the milestones in history from early telegraphy right up to digital and wireless technology and defines basic concepts such as bandwidth, modulation, and signal-to-noise ratio.

Chapter 2 is dedicated to analog communication and covers amplitude modulation, frequency modulation, and phase modulation. Methods of generation and detection, bandwidth efficiency, and practical uses such as radio broadcasting and analog TV are discussed. Limitations of an analog system under the influence of noise and interference are also brought out, setting the stage for the digital techniques to follow.

Chapter 3 covers random processes and noise, introducing the statistical means applied to model and analyze signals under uncertainty. It covers autocorrelation, power spectral density, Gaussian processes, and noise sources such as thermal noise and shot noise. The chapter describes how noise influences system performance and covers how to avoid its effects.

Chapter 4 addresses sampling and pulse modulation, closing the gap between digital and continuous-time signals. The chapter derives the sampling theorem, aliasing, pulse amplitude modulation, pulse code modulation, and delta modulation. Quantization error and signal reconstruction as practical issues are also introduced in the chapter, making it best suited for digital system design.

Chapter 5 provides the fundamentals of digital communication. The chapter talks about binary signaling, decision rules, error probability, and matched filter detection. Chapter 5 also addresses block diagrams of transmitters and receivers for digital signals, synchronization requirements, and the role of timing recovery in signal integrity preservation.

Chapter 6 expands on these fundamentals by focusing on digital modulation schemes. It has detailed exposition of amplitude shift keying (ASK), frequency shift keying (FSK), phase shift keying (PSK), quadrature amplitude modulation (QAM), and orthogonal frequency division multiplexing (OFDM). Performance under noisy environments and bandwidth efficiency are considered for each scheme with examples drawn from cellular, satellite, and Wi-Fi systems.

Chapter 7 covers information theory and coding, detailing Shannon's capacity theorem, entropy, and mutual information. It explains source coding techniques such as Huffman coding and linear predictive coding, and error control techniques such as block codes, convolutional codes, and turbo codes. The chapter centers on the trade-offs between redundancy, efficiency, and reliability in building robust communication systems.

Chapter 8 deals with transmission channels and media, wired and wireless like coaxial cable, optical fibers, radio waves, and satellite links. It explains channel impairments such as attenuation, multipath fading, dispersion, and interference. It also indicates how different media are chosen depending on distance, bandwidth needs, and reliability needs.

Chapter 9 describes spread spectrum and multiple access technologies. It introduces direct sequence and frequency hopping spread spectrum systems, code division multiple access (CDMA), and orthogonal frequency division multiple access (OFDMA). They are described based on cellular communications, military networks, and GPS as their examples of applications to show their usage to achieve secure and efficient transmission.

Chapter 10 addresses the future generation communication systems, including multiple-input multiple-output (MIMO) systems, adaptive modulation, cognitive radio, and cooperative communication. The chapter also touches upon future directions such as 6G, artificial intelligence-based integration in communication networks, and quantum communication systems, giving a futuristic view to students and professionals.

This book will act as an organized and systematically presented introduction to the application and theory of communication systems for graduate and undergraduate electrical and electronics engineering students. Researchers will enjoy the treatment of current developments and integration of mathematical rigor and engineering practice. Telecom experts can utilize it as a design and implementation manual, while professors can utilize it as a core-course teaching text. Policymakers and technologists who deal with digital infrastructure planning will also find the book useful in understanding the scientific background for existing communication systems. Presenting a blend of traditional principles and modern understanding, this book is formulated to equip readers with the competencies and information they need to approach one of the most dynamic and basic fields of engineering.

CHAPTER 1

INTRODUCTION TO COMMUNICATION SYSTEMS

Contents

INTRODUCTION

Communication systems, as the pillars of modern society, are defined as processes and structures through which information is transmitted from receiver to sender. Communication systems can be classified based on their function, medium, and design complexity. In general, communication systems encompass verbal, nonverbal, written, and technological modes. Each mode is designed to serve specific contexts depending on social norms and technological advancements.

Understanding the concept of communication systems enables academics and experts to analyze the effectiveness of various forms of communication, their impact on society, and the potential for innovation and optimization in these systems. On a technical level, communication systems can be described as intricate structures that facilitate the transmission of information across various media, including analog and digital signals. Such systems have a set of components: receivers, transmitters, channels, and message forms. The effectiveness of a communication system is sometimes gauged by its ability to provide clarity, accuracy, and speed of communication of information.

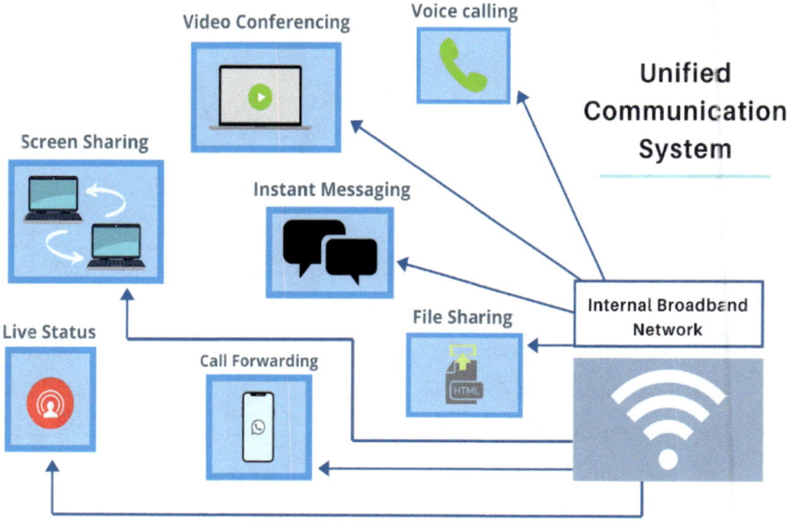

Figure 1.1. *Communication systems.*

Source: https://lh5.googleusercontent.com/uAoG gKNWd gRvWw2Jz-aU2_1N0e_044S_ WnWRNTAFQoo1 kYu4K 4X33OKre 836LEzQVMYL8xnzjxC-Qj9xp01iAoAgxqkmmw0iRq8Ke0TLBFx9QjzgvyJ-UlInYiPU heUPpDuwhucp4.

Furthermore, the development in the digital era has formed a rise in communication systems, generating compound phenomena such as computer-mediated communication and social networking. Thus, the interface of technology with human communication repeatedly redraws the boundaries and processes of communication systems, and highlights the need for ongoing study and adaptation. The value of clarifying communication systems extends beyond mere intellectual interest and into practical utilization, affecting businesses, medicine, and education. With the creation of accurate definitions and models, organizations can clarify their communication methods to enhance productivity and cooperation. For instance, effective organizational team communication mirrors elements of principles of small group communication theory, where cohesion of the group, roles, and norms play an important role in the accomplishment of shared goals. Having an understanding of these dynamics is useful to groups as they enable them to resolve conflicts and solve problems positively. Later, a well-defined concept of communication systems, in theoretical contexts as well as practical realities, becomes a central pillar stone for maximizing interactions in various contexts, testifying to the fluidity of communication in contemporary society (Srivastava et al., 2020; Thomas et al., 2018) (Figure 1.1).

Overview of Communication Systems

In the communication systems field, the emergence of new technologies has revolutionized how information is transmitted and received over long distances. The center of these systems is the range of ways in which data is transmitted, including wired and wireless channels. This binary is not just impacting the speed and reach of communication but also decides the security protocols needed by the environment on which they are being deployed. With the introduction of cutting-edge networks, the need for robust security measures has been higher, particularly for applications under the Internet of Things (IoT). The application of 5G technology is not an exception, with higher bandwidth and lower latency offering higher possibilities of real-time communications across various sectors, such as telemedicine and smart city infrastructure (Vishnu et al., 2024). As a result, the operation of such systems, and their potential vulnerabilities, is basic knowledge for both developers and end users. Further, structural design of communication systems, more specifically network topology, plays a crucial role in determining their effectiveness and efficiency. Networks may be designed in various configurations, with varying strength factors and weaknesses in data transmission and fault tolerance. Physical and logical topologies are critical to understanding how independent computers communicate through networks. The configuration of the systems affects message routing, dimensioning of the network, and even the potential for transmission of viruses, which proves that the engineering of a network sets the stage for its performance (Kaur, 2024). Thus, an accurate analysis of such topological structures allows not just optimal communication flows but also strategic decision-making in the founding or formation of existing networks. The socio-economic aspect of communications systems, together with technological implications, forms the context-dependent complexity of communications systems. When the new formations of communications systems emerge, they would have a tendency to trigger social practice and economic organizational changes. For instance, the shift to more deconcentrated systems, fueled by cheaper and more powerful computing technology, forces organizations to rethink their operating models. In addition, improved communication capability has opened up the possibility of international markets, enabling firms to operate in global markets more readily. Nevertheless, this rapid growth also entails gigantic issues for data security and privacy, especially with the growing dominance of digital communication. The current controversy among stakeholders regarding these matters underscores the importance of understanding not only the technological sides of communication systems but also their overall implications on society as a whole.

Importance in Modern Society

In the contemporary world, communication systems are an essential part in society development and in the formation of economic structures. As the pace of globalization increases day by day, there is a growing demand for efficient channels of communication that enable access to information globally. Efficient communication systems serve not only as channels of access to individual interaction but also as channels of business transactions and foreign trade. The application of newer technologies, such as Artificial Intelligence (AI), suggests the revolutionary nature of such systems. As explained in C. B., et al. (2025), AI technologies are currently being used in marketing campaigns, which reflects a paradigmatic shift in how companies are studying consumer behavior. It demonstrates the pivotal role played by communication systems in constructing interpersonal relationships as well as promoting economic growth in a world of globalization. On top of this, the development of digital technologies, particularly in communication systems, addresses the challenges that come with rising demands for real-time and effective provision of information. In light of the rapid advancement of digital communications, traditional methods are being increasingly complemented by emerging methods that enhance performance and improve user interaction. The role of Deep Learning (DL) technology cannot be overestimated in this regard. As recognized in Guo et al. (2024), DL models are central to the enhancement of the performance of communication systems through optimization of processes such as network security and signal processing. These enhancements facilitate more accurate resource allocation and estimation of the channel to allow systems to handle the enhanced quantity of data effectively. Thus, modern communication systems are fundamental in regulating the intricate web of information that characterizes the modern digital world. Lastly, the social effects of robust communication networks extend beyond service performance to security and resilience. In a world marked by heightened exposure to cyber-attacks and malicious threats, capacity for securing communication networks is at a premium. As described in Guo et al. (2024), DL technology has better capacities for the detection of network intrusions compared to traditional means in reliability and effectiveness. This capacity not only protects confidential information but also ensures the integrity of larger systems of communication upon which economic security relies. The continued growth in the development of communication systems signifies a broader societal willingness to create communities where information moves efficiently and securely. To this extent, their function in modern society far exceeds the encouragement of connectivity to the creation of trust and security in our increasingly digital worlds (Figure 1.2).

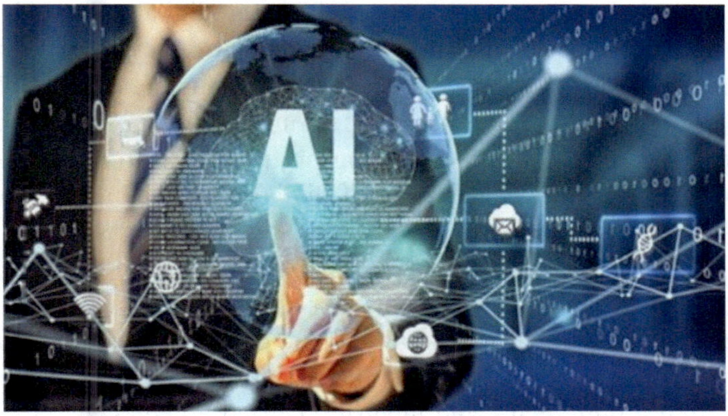

Figure 1.2. *Importance in modern society as artificial intelligence (AI).*

Source: https://external-content.duckduckgo.com/iu/?u=https%3A%2F%2Ftse3.mm.bing.net%2Fth%2Fid%2FOIP.9yEgWBAuhDuAKHy8I80qUQHaEK%3Fpid%3DApi&f=1&ipt=20d9f73680e0cc16efa1419293f2e393a2a32973d61aa68b8ae64009d3e13dca.

Types of Communication Systems

The communication system environment is diverse and multidimensional, such as those that support various channels of information exchange. Circuit-switched systems, for example, establish a dedicated communication channel between two ends during the call, a scheme that has been dominant in classical landline telephony. Such a system offers guaranteed bandwidth and low latency, which is well adapted to voice communication, wherein real-time interaction is highly important. However, development in data-intensive applications has promoted a shift from circuit-switched to packet-switched technology. Unlike circuit-switching mechanisms, packet-switching dissects messages into packets transmitted individually, which maximizes channel utilization. By the advent of packet-switched technology, the throughput of communication systems is greatly enhanced, as well as being able to transmit numerous data flows simultaneously, something that has become crucial with growing demands for digital services (Hasegawa et al., 1983). In recent years, digital communication systems have gained more significance due to their ability to meet the rising need for data transmission services. Aside from the problem of mere voice communication, these systems have services like video conferencing, file transfer, and streaming. Incorporating digital technology into communication infrastructure improves them by making them efficient and scalable. Especially, maritime communication has begun to become digital and address these developments, addressing current analog systems' deficiencies that lack the capacity to cater to today's user demands. The transition to digital systems allows for the use of the frequency spectrum in an efficient manner with more users catered to without sacrificing on quality. With the growing need for data services, strategic integration of digital communication technologies is responsible for guaranteeing that there is quality communication between different industries, for instance, maritime applications (Val Sčić et al., 2012). Hybrid communication systems also enter the scene as a way of bridging the gap between traditional analog systems and modern digital networks. By allowing both modalities to coexist, these systems facilitate heterogeneous communication requirements and integrity. During emergencies, for instance, analog systems allow for reliability and toughness. Under regular operation modes, digital systems, on the other hand, can facilitate improved spectral efficiency and data processing. The convergence of all these strategies not only improves system resilience but also provides a clear trajectory for phasing into fully digitized communications. Ultimately, understanding the various types of communication systems—circuit-switched, packet-switched, digital, and hybrid—enables stakeholders to make informed decisions on embracing and deploying technology that best responds to the users' requirements (Val Sčić et al., 2012).

1.1. COMPONENTS OF COMMUNICATION SYSTEMS

To learn the components of communication systems is to learn both the organizational and technological aspects that make up effective information transfer. In essence, a communication system consists of transmitters, receivers, and the media through which signals are transmitted by, including electromagnetic waves in wireless networks or fibers in wired networks. Each of these is critical in the process of transmission. For instance, transmitters convert information into transmission format and convert signals back to usable information. Furthermore, efficiency and quality of such components determine overall system performance. Such entangled interaction among components further emphasizes the need for a robust infrastructure to ensure free flow of communication. In modern telecommunications, most of the issues arise due to competition intensification and integrated services requirements. With changing globalization and technological evolution, the sector keeps on being transformed. Communication networks are forced to adapt in order to meet these altering demands. New business models have to deal with the complexity resulting from converging services and technologies (Al-Debei et al., 2009). Convergence of various telecommunication service offerings presents challenges that have to be effectively managed.

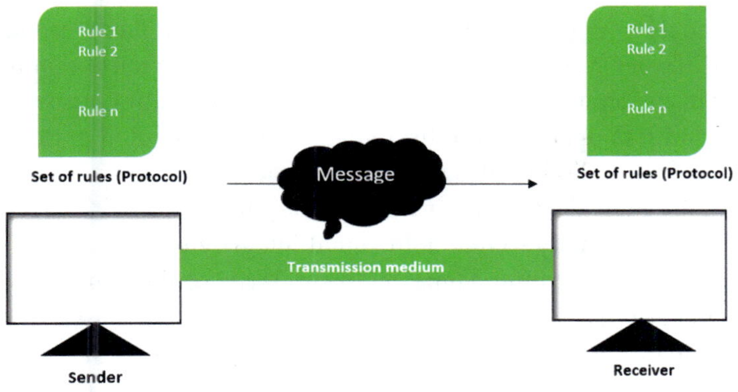

Figure 1.3. *Components of communication systems.*

Source: https://media.geeksforgeeks.org/wp-content/cdn-uploads/20200625134316/1118.png.

As pointed out, telecommunications operators need business models that ensure flexibility in a constantly changing environment. Flexibility is particularly crucial as organizations struggle to stay competitive and profitable amidst deregulation and liberalization trends in the market. The evolution of communications systems also demands close examination of integration issues confronting telecommunications developers. The likelihood of creating and managing integrated services expands with the objective of offering streamlined and efficient communication solutions. Seamless integration among various organizational domains is imperative in an attempt to offer an integrated user experience. The ACTS project Prospect illustrates how collaboration can enable the development of integrated telecommunication services tailored to the needs of different users (Bracht et al., 1997). Through the attack on the application of technological standards and architectural approaches, business operators enhance the interoperability and efficiency of their communication systems. Such initiatives are critical to fostering a robust environment for the delivery of innovative services in an increasingly competitive market (Figure 1.3).

1.1.1. Transmitter Functions and Types

In the area of communications systems, transmitters are essential as they convert and send information via various channels. Transmitters' fundamental function is to receive input signals in the forms of audio, video, or data and modulate them into a suitable format that can be communicated effectively via a medium. Different transmitters are classified based on their function and application, including analog, digital, and molecular transmitters. Each group operates based on different principles that are best suited for specific communication needs. For instance, analog transmitters operate with continuous signals, while digital transmitters operate with discrete data, thus providing extensive sets of applications from radio broadcasting to computer networking. Their efficiency is regulated by their design and the technology involved, with the technology controlling their capacity to maintain signal integrity over distance. Molecular communication is an intriguing type of transmitter, especially in nanotechnology and bioengineering. Signaling molecules serve as the media used in conveying information in this novel system, which allows the possibility of communication on microscopic levels. The most important job of molecular transmitters is to disseminate these signal molecules, which attach to receivers that are sensitive to the presence of the molecules. This attachment is usually a sophisticated series of chemical reactions that generates an output signal proportional to the timing and concentration of the signal molecules. The random character of chemical reactions and diffusion introduces noise into the output signal and thus requires sophisticated analysis to be able to read transmission effectiveness and reliability. Understandings of molecular transmitters' characteristics including response dynamics, provide gateway to use in a wide range of products from medicine to environmental monitoring (Chou et al., 2013). The operation and structure of transmitters can also influence the use in applications within communication studies. A case in point is in vehicular technology where electronic fuel injectors are a form of specialized transmitter. These injectors work by modulating the amount of fuel to be injected into an engine depending on actual conditions, thus optimizing performance parameters such as power and fuel economy. The air-fuel interaction enables a complex response mechanism like traditional communication systems where data is modulated in order to transmit. This new application reflects the extent with which transmitter functions can be maximally stretched from traditional boundaries to optimize system performance. In-depth studies of variables affecting such injectors, such as plenum length, yield insight that can be converted into improved designs and efficiencies in various applications. Hence, the understanding of transmitter operations covers a broad spectrum of technologies and uses, further justifying the necessity of innovative communication systems (Bakar et al., 2020).

1.1.2. Channel Characteristics and Types

Understanding the various attributes and types of communication channels is required for reconciling the complexities of modern communication systems. Channels are largely differentiated based on their physical attributes, i.e., if they are wired or wireless. Wired channels, which consist of coaxial cables and fiber optics, offer larger bandwidth and lower latency and thus are best suited for high-speed data transfer. Contrary to this, wireless channels like radio waves and microwaves are characterized by their ability to travel through the air but for their vulnerability to interference and noise. The latter calls for a deep understanding of the strengths and limitations of each channel, particularly as communication demands grow in scope and complexity. A thorough analysis of these shapes and characteristics forms the basis of exploring their usability in application within communication systems, leading to greater theoretical analysis and technological development (Tang, 2023). Progression through the characteristics of channels introduces consideration of noise action, bandwidth, and signal transmission. Noise, ever-present enemy of communication systems, has the ability to disrupt signals, leading to increased rates of error. Channels can vary in their immunity to noise, which is crucial in maintaining signal integrity during transmission. Channel characterization of its capability to transmit information error-free is encapsulated in parameters such as signal-to-noise ratio (SNR) and bandwidth efficiency. For example, a larger bandwidth channel will carry more information simultaneously. Moreover, channels may possess characteristics such as fading and shadowing, particularly in wireless transmissions, as environmental issues are likely to impact signal strength. Taking this into consideration becomes very essential while planning systems that can transmit data in a manner that prevents losses due to noise and other types of interference (Pavlov, 2022). Channel modeling is becoming more essential to support the optimal transmission and operation of the system. Paths such as Additive White Gaussian Noise (AWGN) paths and fading paths are crucial models to achieve the effects that various perturbations impose on signals. Models creation is a ongoing process from empiricism to sophisticated statistical, physical, and evolutionary descriptions that can properly model channel performance under different conditions. Each modeling method has unique insights and aids in the formulation of communication systems with the capability to cope with the inherent uncertainties of real-world environments The development in channel modeling emphasizes continuous R&D work, thereby equipping future communication systems with the needs of today as well as with the potential to cope with emerging challenges in technology and application environments (Tang, 2023).

1.1.3. Receiver Roles and Technologies

In communication systems, receivers take a pivotal role in ensuring signal detection and effectiveness of processing. Receivers sense transmitted signals, which usually travel through various media and are susceptible to noise and interference. A good receiver enhances the fidelity and intelligibility of the signals, which is critical in ensuring effective communication. Importantly, receiver devices like sub-harmonic mixers have been of great promise in improving the performance of receivers. They are increasingly used because they can handle the sub-multiple frequencies in relation to the local oscillator, thus giving improved performance with lower power consumption. These technologies address pressing requirements for low-power and high-bandwidth technologies for modern communication systems, hence proving the widespread relationship between receiver technology and overall system performance (Ali et al., 2025). With every advancement in communication technologies, the integration of dynamic wireless charging (DWC) into electric vehicles (EVs) is a good example of receiver function within a greater system. Here, the receiver is not a typical radio signal receiver but an inbuilt device within EVs designed to talk to magnetic transmitters that provide power dynamically as the vehicle moves. This is an innovative approach to surmount the issue of EV range anxiety and render it more convenient and viable. The technology is based on advanced communication protocols in order to make the transaction between the vehicle and charging infrastructure smooth. Moreover, safety and efficacy take center stage, hence the thought about newer technologies such as blockchain to authenticate transactions, signifying the different roles played by receivers in the new communication environment (Rajamanickam et al., 2024). Despite the colossal increase in receiver technologies, considerable challenges remain that still need research and development. Issues such as guaranteeing linearity, optimizing noise figures, and minimum power consumption are critical considerations in designing efficient receivers. Future studies in these subject areas are inevitable as the industry strives to provide receivers that will keep up with the increasing performance requirements brought about by advances in wireless communication technologies. The consistent evolution of receivers will, in time, make the communication systems more stable and robust, allowing for the deployment of high-performance applications in various fields, including IoT and satellite communications. As there is growing demand for glitch-free connectivity, so also will there be the requirement to evolve receiver technologies toward enabling the next generation of communication systems (Ali et al., 2025) (Figure 1.4).

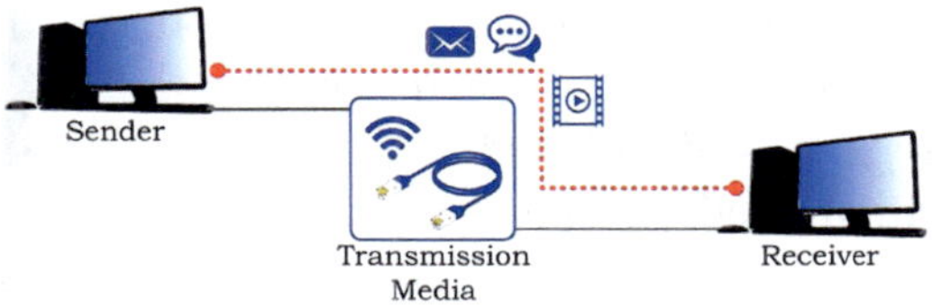

Figure 1.4. *Receiver roles and technologies.*

Source: https://cuet.iitk.ac.in/sathee-cuet/student-corner/ncert-books/class-12/computer-science/chapter-11-data-communication/.

1.2. TYPES OF COMMUNICATION

Effective communication is at the heart of the context of diverse communication systems, with many types and modalities of communications such as written, verbal, non-verbal, and visual communication. Each serves distinct purposes and functions in various frameworks, making it possible for individuals to communicate more effectively. Verbal communication, for instance, involves spoken interactions that are capable of providing immediate feedback, fostering lively and dynamic interactions. In contrast, nonverbal communication, as body language and facial expressions, conveys emotions and attitudes that are difficult to put into words. Thus, awareness of the modalities enables effective participation in any communications system as well as the development of enhanced interpersonal relationships. Written communication is vital in providing a permanent record of transactions, particularly in business and academic settings. Compared to verbal communication, written forms can be explicitly developed and refined to guarantee clarity and accuracy in the conveyance of complex concepts. Written communication is in the form of emails, reports, and research papers that call for a formalistic stance to maintain the message conveyed. Written communication as a vehicle in the field of communication systems allows the information to be carried beyond time and space, such that people can record and disseminate valuable information. The combination of various modes of communication improves discourse and invites substantial understanding among the communicators, enhancing communication effectiveness overall. Besides this, visual communication has emerged as an integral component of modern systems of communication, employing images and graphics to convey information concisely and efficiently. This type encompasses anything from infographics to presentations based on the design elements employed to improve understanding and memory of the information. In sophisticated communication environments such as software-defined networks, visualizations can shed light on esoteric concepts and facilitate the description of complex systems, as proposed in Kong et al. (2017). Additionally, heterogeneous distributed sensor networks (HDSNs) demonstrate concurrent verification needs within visualizing environments, as they offer dynamic interaction between various sensors within a defined area, offering group functionality and security, as presented in Heo et al. (2007). Together, these forms of communication interact to form an integrated understanding necessary for efficient communication systems (Figure 1.5).

Figure 1.5. *Types of communication.*

Source: https://getuplearn.com/wp-content/uploads/2021/10/Types-of-Communication-scaled.jpg.

1.2.1. Analog vs. Digital Communication

The distinction between analog and digital communication systems forms the foundation of understanding the broader selection of communication technologies. Analog communication employs continuous signals that measure physical quantities, such as sound waves in a voice. The process allows direct representation of the amplitude and frequency of a wave, retaining nuances of the original signal. Traditional analog signal examples are gramophone records and radio transmission, where an audio wave of the original form as much as possible is played back. Analog signals are noise-prone in their nature, which over long-distance transmission or transmission through a medium of transmission corrupts the signal ultimately rendering the received data in poor quality. This vulnerability suggests an inherent lack in analog systems that calls for a shift towards more robust digital methods that will offer higher fidelity in signal transmission (Reimeir et al., 2024). Digital communication, however, translates information into binary form—sequences of 0s and 1s—facilitating improved processing and transmission over networks. Digital signals can be easily compressed, encrypted, and manipulated, enabling a vast range of applications from data security transmission to streaming over the internet. The primary benefits of digital communication are greater noise resistance, which greatly enhances clarity and reliability throughout transmission. Additionally, digital systems enable advanced signal processing algorithms, such as error-correcting algorithms, which contribute to the integrity of the process of communication. Such advancements signify an evolutionary leap in communication technology, characterized by the ability to transmit enormous amounts of information quickly and securely, a feature typically out of the question with analog technology (Anders, 2024). In the midst of the growing hegemony of digital communication, analog systems are still relevant in some situations. For instance, in certain medical and audio applications, analog gear has the ability to record sound in a way that is rich and warm and that sometimes digital gear can't. Also, the straightforwardness of analog signals can be an advantage in certain real-time applications where latency isn't just an important consideration, but the most important. But as telecommunication continues to evolve, the integration of both analog and digital technologies presents interesting opportunities and challenges. It is essential for future innovation to realize how these two paradigms of communication subtly interact with each other because new technologies are constantly looking to bridge gaps and enhance the efficiency of communication systems (Reimeir et al., 2024).

1.2.2. Unidirectional vs. Bidirectional Communication

In exploring the communication dynamics of systems, a crucial distinction is made between unidirectional and bidirectional communication. Unidirectional communication demonstrates a one-to-one direction transmission of information where data is sent from a sender to a receiver in the expectation of neither receiving nor sending feedback. It is normally used in broadcasting, such as radio and television, that allows the dissemination of information to large masses of people at low costs. Although unidirectional communication is effective in passing messages to many people, it lacks interaction that could lead to misunderstanding or alienation. Subsequent use of this model then limits the full exchange that often occurs in more interactive settings, and thus contracts the richness of the communication process. Bidirectional communication, instead, promotes a two-way exchange of information with space for feedback and interaction between the participants. This model is more common in situations where dialogue is indispensable, including interpersonal communication, classroom settings, and collaborative workplaces. The interaction promoted by two-way communication improves clarity and comprehension, as parties are able to clarify misunderstandings and modify messages according to real-time feedback. More importantly, this interactivity promotes bond strengthening and facilitates trust development among the parties, leading to a more integral experience of communication. The interdependency present in two-way communication successfully enhances the flow of information, showing how it can contribute to increased participation and a solution-focused conversation. Moreover, the implications of these communicative modalities reach beyond interhuman encounters into other areas of technology and learning systems. For instance, one-way communication is prevalent in novel technologies, e.g., those implementing

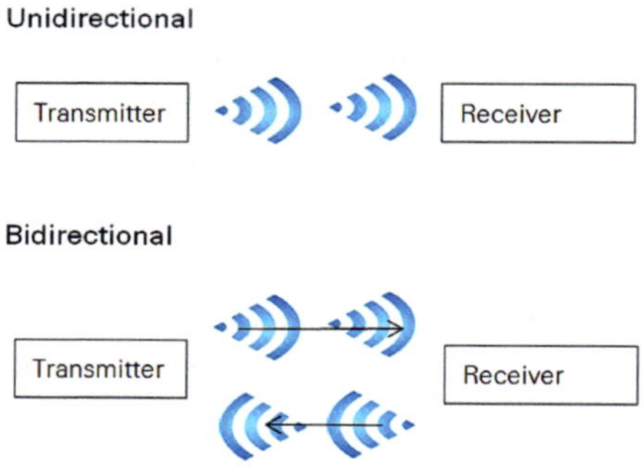

Figure 1.6. *Unidirectional vs. bidirectional communication.*

Source: https://res.cloudinary.com/dcvgho2zc/image/upload/v1640569831 Tech%20Blog/unidirectional-and-bidirectional-communication.png.

Federated Learning, where the data remains locally resident without explicit sharing between entities (Kairouz et al., 2021). Feedback mechanisms in such system types are restricted, which can impact collaborative problem-solving. Conversely, bidirectional approaches within learning environments, particularly in research on neural circuits in fruit flies, emphasize the significance of feedback when unraveling complex systems (Louis et al., 2020). This emphasizes the way in which bidirectional communication not just has an indispensable role to play in human interactions but is also the foundation of innovation in machine learning and neurobiological studies, highlighting its widespread applicability within contemporary communication systems (Figure 1.6).

1.2.3. Wired vs. Wireless Communication

Development of communication technology has generated two dominant paradigms: wired and wireless communication systems. Wired communication is the transfer of data through physical mediums such as copper wires, fiber optics, or coaxial cables. Wired communication is a form of communication that typically offers high bandwidth, low delay, and more security and is hence appropriate for applications requiring trusted links, such as enterprise networks and data centers. However, wired networks are prone to cost a lot to implement in terms of investment and physical realignment of existing spaces. The implementation of wired solutions would therefore not always be cost-effective, especially in geographically dispersed locations where the physical cabling would be difficult to install. Wireless communication, on the other hand, has become extremely popular due to its convenience and flexibility. It eliminates the need for extensive wiring by transmitting data through radio waves, infrared signals, or satellite connections.

This adaptability makes wireless communication systems particularly attractive for mobile devices and locations where establishing a wired network is impractical. As wireless technologies evolve, potential applications in sectors like healthcare and automotive have expanded significantly. Nonetheless, wireless communication comes with a range of its own vulnerabilities in terms of susceptibility to interference, security compromise, and reduced bandwidth by nature compared to hardwired systems. Furthermore, the resultant demand for more multimedia services exhibits an overall trend requiring the utilization of both wireline and wireless technology to meet existing traffic needs. This merging into wire and wireless technology is a step in the direction of hybrid communication systems that leverage the advantages of both modalities. For example, wire infrastructures can be merged with wireless technology to make connectivity and performance straightforward, particularly in high-complexity applications like the internet of things (IoT). This merging can facilitate first-class data transmission, removing bottlenecks and accommodating a growing variety of devices and services. Through guidelines on optical and wireless communications research, the combination of the two technologies can adequately address future demands for interactive video, voice, and data service demand, leading to more efficient communications solutions (Abdollahi et al., 2010). Hybrid systems also have the advantage of enhancing network scalability, especially in multicasting and broadcasting applications, as observed in wireless network-on-chip architecture comparisons (Cavall et al., 2016). By embracing such integration, communications systems can evolve to address the increasing needs of an increasingly globalized world.

1.3. COMMUNICATION PROTOCOLS

Protocols in communication systems are the rules of operation behind data transmission between devices. Protocols specify the structure, timing, ordering and error-checking functions required for successful communication so that systems of different types may communicate smoothly with each other. Protocols differ based on the requirements of an application, with some prioritizing speed, others prioritizing accuracy, or security. For instance, Transmission Control Protocol (TCP) values reliable data delivery, while User Datagram Protocol (UDP) values speed at the expense of reliability. This choice of protocol has far-reaching implications on system performance and user satisfaction from the perspective of system design, hence the need for careful consideration during system design. Understanding the nuances of these communication protocols is critical to constructing good systems that can address the more complex requirements of modern-day applications, including telematics and other emerging technologies. The emergence of two dominant paradigms in telematics systems—protocol-based and object-based—demonstrates the dynamicity of communication protocols. The paradigm of protocols is concerned with the standardization of interdevice communication in order to facilitate interoperability in a variety of different systems, a vital consideration in the continuously increasing realm of networked devices. By contrast, the object-based approach is concerned with the data itself, allowing for managed structuring and access to information and still maintaining the principles of communication protocols. The coexistence of these paradigms not only adds richness to the potential of telematics system design but also brings some issues, primarily associated with the proper integration of the two paradigms. As identified in recent literature, this synergy raises crucial system architecture concerns that require continued research to realize the utmost potential of communication technologies in various applications (Pires et al., 1997). In meeting data handling and accessibility needs of communication systems, the development of applications data services (ADS) is a significant development. The ADS addresses data integration and access issues, providing an easy way for users to be able to conveniently locate, gain access to, and utilize different datasets by numerous various applications. Through the provision of a common service layer, the ADS enhances availability and visibility of data, promoting improved coordination among users with diverse needs. This approach not only streamlines the integration of data but also allows for faster delivery of usable information, a necessity in timely settings. With more advanced communication systems, services such as these will become ever more central to enhancing the effectiveness and efficiency of data exchange, validating the pertinence of their role against the backdrop of communication protocols (Kuch et al., 2025) (Figure 1.7).

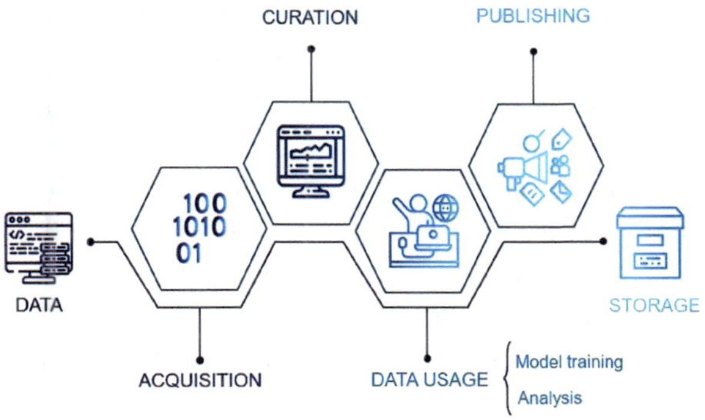

Figure 1.7. *Communication protocols.*

Source: https://pub.mdpi-res.com/sensors/sensors-23-05694/article_deploy/html/images/sensors-23-05694-g001.png?1687143196.

1.3.1. Definition and Purpose of Protocols

In the context of communication systems, protocols are essential frameworks that outline the conventions and rules for data exchange. As a structured set of regulations, protocols enable heterogenous systems to communicate in a timely and effective way, ensuring that the sent information is accurate and reliable. The significance is increased by the requirement for compatibility between various hardware and software configurations. Without protocols, different systems would not know how to interpret each other's data being communicated, which may lead to possible errors and miscommunications. Designing communication protocols is not just about syntactic rules, but also semantics—clarifying communicative intent. Such a complexity determines the role of protocols in facilitating seamless communication among various networks and devices towards a shared environment of computer communication within the overall framework of open distributed systems and their technology infrastructure (Pires et al., 1994). Protocols do not perform just the direct transfer of information; they determine the process by which inter-operability among various systems is achieved under the scenario of a communications network. Protocols allow for many functions like the detection and correction of errors, data compression, and flow control that ultimately reinforce communication systems. By standardizing how data is packaged, sent, and received, protocols are at the forefront of ensuring the integrity and efficiency of communications. This systematic approach is particularly critical in environments where timely data exchange is paramount, such as telecommunications and internet services, where delayed loss of time or data can have significant effects. Implementation of robust communication protocols has a critical influence on the success of technology deployment because they enable various constituents to complement one another without a hitch while adopting a common standard (Desvaux et al., 2005). In addition, the constant change in the protocols of communication reflects the dynamic nature of technology advancements and the continuously increasing level of sophistication in communications systems. With emerging technologies—such as the Internet of Things (IoT) and advanced wireless networking—protocols tend to be updated to address emerging problems and enhance performance. This sort of development necessitates a constant dialogue between stakeholders like operators, designers, and users to create protocols that are responsive enough to adjust to changing needs and innovations. Setting up robust communication protocols is thus not an irreversible process but an adaptive process to enhance the usability and security of networks. In general, the use and meaning of protocols are building blocks in being aware of and improving communication systems, with their pivotal role in digital interaction across platforms and devices highlighted (Pires et al., 1994).

1.3.2. Common Communication Protocols (E.g., TCP/IP, HTTP)

The building blocks of new communication systems are common communication protocols that allow data exchange on different networks. Of these, the Transmission Control Protocol/Internet Protocol (TCP/IP) suite is of greatest significance. TCP/IP is a suite of protocols employed to achieve reliable internet communication. The exchange between TCP and IP allows connections to be established and the efficient routing of packets of data. Being at the very center of the internet, TCP/IP facilitates numerous programs and is therefore versatile enough to function in various network topologies. Also, its versatility both with regard to IPv4 as well as the new IPv6 protocols speaks volumes about its value as well as applicability in addressing future networking challenges further establishing its significance in ongoing advancements in communication systems (Ferretti et al., 2015). In addition to TCP/IP, the Hypertext Transfer Protocol (HTTP) is another important protocol of data communication within web environments. As a hypertext request transfer protocol and internet information, HTTP provides the means to transfer information between web browsers and web servers without interruption. The fact that HTTP is a stateless protocol renders it simple to interact, with an option of obtaining data swiftly and displaying it without the need to hold open connections, thereby improving user experience immensely. Furthermore, HTTP enhancements, such as HTTP/2, target the increase of speed and efficiency in an endeavor to satisfy the growing demands of web applications. Given that communication systems advance to accommodate complex data interactions, the task of HTTP is further enhanced, indicating its significance in developing the web architecture robust and user-friendly (Eggert et al., 2008). The efficiency of the above protocols is radically tied to ongoing innovation in mobility architectures that facilitate user experiences in different environments. For instance, mobility management protocols address the issue of handovers to ensure effortless service during user switching between networks. This capability is not only vital for quality of service (QoS) support but also for legacy system support in current internet environments. Studies have indicated that there are new approaches unique to IPv4 and IPv6 networks that enhance the handover process, such as solutions to issues such as network address translation (NAT) and firewall compatibility. These advancements provide precious guidelines towards the development of future communication systems and make them able to incorporate changing user requirements and the emerging digital world (Ferretti et al., 2015) (Figure 1.8).

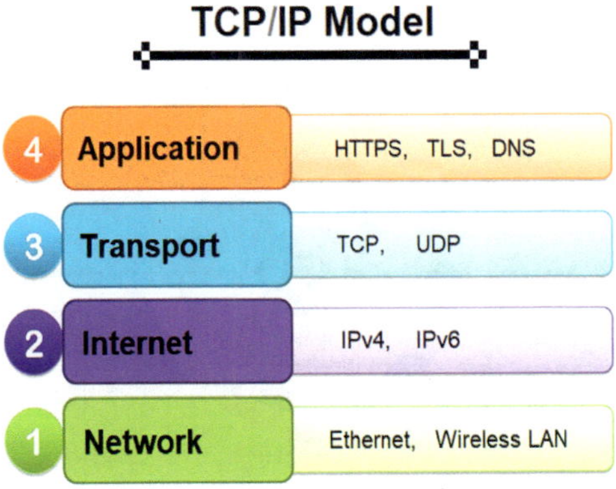

Figure 1.8. *Common communication protocols (e.g., TCP/IP, HTTP).*

Source: https://tecadmin.net/wp-content/uploads/2023/06/tcp-ip-model.png.

1.3.3. Role of Protocols in Data Transmission

Protocols are the underlying rules or agreements that allow devices to exchange information with one another when data is being sent. They define the format, timing, sequence, and error-checking needs that are necessary for successful data exchange. Without protocols, communications systems would experience enormous issues, including data misinterpretation and loss of data integrity. For example, Transmission Control Protocol (TCP) employs a connection-oriented approach to ensure guaranteed data transfer through employing flow control and error recovery. In essence, protocols manage how data is framed, sent, and received, providing a setting that enhances communication systems' performance and efficiency. This orchestration is important in facilitating various applications and devices to communicate without hitches, laying the groundwork for information exchange across networks seamlessly. The necessity of protocols becomes even more apparent with consideration of specific categories of protocols, such as routing protocols and application-layer protocols. Routing protocols, for instance, are important in directing data packets through complex networks so that they reach their destinations. Technological advancements in routing protocols like the proposed Dovetail system are best examples of how data anonymity and privacy in transit may be enhanced without compromising peerless operational performance. As identified in recent research, well-intentioned routing will be capable of repelling unauthorized access by concealing users' identities without compromising data integrity (Acquisti et al., 2014).

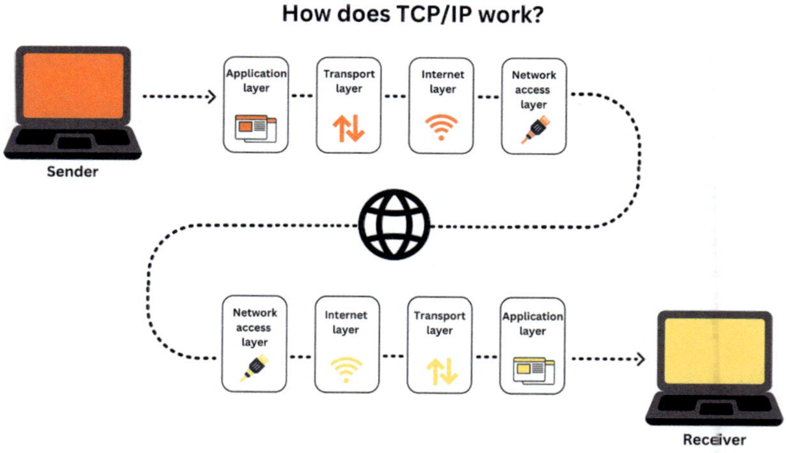

Figure 1.9. *Role of protocols in data transmission.*

Source: https://www.cloudns.net/blog/wp-content/uploads/2022/12/ClouDNS-TCPIP-5-7.png.

Application-layer protocols, on the other hand, focus on end-user experience and provide application-specific rules for protocols such as email or web browsing. Together, these varied protocols not only ensure secure and correct transmission of information but also that information meets the functional specifications of users in a broad spectrum of applications. In particular, protocols managing wireless sensor networks demonstrate the adaptive role of protocols in evolving communications environments. The use of more advanced routing strategies, such as the modified LEACH (MODLEACH), shows how protocols can be tailored for better performance in niche networks. MODLEACH introduces new cluster structure and double power levels and has significantly improved throughput with increased network lifetime (Javaid et al., 2013). This adaptability is crucial in use cases like environmental monitoring or smart cities, where sensor networks must be backed by robust and flexible communication protocols that can handle large amounts of data efficiently. These are symptomatic of a larger movement in the design of protocols that prioritize not only efficiency and dependability but also scalability and resilience in heterogeneous uses. By and large, the design and deployment of various data transmission protocols are the backbone of modern communications systems (Figure 1.9).

1.4. MODULATION TECHNIQUES

The field of communication systems extensively uses modulation methods for successful transfer of information through various channels. Modulation, the process whereby the carrier signal is altered in accordance with the message input signal, is vital for successful data transmission. Various techniques such as Amplitude Modulation (AM), Frequency Modulation (FM), and Phase Shift Keying (PSK) each possess different strengths and applications. For instance, FM is used because it is immune to noise and thus ideal for transmitting high-fidelity sound. In digital communication, on the other hand, PSK is preferred due to its capacity to utilize bandwidth efficiently. It is thus important to understand these basic modulation techniques so as to come up with stable communication systems capable of transferring information without much loss and with maximum fidelity. The recent advancements in modulation techniques have been strongly influenced by the convergence of emerging technologies. One new breakthrough is the Fluid Antenna Index Modulation (FAIM), which integrates fluid antenna systems and index modulation techniques. This technique facilitates the swift adaptation of channel conditions through the utilization of software-programmable fluidic elements, thereby enhancing the general performance of wireless communication systems. The success of FAIM lies in its ability to exploit channel-independent variations to convey additional information without involving other data transmission resources. As illustrated by research, with this approach, it is possible to have enhanced classification of index patterns using wavelet scattering neural networks (WSNN), which is a feasible solution in the optimization of communication networks in noisy channels (Chen et al., 2024). Thus, FAIM signifies the ongoing evolution of modulation techniques that meet the increased expectations of high-capacity and reliable communications systems. Further, the finding of nonlinear analytic techniques has emerged as a revolutionary factor in modulation techniques. By going beyond traditional linear forms, the nonlinear techniques expose new possibilities of system optimization and robustness in communication networks. They enable the creation of robust signal processing algorithms and adaptive modulation techniques that are capable of coping with the dynamicity of communication links. Recent studies point out that incorporation of nonlinear optimization mechanisms leads to enhanced error correction schemes, which result in more robust communications under interference and signal distortion (Et al. Upadhye, 2023). Although questions regarding real-time implementation and computational complexity remain open, the integration of nonlinear approaches is on the cusp of propelling the next generation of adaptive networks. In effect, these advances are a confirmation of ongoing efforts toward maximizing modulation techniques in order to keep up with ever-changing communication technologies (Figure 1.10).

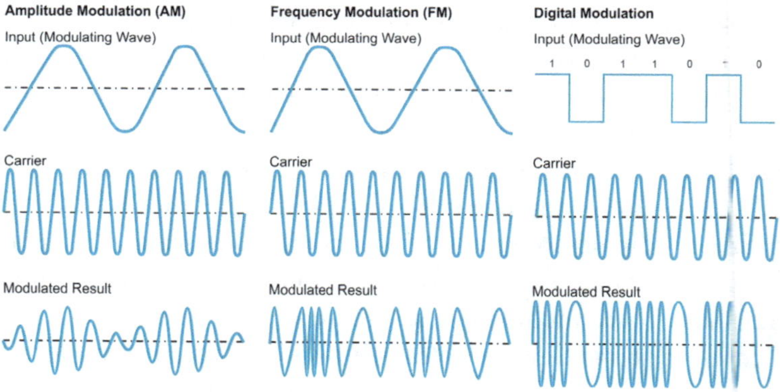

Figure 1.10. *Modulation techniques.*

Source: https://www.taitradioacademy.com/wp-content/uploads/2014/10/Image-8.png.

1.4.1. Definition and Significance of Modulation

Modulation is valuable in communications systems through making information transfer over media possible. Modulation actually refers to the process of how a base signal (usually from audio, video, or data) is converted to impress information on a carrier wave. It involves the alteration of one or more characteristics of the carrier signal—amplitude, frequency, or phase—to effectively carry the intended information. Direct transmission of baseband signals would be impractical in the absence of modulation, particularly due to signal attenuation over long distances and due to the need for several signals to pack into a thin bandwidth. Modulation is thus an underlying method that enhances data transmission efficiency and accuracy within communication systems. The significance of modulation extends beyond transmission itself; it is also bandwidth consumption and overall quality of communication optimization. Different modulation techniques such as amplitude modulation (AM), frequency modulation (FM), and phase shift keying (PSK) achieve different trade-offs in terms of complexity, interference immunity, and spectral efficiency. Frequency modulation, for instance, is extremely renowned for interference immunity and therefore nicely employed in high-fidelity audio broadcasting. On the other hand, modulation methods like quadrature amplitude modulation (QAM) are at the heart of modern digital communication, as they efficiently transfer high data rates over narrow bandwidth. With all these various techniques, modulation not only enables efficient information transmission but also accommodates the evolving needs of communication technology, particularly in the case of future wireless standards. Moreover, modulation integration in communication systems is essential in bridging current challenges, especially with the growing demand for high-speed data transfer. The new technologies, such as the 5G network, are transforming the communication environment, putting additional pressure on modulation techniques to accommodate more traffic volumes and real-time data processing needs. As highlighted in current research, the heterogeneity of mobile environments necessitates innovative solutions, including the application of advanced algorithms and machine learning techniques to optimize modulation schemes. These advancements will not only make transmission systems capable of accommodating the flood of data traffic but also enhance user experiences significantly. With advancements in the field, the ongoing discovery of modulations capabilities is crucial, ultimately spearheading advancements in the efficiency and effectiveness of current communication systems (John et al., 2019; Zhang et al., 2019).

1.4.2. Types of Modulation (AM, FM, PM)

The landscape of communications networks is ultimately characterized by modulation techniques, which are crucial to transmit information through various media. Amongst these techniques, Amplitude Modulation (AM) is one of the oldest forms used, founded on echoing alterations in the amplitude of the carrier wave to show data. AM is particularly pervasive in broadcasting due to the fact that it makes it possible to be compatible with common amplitude-type receivers. Despite its extensive application, AM is more prone to noise primarily based on how easy it is for amplitude fluctuations to be overwhelmed by unwanted noise, undermining signal integrity and clarity. The simplicity of AM implementation and, particularly, in its use in long-wave radio transmission generally compensates for its detriments to deliver long range at the expense of degraded audio quality compared to other advanced modulation methods (Ferrel 2018). In contrast to AM, Frequency Modulation (FM) varies the frequency of the carrier wave to encode the data to be transmitted. The method makes the signal more resilient to noise, giving clearer and more accurate sound reproduction and is why it is superior in music broadcasting and communication systems. FM's immunity to signal degradation even in the presence of interference is primarily due to its insensitivity to amplitude variations, hence providing higher signal-to-noise ratios.

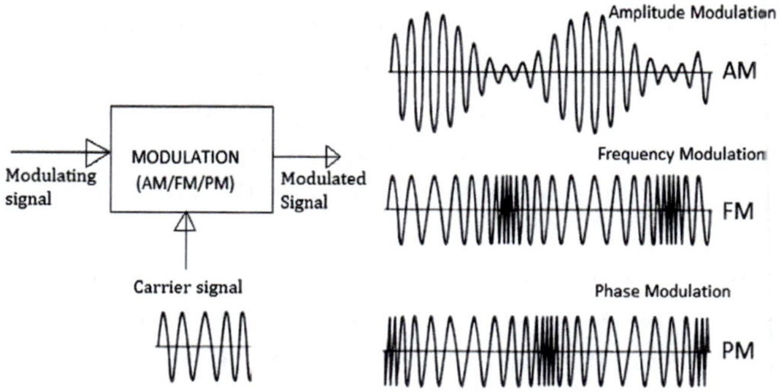

Figure 1.11. *Types of modulation (AM, FM, PM).*

Source: https://thestudyias.com/blogs/wp-content/uploads/2024/09/unnamed-1-4 png.

Also, techniques such as Wideband FM enable the transmission of high-fidelity audio, hence making FM appealing to music broadcasters. Also, the introduction of sophisticated FM demodulation techniques supports cost-effective frequency discrimination in precise accordance with the nature of sound signals, further optimizing listening experience through various delivery modes (Ferrel, 2018). Phase Modulation (PM) is another crucial technique of modulation. During this process, information is encoded by adjusting the phase of the carrier wave, making it particularly effective in digital communication networks. PM is closely related to FM but differs in that it processes information. The principal advantage of PM is its efficiency and compatibility with digital data streams wherein discrete phases unambiguously convey binary information. Moreover, advanced applications of PM, for instance, in quadrature amplitude modulation (QAM), enable the transmission of several bits simultaneously by superimposing phase and amplitude variations. This multiplexing capability is critical for high data rates in modern communications systems, including wireless communications. The applications of PM and its variants serve to show the evolution of modulation techniques necessary for modern communication requirements (Figure 1.11).

1.4.3. Applications of Modulation in Communication Systems

Applications of modulation in communication systems are numerous and are at the core of modern telecommunications. Modulation is essential to facilitate information transfer in various media, including radio waves, optical fiber, and satellite communications. One of the significant applications is in the utilization of spread spectrum techniques, which provide multiple access systems for an extensive number of low-data-rate terminals. This is particularly relevant in Very Small Aperture Terminal (VSAT) satellite networks, where various modulation methods, like M-Ary Frequency Shift Keying (MFSK), are used. The performance of these systems considers significant parameters, like satellite power limitations and the presence of proximate satellite interference. In addition, techniques like chirp modulation have been shown highly useful in achieving maximum robustness against interference, ultimately maximizing the data transmission efficiency in hostile environments, thereby accentuating the significant contribution of modulation in achieving maximum communication systems (Ha et al., 2025). Another significant application of modulation appears in the context of wireless communication systems. Here, the modulation schemes are tasked with making signals pass through various transmission conditions and yet maintain data integrity. A good example is the application of amplitude modulation (AM) and frequency modulation (FM) in broadcasting to facilitate the transmission of audio signals over long distances. In addition, advanced digital modulation techniques such as Quadrature Amplitude Modulation (QAM) have revolutionized the capabilities of today's mobile networks to accommodate higher data rates necessary for video streaming and real-time applications. These techniques enable several bits to be transmitted per symbol, fully utilizing bandwidth. Evolution of modulation techniques always adapts to meet increasing demand for efficient and portable communications in terrestrial and satellite communications (Buswig et al., 2011). Modulation in communication systems extends beyond use for general-purpose data transmission; it also has a vital role in power stabilization and power conversion in renewable energy systems, for example, wind energy systems. In such applications, modulation is employed to convert direct current (DC) generated by wind turbines into alternating current (AC), which is grid compatible. Use of artificial neural network controllers in managing this process of conversion is a demonstration of the merging of modulation and intelligent control systems. With the application of advanced modulation strategies, the output voltage can be stabilized adequately, ensuring effective performance even with varying input conditions. This reflects how modulation not only helps towards communication efficiency but also plays a key role in optimizing energy systems, illustrating its broad application in the grand scheme of technological advancement (Buswig et al., 2011).

1.5. NOISE AND INTERFERENCE

The interference and noise phenomenon are significant issues in communication systems because they degrade the integrity of signals and the reliability of information transmission. The noise can be categorized into several types including the thermal noise produced by electron agitation within conductors and the shot noise caused by the quantized nature of electric charge. Further, environmental conditions may add extra interference to the communication environment and make things more complex. For instance, electromagnetic interference in optical communication systems can confuse signals, leading to reduced clarity and loss of data. Understanding the sources and consequences of such noise is important for engineers seeking to optimize communication protocols and ensure smooth data exchange with minimal distortion. One of many methods used to combat noise, probabilistic shaping has gained popularity due to its ability to lower transmit power without sacrificing data integrity. The approach is founded on designing the transmitted symbol distribution to mitigate the effects of noise, particularly in linear interference characterized channels. Maximizing the input probability distribution of a discrete-time channel has been shown, as dictated by existing research, to be a central topic in the study of digital communication systems. This optimization is especially applicable to optical fiber communication, in which the probabilistic amplitude shaping can lead to substantial performance gains through carefully tuning data rates and the influence of nonlinearity noise, an issue that arises with increasing bandwidth demands. These tailored solutions underscore the principal relationship between noise control and system performance enhancement (Askari et al., 2024). Interference and noise in underwater wireless optical communication (UWOC) systems manifest uniquely, primarily as a consequence of environmental factors on signal propagation. Solar noise, for example, manifests as a strong source of interference, typically rendering the communication process tedious as it results in waveform distortion during daytime. One of the issues is simulating the channel impulse response (CIR) in this type of environment, one which has not been dealt with well in the past. New research with the Monte Carlo method is now modeling these effects and discovering consequences to signal quality and clarity. With breakthroughs such as the composite exponential function model, communication system designers can better anticipate performance bounds in turbid mediums, paving the way for further theoretical understanding and practical application in future UWOC system development (Li et al., 2023) (Figure 1.12).

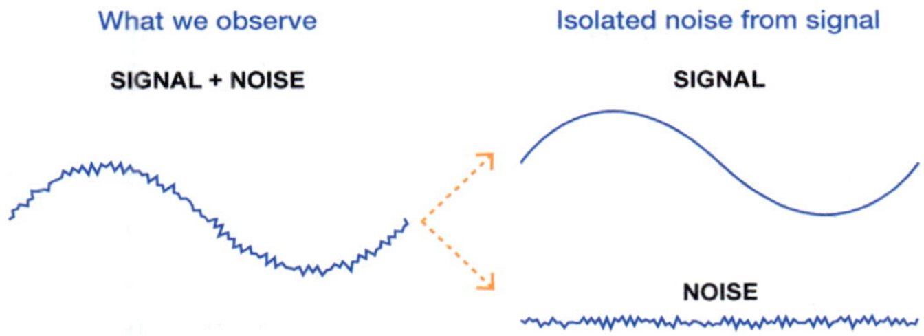

Figure 1.12. *Noise and interference.*

Source: https://external-content.duckduckgo.com/iu/?u=https%3A%2F%2Fwww.predig.com%2Fsites%2Fdefault%2Ffiles%2Fimages%2FIndicator%2F0216%2Fsig_noise_figure_2.jpg&f=1&nofb=1&ipt=39c1c3e8f311a65626300f8b0cf8c1d8e3c6a757786e27c0a647a265bad5dac2.

1.5.1. Types of Noise Affecting Communication

In any communication process, there are a number of classes of noise that can significantly hinder the effectiveness of message transmission, and indirectly affect clarity and comprehension. One of the major classes of noise is physical noise, where external noises interfere with information transmission. One instance of environmental sources such as background conversation, cars, or equipment making noisy distractions, thus it becomes hard for receivers to focus on the desired message. These distractions can lead to misunderstandings or incomplete exchanges, since important information can be lost amidst the noise of ambient sounds. Identification and minimization of such physical sound sources are necessary in an attempt to enhance communication effectiveness, particularly in environments like classrooms or crowded public spaces where effective information transfer is critical for effective interaction and understanding. Another dominant category of noise in communication is semantic noise, which is caused by misunderstandings or misinterpretations due to meaning and language. This can occur when the receiver and sender possess different cultures, degrees of experience, or knowledge of technical jargon, leading to ambiguities or misunderstandings. For example, technical jargon that is well understood by professionals might be ambiguous to non-professionals, creating enormous communication barriers. Furthermore, contextual conversation plays a key part; without a common frame of reference, messages have the tendency to lose their intended meaning. It is imperative that communicators seek to be clear and to frame messages around the audience's comprehension level so as to reduce semantic noise and to facilitate a more effective information exchange. Finally, network interference is another paramount element of noise affecting communication systems, especially in wireless environments. As has been demonstrated in communication network research, network interference can be from a variety of sources including overlapping frequencies, congestion in the signals, and environmental obstructions. Having a number of interferers that operate asynchronously in a wireless environment can complicate signal clarity and have significant effects on overall performance in communication (Moe et al., 2007). All such network noise causes issues such as dropped calls, delayed data transfer, and increased error rates, making communications of lower quality. It is essential to comprehend the actions of interferers so as to improve communication technologies, as scholars strive to advance ways of minimizing noise and maximizing signal fidelity within increasingly complex networks (Moe et al., 2007) (Figure 1.13).

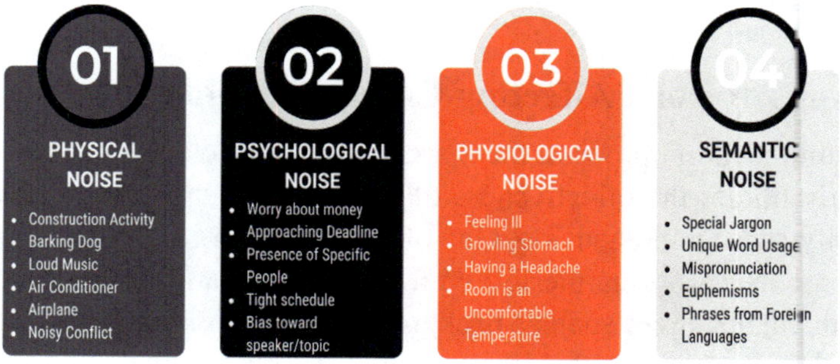

Figure 1.13. *Types of noise affecting communication.*

Source: https://open.library.okstate.edu/app/uploads/sites/154/2016/06/Types-of-Noise.svg.

1.5.2. Impact of Interference on Signal Quality

Interference effect on signal quality is a critical problem in communication systems, where so many varied factors can diminish performance. Interference is generated by a lot of different sources, from electromagnetic radiation from other devices to environmental factors and inherent noise in the communication medium itself. The resulting degradation due to these interferences makes signal clarity lessen, contributing to increased error rates and corrupted data integrity. Among the significant challenges is the existence of various wireless systems operating in overlapping frequency bands, which throws into sharp relief the need for effective ways of minimizing interference. New advances in statistical methods, especially those based on amplitude probability distribution, have shown promising capabilities to manage the challenges more effectively than traditional pass/fail thresholds, as seen with modern standards in electromagnetic compatibility (EMC) (Azpúrua et al., 2019). Understanding these dynamics sets the groundwork for improving signal quality in more dense communication environments. In addition to extrinsic interference, the intricacies of signal processing methodologies must also be taken into consideration when evaluating their influence on signal quality. As a case in point, for large Multiple Input Multiple Output (MIMO) systems, phase noise can impact overall performance severely, particularly for scenarios with passive eavesdroppers. Phase noise, being one of the types of interference created by timing errors in local oscillators, may add uncertainties to the received signal, making it oscillate and introduce additional uncertainties. Studies have shown that distributed locally deployed oscillators can optimize the secrecy rate of the communication system to perform better than systems utilizing a single shared oscillator (Bhargava et al., 2016). This illustrates how the design and form of communication hardware can neutralize the degrading effects of interference, ultimately resulting in more robust and secure signal transmission. Interference is the problem that calls for continued research and the formulation of standards to enable dependable communication systems. With further developments in wireless technologies and their growing interdependence, the effect of interference on signal quality will only become more pronounced. Thus, global standardization bodies must put highest priority on redefining EMC compliance requirements in such a way that new tests reflect today's realities of communications systems. Employing extensive statistical means of interference effect assessment, the communications industry can adapt to these advanced environments. This forward-looking solution will not only enhance the quality of signals but will also pave the way for innovation within communication systems as they evolve to meet the demands of an increasingly networked world (Azpúrua et al., 2019).

1.5.3. Techniques for Noise Reduction

In communications systems, the significance of effective noise-reduction techniques cannot be overstated. Noise, which could cause signal integrity as well as the overall system performance to become severely degraded necessitates systematic techniques for mitigation. Optical phase conjugation is an important technique in reducing issues of group-velocity dispersion as well as self-phase modulation. The process has been shown to make significant contributions toward reducing noise figures in optical systems. For instance, a comparison between dispersion-shifted fiber (DSF) and semiconductor optical amplifier (SOA)-based phase conjugators revealed that while both technologies share comparable noise figures, fiber grating utilization enhances signal retrieval and diminishes amplified spontaneous emission (ASE) noise, ultimately allowing for closer wavelength spacing and superior transmission quality (Cole et al., 1997).

Figure 1.14. *Techniques for noise reduction of communication systems.*

Source: https://soundproofcentral.com/wp-content/uploads/2020/06/Cubicle-Noise-Reduction-1.jpg.

Again, another new method of noise reduction has been reported in multicarrier transmission systems, and that is by employing nonlinear companding transformations. Such transformations are primarily responsible for the reduction of the usual high peak-to-average power ratio (PAPR) that usual such systems suffer. Although the application of companding significantly lowers PAPR, as a pity, it introduces companding noise that spoils the bit-error-rate (BER) performance. New advancements propose a simpler scheme for noise estimation and cancellation and thus improving the overall system efficiency. Through the view of companded signals as the sum of original signals and added noise, systems can maximize BER performance through effective noise cancellation techniques (Jian et al., 2015). Complemented with companding and optical methods, the application of filtering processes plays a crucial role in noise reduction strategies in communication systems. In SOA-based systems, for example, bandpass and notch filtering processes applied strategically can easily minimize ASE noise such that only cleared signals pass through the communication channel. Filtering not only maintains signal integrity but also significantly improves system performance under most operating conditions. Also, integrating these filtering techniques with advanced modulation techniques has the potential to provide still better-quality noise reduction. Advances in communications, recurrent progress in the development of noise reduction techniques will be demanded to fit the increasing demands for high-fidelity signal transmission and reception, which will ultimately establish more trustworthy communications networks (Cole et al., 1997) (Figure 1.14).

1.6. EMERGING TECHNOLOGIES IN COMMUNICATION

The communication system landscape is being revolutionized by advancing technologies that maximize connectivity and efficiency. The most significant revolution is the emergence of 5G technology, which dramatically enhances data transfer rates, minimizes latency, and enables a higher concentration of connected devices per communication network. This new capacity is crucial as the Internet of Things (IoT) expands, enabling seamless sharing of information amongst devices. Furthermore, technologies such as digital twins—computerized replicas of physical objects—are revolutionizing sectors by enabling real-time monitoring and analysis of operations. These technologies not only enhance communication but also create avenues for improved decision-making, thereby increasing the overall level of operational efficiency in communications systems. Besides, the sustainability of communication technology is a major issue today, particularly concerning their impacts on the environment. Techniques for analyzing life cycle emissions of digital information and communication technology (ICT) are at the forefront of such an issue. More so, the simplified Consequential LCA (SCLCA) method yields useful conclusions on the environmental track record of ICT solutions since it combines the assessment of positive impacts with traditional analytical approaches. This holistic evaluation is important to organizations that want to implement greener technology in communications systems. As ecologically sound practices are integrated, they can offset adverse environmental impacts, and in turn, the organization has a way of balancing communication systems with environmental sustainability, thereby mitigating both the pressure of regulation and consumer demand for more sustainable practices. In agri-environmental settings, emerging technologies are causing remarkable advancements in communications and operating practices. The application of digital twin technologies represents a paradigm change in business areas like animal husbandry by offering virtual copies of physical places that are able to replicate and optimize performance. The innovation makes communication between systems easier, allowing the farmer to administer and track resources efficiently. For instance, in research into HVAC systems related to animal welfare, digital twins can simulate energy needs and optimize efficiencies. The effective use of digital twins in agriculture not only enhances productivity but also promotes sustainable agriculture through the assistance of making informed decisions and resource management. This kind of technology and communication convergence in agriculture is a demonstration of new technologies transforming the communication system in prime sectors.

1.6.1. Role of 5G Technology

The advent of 5G technology marks a turning point in communication systems, transforming not only mobile internet access but also the entire nature of a lot of industries. With its unparalleled speed, significantly lower latency, and higher capacity, 5G makes real-time data transmission and connectivity possible, which is critical for technologies such as autonomous vehicles and telemedicine. This is made possible by the technologys ability to support multiple devices simultaneously, hence propelling the growth of the Internet of Things (IoT). As there's a growing demand for smart devices, 5G offers a critical infrastructure that supports pervasive connectivity, propelling productivity and efficiency across numerous industries. This wider scope of 5G technology makes it the cornerstone of modern communication systems, fueling innovations that were unattainable earlier. The launch of 5G also provides a foundation for utilizing a digital twin, the virtual replica that delivers maximum operating efficiency across all industries. With increased business adoption of digital twins for real-time monitoring and simulation, its integration of 5G technology becomes paramount. With the aid of 5G's high-speed data transfer, these digital twins can provide instant feedback and predictive analytics, which will allow for data-driven decision-making. All this is critical in engineering fields, where simulation and performance monitoring are central to success. The collaboration between 5G and digital twin technology is the best the revolutionizing capability of communication systems has to offer, as they enable organizations to make operations easier while concurrently revolutionizing practice in product development, manufacturing, and so forth, thereby addressing issues related to managing complicated systems (Diego et al., 2022). Future development of wireless communication systems will extend beyond 5G, as revealed by the studies currently being conducted on 6G technology. Presenting exploration into this next generation looks towards revolutionary advancements that will surpass 5G limitations such as latency and bandwidth constraint. Of special attention is the realization that 6G will offer a huge throughput improvement and support a variety of diverse applications along societal dimensions. With this transformation, technical problems related to spectrum use and network design must be addressed carefully to allow integration and performance enhancement with ease. Emphasizing an end-to-end approach, the construction of 6G systems will rely on redesigning core network architectures, facilitating the connectivity for the digital futures, thus proving that advances in communication technology are interwoven with the basic needs of society (Tataria et al., 2021) (Figure 1.15).

Figure 1.15. *Role of 5G technology systems.*

Source: https://cloudgate.ie/wp-content/uploads/2024/07/5g-Connectivity-And-Communication-scaled.jpg.

1.6.2. Impact of Satellite Communication

Satellite communication has changed the global communication systems world by offering real-time connectivity over extended distances. Being capable of transmitting data, voice, and video signals via satellites, communication free from geographical locations has become possible, hence facilitating global discourse, commerce, and emergency services. Such a breakthrough revolution is enabled by the developments in Very Large-Scale Integration (VLSI) technology, which has significantly enhanced the performance and capability of communication systems. With miniaturization and greater integration density, satellites have more power and efficiency that allow for sophisticated modulation techniques and error correction mechanisms. Consequently, satellite technology development not only improves communications quality but also extends services reach into far-flung areas where conventional communications infrastructure may be non-existent, hence increasing information and technology access across various groups (Guvvala, 2025). Moreover, the advent of high-throughput satellites and Low Earth Orbit (LEO) satellite constellations brings with it a new satellite communication paradigm, particularly bandwidth distribution and latency. These new systems allow for increased levels of augmented data transmission rates to accommodate growing demand for high-speed internet. A number of access techniques, like Frequency Division Multiple Access (FDMA) and Code Division Multiple Access (CDMA), play a crucial role in optimizing resource utilization in satellite networks, thereby enhancing overall system efficiency. As satellite communications continue to be integrated into next-generation networks such as 5G, technology such as this is critical in facilitating seamless connections and making possible paramount applications, including IoT and smart city initiatives. This development in the satellite system is a reflection of the broader advancements in communication technology, which are striving to meet the requirements of an increasingly networked world (Kamarunisha et al., 2025). With satellite communication improving, its influence in various areas of life, including disaster management, global positioning, and environmental monitoring, cannot be overestimated. The ability of satellites to provide timely and precise information during natural disasters allows for rapid response and relief interventions, which equal lives saved and economic loss averted. Satellite communications are also an essential element in global positioning and navigation services, which play a critical role in the transport and logistics industries. With increasing environmental concerns, satellite technology also has a major contribution to monitoring climate changes and natural resources and facilitates large-scale analysis of data to inform policies. Thus, the impact of satellite communication far exceeds basic connectivity; it is a critical component of managing today's challenges and fostering sustainable development in a networked global society. The ongoing innovation in this area is certain to introduce yet even greater benefits, paving the way for innovative solutions to difficult global issues.

1.6.3. Future Trends in Communication Systems

With the continuous advancements in communication systems, future trends will be increasingly dependent on convergence of technologies with an open policy of integration across platforms. This is significantly driven by the rapid advancement of artificial intelligence (AI), machine learning, and the increasing potency of the Internet of Things (IoT). AI will play a central role in decomposing large data sets, maximizing network efficiency, and personalizing user interfaces. The direction of ubiquitous connectivity, where machines remain online all the time and interactively conversing, will revolutionize communication systems in a dramatic way. Such assimilation not only will increase efficiency but also will enable creative applications across domains like intelligent cities, driverless cars, and medicine, further demonstrating the possibility of integrated systems in transforming communication ecosystems (Crolla et al., 2015). Another significant trend is the ongoing development towards 5G and beyond 6G networks, which will provide unprecedented speed, lower latency, and increased capacity. It is expected that 5G technology will be a key driver of next-generation communications by enabling the deployment of advanced applications such as AR and VR. These are applications that require high-bandwidth and low-latency support, which will be provided by 5G and 6G networks. Further, such networks will enable an explosion of connected devices with the demand for robust solutions to data management, security, and interoperability. Furthermore, advances in satellite communications technologies will extend connectivity to rural and underserved communities, thereby democratizing access to high-speed communication and empowering global participation in the digital economy (Li et al., 2015). Lastly, the future of communications systems will emphasize a robust emphasis on security and privacy. As cyber-attacks increase, having robust cybersecurity measures will be the solution to keeping data exchanged across networks safe. Encryption protocols, blockchain, and decentralized communication platforms will gain traction as businesses will want to keep confidential information secure. Policy and regulations for maintaining data privacy will also become stricter, having effects on communication technology deployment and design. This holistic security mechanism coupled with the implementation of advanced technology will ensure that next-generation communication systems not only meet the demands of current connectivity but also esteem the users' trust as well as data protection (Crolla et al., 2015).

SUMMARY

In summarizing our research into communication systems, it is important to synthesize the intricate interactions that mark successful communication both within the biological and technological paradigms. The intricate networks attested in human awareness indicate the way the brain processes and synthesizes information through various modes of transmission, such as electrical signals and chemical signals. Such processes can be described based on the model proposed in Gaiseanu et al. (2019), in which the informational structures of the brain are categorized and their influence on consciousness accounted for. Such an abundance of insight into communication reveals the importance of connectivity, not only between neurons but even between subjects and the external world. With the ongoing developments in communication systems, knowledge of the interaction among different models of information will become central to advancing our understanding of human consciousness and technology development. Furthermore, findings of research on continuous electroencephalography (cEEG) in intensive care units point out the working aspects of communication systems. The above-described disparities in different healthcare centers, as claimed in Fernandez et al. (2020), point towards the importance of creating standardized protocols. This disparity affects the outcomes of patients that directly correlate with the communication pathways used in intensive care. With the implementation of standardized parameters and designs across cEEG programs, health workers have been able to enhance improved communication, ultimately bettering clinical decision-making. Such integration not only benefits individual institutions but also contributes to the general discourse regarding health equity in the critical care environment, showing the significance of robust communication infrastructures in high-risk environments. Finally, the overarching theme in the study of communication systems illustrates that there must be an adaptive strategy in both the biological as well as technological contexts. The constantly developing nature of such systems requires constant evaluation and adaptation in order to keep up with challenges based on changing situations. Seeing through the information acquisition, processing, and affective response dynamics in the human brain portrayed in Gaiseanu et al. (2019) how this can yield benefits increasing user satisfaction and operation effectiveness in the technological environment as well. The conclusions reached through this comparative review pave the way for future methods of communication that are not only effective but also equitable and participatory. Ultimately, it is our ability to learn from diverse sources and transform them into tools supporting nascent needs that governs the development of integrated communication systems.

Communication systems are an inter-disciplinary area that encompasses a broad range of core concepts, technologies, and frameworks that facilitate effective information sharing across various media. Knowledge of communication protocols is one of the core aspects of this discipline that facilitates devices and users communicating freely. In the context of modern communication systems, protocols dictate how data is structured, transmitted, and processed, hence making different systems interoperable. Furthermore, the development of communication technologies has led to tremendous improvement in transmission means, for instance, transitioning from traditional wired connections to wireless systems that improve on convenience and effectiveness. This transition points to the realization of adapting strategies in communication to cater for the demands of today, focusing on the fundamental knowledge entailed in performing effectively in the complexity of modern and emerging communication systems. Apart from protocols, system architecture is also highly significant in communication. An architecture decides the physical or structural layout of components of a communication system as well as how the various components interact and operate while sending information. Using a methodical approach, data transmission can be optimized and latency and errors minimized through designing robust systems. Furthermore, bandwidth, signal quality, and noise issues must be addressed in architectural design so that communication paths are able to handle increasing levels of data. This kind of sophistication evidences the need for careful planning and foresight during system design so that it forms a basis for effective communication systems capable of withstanding changing technology and user needs. The value of well-designed architecture is self-evident in a number of applications, ranging from telephony to data networks, where performance and reliability are paramount. Ultimately, analysis of communications systems transcends technology concerns and encompasses issues of immense importance regarding policy and regulation. The installation of communications infrastructures is influenced significantly by governmental and institutional frameworks governing their use, privacy, and security. For instance, accurate understanding of the policies is crucial for addressing issues of data protection, intellectual property, and ethicality in communication. Organizations must steer through such regulatory landscapes to ensure compliance and gain trust among users, which ultimately decides the success of communication endeavors. As is the case with rapid evolution in response to issues like public health emergencies, effective communication remains at the forefront of organizing responses and passing information. The experience acquired through such projects highlights the continued need for enhancing the refinement in communication strategy, as evident in the interdependent relationship between technology, policy, and human interaction in the evolving landscape of communication systems (Pomer et al., 2024; Park et al., 2023).

REFERENCES

1. Abadal Cavallé, S., Alarcón Cot, E. J., Cabellos Aparicio, A., et al. (2016). Scalability of broadcast performance in wireless network-on-chip. https://core.ac.uk/download/81578757.pdf (accessed on 12th September 2025).
2. Abdollahi, S. R., Al-Raweshidy, H., Darzi, A., et al. (2010). Future broadband access network challenges. https://core.ac.uk/download/337968.pdf (accessed on 12th September 2025).
3. Abu Bakar, M. A., Ithsan, M. Z., Khalid, A., et al. (2020). Engine performance characteristics and evaluation of variation in the length of intake plenum. https://core.ac.uk/download/12007819.pdf (accessed on 12th September 2025).
4. Acquisti, B. N., Levine, P., Eckersley, P., Syverson, S., Chen, S., & Murdoch, S. J. (2014). Dovetail: Stronger anonymity in next-generation internet routing. http://arxiv.org/abs/1405.0351 (accessed on 12th September 2025).
5. Al-Debei, M. M., & Avison, D. (2009). Business model requirements and challenges in the mobile telecommunication sector. https://core.ac.uk/download/336341.pdf (accessed on 12th September 2025).
6. Ali, M. R., & Muttlak, S. G. (2025). Advancements and challenges in sub-harmonic mixer technology for high-speed, low-power communication systems. International Research Journal of Innovations in Engineering and Technology. https://www.semanticscholar.org/paper/7c4a219aaae77be457443c3a2dc0ff1c7bc18ae5 (accessed on 12th September 2025).
7. Andrae, A. S. G. (2024). Method for calculating the avoided impact of specific information and communication technology services. International Journal of Environmental Engineering and Development. https://www.semanticscholar.org/paper/b62fd9c246c67be43b271b3661ac3b7429f99939 (accessed on 12th September 2025).
8. Askari, M. T., & Lampe, L. (2024). Probabilistic shaping for nonlinearity tolerance. Journal of Lightwave Technology, 43, 1565–1580. https://www.semanticscholar.org/paper/68e42b0c761b9a27efdc1ffde23b7cde48dda97f (accessed on 12th September 2025).
9. Azpúrua Auyanet, M. A., Pous Solà, M., & Silva Martínez, F. (2019). Improved electromagnetic compatibility standards for the interconnected wireless world. https://core.ac.uk/download/288625020.pdf (accessed on 12th September 2025).
10. Bhargava, V. K., Schober, R., & Zhu, J. (2016). Physical layer security for massive MIMO systems impaired by phase noise. http://arxiv.org/abs/1603.01869 (accessed on 12th September 2025).

11. Botín-Sanabria, D. M., Mihăiţă, A. S., Peimbert-García, R. E., Ramírez-Moreno, M. A., Ramírez-Mendoza, R. A., & Lozoya-Santos, J. J. (2022). Digital twin technology challenges and applications: A comprehensive review. Remote Sensing, 14, 1335–1335. https://doi.org/10.3390/rs14061335 (accessed on 12th September 2025).
12. Bracht, R., Lewis, D., McEwan, A., et al. (1997). Inter-domain integration of services and service management.
13. Buswig, Y., & Mohammed, Y. (2011). Development of a DC-AC power conditioner for wind generator by using neural network. https://core.ac.uk/download/12007862.pdf (accessed on 12th September 2025).
14. Chen, Y., & Xu, T. (2024). Fluid antenna index modulation communications. IEEE Wireless Communications Letters, 13, 1203–1207. https://www.semanticscholar.org/paper/cdc0ce830a9bca92a6f2cd4157600a616ef41374 (accessed on 12th September 2025).
15. Chiranjeevi, V. B., & Giridhar, K. V. (2025). Role and importance of artificial intelligence in modern marketing–A study. International Journal for Multidisciplinary Research. https://www.semanticscholar.org/paper/ef340c1dd3e5dd1321fb3e6051b965ae9261b49d (accessed on 12th September 2025).
16. Chou, C. T. (2013). Impact of receiver reaction mechanisms on the performance of molecular communication networks. http://arxiv.org/abs/1312.1375 (accessed on 12th September 2025).
17. Cole, M. J., Geiger, H., Laming, R. I., et al. (1997). Optimization of DSF and SOA based phase conjugators by incorporating noise-suppressing fiber gratings.
18. Crolla, D. A., Foster, D. E., 小林, 敏雄, & Vaughan, N. D. (2015). Encyclopedia of Automotive Engineering, 53, 53–1072. Choice Reviews Online. https://doi.org/10.5860/choice.191861 (accessed on 12th September 2025).
19. Cryan, J. F., O'Riordan, K. J., Cowan, C. S. M., Sandhu, K. V., Bastiaanssen, T. F. S., Boehme, M., Codagnone, M. G., et al. (2019). The microbiota-gut-brain axis. Physiological Reviews, 99, 1877–2013. https://doi.org/10.1152/physrev.00018.2018 (accessed on 12th September 2025).
20. Desvaux, S., Goutard, F., & Roger, F. (2005). Workshop to deliver epidemiology and laboratory training for HPAI diagnostic network. http://agritrop.cirad.fr/529308/1/ID529308.pdf (accessed on 12th September 2025).

21. Eggert, L., & Fairhurst, G. (2008). Unicast UDP usage guidelines for application designers. https://core.ac.uk/download/19915619.pdf (accessed on 12th September 2025).

22. Fernandez, M. D., Andres, M. S., Frasso, R., et al. (2020). Characteristics of continuous EEG programs in the critical care setting. https://core.ac.uk/download/322850220.pdf (accessed on 12th September 2025).

23. Ferreira Pires, L., & van Sinderen, M. (1997). Protocols versus objects: Can models for telecommunications and distributed processing coexist? https://core.ac.uk/download/pdf/11454469.pdf (accessed on 12th September 2025).

24. Ferreira Pires, L., Katoen, J. P., van Sinderen, M., et al. (1994). A design model for open distributed processing systems. https://core.ac.uk/download/pdf/11454281.pdf (accessed on 12th September 2025).

25. Ferretti, S., Ghini, V., & Panzieri, F. (2015). A survey on handover management in mobility architectures. http://arxiv.org/abs/1509.01395 (accessed on 12th September 2025).

26. Fields, J. M., & Powell, C. A. (1991). Human response to aircraft noise. https://core.ac.uk/download/pdf/10437095.pdf (accessed on 12th September 2025).

27. Gaiseanu, F. (2019). Informational mode of the brain operation and consciousness as an informational related system. https://core.ac.uk/download/210995796.pdf (accessed on 12th September 2025).

28. Guo, F., & Guo, J. (2024). Application of deep learning technology in performance optimization of communication system. 13445, 1344503-1344503-8. https://www.semanticscholar.org/paper/1a761f6011c77f47a6 96a7c4cba717dac1ffcd5f (accessed on 12th September 2025).

29. Guvvala, B. (2025). VLSI technology: Revolutionizing modern communication systems. International Journal of Scientific Research in Computer Science, Engineering and Information Technology. https://www.semanticscholar.org/paper/81bbe039a958d22b93a771fb586eebef94e862f4 (accessed on 12th September 2025).

30. Ha, T. T., & Pratt, T. (2025). Study of spread spectrum multiple access systems for satellite communications with overlay on current services. https://core.ac.uk/download/pdf/42828047.pdf (accessed on 12th September 2025).

31. Harris, T. E., & Nelson, M. D. (2018). Small Group and Team Communication. Routledge eBooks. https://doi.org/10.4324/9781351019583-10 (accessed on 12th September 2025).

32. Hasegawa, T., Miyahara, H., & Suda, T. (1983). Performance evaluation of a packetized voice system simulation study. https://core.ac.uk/download/161438864.pdf (accessed on 12th September 2025).

33. Heo, G., Hong, C. S., & Pathan, A. S. K. (2007). A secure lightweight approach of node membership verification in dense HDSN. http://arxiv.org/abs/0712.4177 (accessed on 12th September 2025).

34. Javaid, N., Mahmood, D., Mahmood, S., et al. (2013). MODLEACH: A variant of LEACH for WSNs. http://arxiv.org/abs/1307.7059 (accessed on 12th September 2025).

35. Jeong, D. Y., Jo, S. K., Lee, I. B., Shin, H., & Kim, J. G. (2023). Digital twin application: Making a virtual pig house toward digital livestock farming. IEEE Access, 11, 121592–121602. https://www.semanticscholar.org/paper/6c9adfecee2ba8ac0036447c1b62005d64a8c08b (accessed on 12th September 2025).

36. Jian, W., Peng, S., Shen, Y., et al. (2015). A simplified scheme of estimation and cancellation of companding noise for companded multicarrier transmission systems. https://core.ac.uk/download/30312259.pdf (accessed on 12th September 2025).

37. Kairouz, P., McMahan, H. B., Avent, B., Bellet, A., Bennis, M., Bhagoji, A. N., Bonawitz, K., et al. (2021). Advances and open problems in federated learning. https://doi.org/10.1561/9781680837896 (accessed on 12th September 2025).

38. Kamarunisha, M., Vimalanand, S., & Dhamodharan, T. (2025). Multiple access techniques in communication systems: Enabling efficient resource sharing. EPRA International Journal of Multidisciplinary Research (IJMR). https://www.semanticscholar.org/paper/0997eb15b7e984c392806acd2a39ac2fa673ea66 (accessed on 12th September 2025).

39. Kaur, K. (2024). Overview of network topology. International Journal for Multidisciplinary Research. https://www.semanticscholar.org/paper/ffa9db12edb7f40eabdb9b6d5264017b632e5331 (accessed on 12th September 2025).

40. Kong, N. (2017). Design concept for a failover mechanism in distributed SDN controllers. https://core.ac.uk/download/153450597.pdf (accessed on 12th September 2025).

41. Kuch, T., & Sakamoto, R. (2025). Survey of federal, national, and international standards applicable to the NASA applications data services. https://core.ac.uk/download/pdf/42860346.pdf (accessed on 12th September 2025).

42. Li, X., Wei, B., Ge, T., Li, H., & Xu, N. (2023). Investigation of solar noise in underwater wireless optical communication systems. 2023 International Conference on Intelligent Communication and Networking (ICN). https://www.semanticscholar.org/paper/cd75a3fe18d72fa0facaeb55aecd726c1e1833a3 (accessed on 12th September 2025).

43. Li, Y., Lee, C. H., & Gao, J. (2015). From computer-aided to intelligent machining: Recent advances in computer numerical control machining research. Proceedings of the Institution of Mechanical Engineers Part B: Journal of Engineering Manufacture, 229, 1087–1103. https://doi.org/10.1177/0954405414560622 (accessed on 12th September 2025).

44. Park, Y., Vollmuth, P., Foltyn-Dumitru, M., Sahm, F., Ahn, S., Chang, J. H., & Kim, S. H. (2023). The 2021 WHO classification for gliomas and implications on imaging diagnosis: Part 1—Key points of the fifth edition and summary of imaging findings on adult-type diffuse gliomas. Journal of Magnetic Resonance Imaging, 58. https://www.semanticscholar.org/paper/2ba3e44bfa288f4df1a088672a9424e9cf9e5f36 (accessed on 12th September 2025).

45. Pavlov. (2022). Analysis of the theory and practice of existing invariant communication systems. H&ES Research. https://www.semanticscholar.org/paper/b3db70c5db8541ccf35f68ec42601d4b90c4fd8c (accessed on 12th September 2025).

46. Pomer, A., Munigala, S., Coles, C. L., Pope Mitro, J., Schoenfeld, A. J. Weissman, J. S., & Koehlmoos, T. (2024). The response of the Military Health System (MHS) to the COVID-19 pandemic: A summary of findings from MHS reviews. Health Research Policy and Systems, 22. https://www.semanticscholar.org/paper/803ddb2387850e1edb3fe9e9114d3c2aa716f8d (accessed on 12th September 2025).

47. Rajamanickam, N., Vishnuram, P., Abraham, D. S., Goňo, M., Kačor, P., & Mlcak, T. (2024). Review of authentication, blockchain, driver ID systems, economic aspects, and communication technologies in DWC for EVs in smart cities applications. Smart Cities. https://www.semanticscholar.org/paper/52f7bba34d603d9c0869c5d38d4227e11c018e48 (accessed on 12th September 2025).

48. Reimeir, B., Wargel, A., Riedl, F., Maach, S., Weidner, R., Grömer, C., & Federolf, P. A. (2024). Neil Armstrong's digital twin: An integrative approach for movement analysis in simulated space missions. Current Issues in Sport Science (CISS). https://www.semanticscholar.org/paper/e7e6512e0da3220820814b1472b5d17075cfbe97 (accessed on 12th September 2025).

49. Roy, N., Hassanieh, H., & Roy Choudhury, R. (2017). BackDoor. https://doi.org/10.1145/3081333.3081366 (accessed on 12th September 2025).

50. Scheffer, L. K., Xu, C. S., Januszewski, M., Lu, Z., Takemura, S. Y., Hayworth, K. J., Huang, G. B., et al. (2020). A connectome and analysis of the adult Drosophila central brain. eLife, 9. https://doi.org/10.7554/elife.57443 (accessed on 12th September 2025).

51. Srivastava, A., Santagostino, E., Dougall, A., Kitchen, S., Sutherland, M. L., Pipe, S. W., Carção, M., et al. (2020). WFH guidelines for the management of hemophilia, 3rd edition. Hemophilia, 26, 1–158. https://doi.org/10.1111/hae.14046 (accessed on 12th September 2025).

52. Stremler, F. G. (2018). Introduction to Communication Systems. CRC Press eBooks. https://doi.org/10.1201/9781315218533-17 (accessed on 12th September 2025).

53. Tang, A. (2023). Analysis of modeling methods for different types of channels. Highlights in Science, Engineering and Technology. https://www.semanticscholar.org/paper/04aebd385a10a4d7e2153d67b7db686b94d3786a (accessed on 12th September 2025).

54. Tataria, H., Shafi, M., Molisch, A. F., Döhler, M., Sjöland, H., & Tufvesson, F. (2021). 6G wireless systems: Vision, requirements, challenges, insights, and opportunities. Proceedings of the IEEE, 109, 1166–1199. https://doi.org/10.1109/jproc.2021.3061701 (accessed on 12th September 2025).

55. Upadhye, P. A. (2023). Optimizing communication systems with applied nonlinear analysis techniques. Communications on Applied Nonlinear Analysis. https://www.semanticscholar.org/paper/1f25ac0b834496a9269ce7a2a124dc9f23cbb5f8 (accessed on 12th September 2025).

56. Valčić, S., Tomas, V., & Mrak, Z. (2012). Analysis of land digital communication systems with emphasis on possible applications in the maritime VHF band. https://core.ac.uk/download/14458880.pdf (accessed on 12th September 2025).

57. Vishnu, P., Manjunath, L., Shalini, P., Amalraj, P., Krishnaveni, S., Subramaniam, N., & Loganathan, S. (2024). Analysis of Internet of Things cybersecurity and communication systems and its applications. 2024 Third International Conference on Electrical, Electronics, Information and Communication Technologies (ICEEICT). https://www.semanticscholar.org/paper/706beeb1979a7d0f18aeb82265e19ba7747fdcfb (accessed on 12th September 2025).

58. Win, M. Z., & Pinto, P. C. (2007). Communication in a Poisson field of interferers – Part I: Interference distribution and error probability. http://arxiv.org/abs/1001.4519 (accessed on 12th September 2025).

59. Zhang, C., Patras, P., & Haddadi, H. (2019). Deep learning in mobile and wireless networking: A survey. IEEE Communications Surveys & Tutorials, 21, 2224–2287. https://doi.org/10.1109/comst.2019.2904897 (accessed on 12th September 2025).

CHAPTER 2

ANALOG COMMUNICATION

Contents

INTRODUCTION

The domain of analog communication has had a significant influence on the evolution of information transfer systems and is the basis for initial telecommunications systems. Analog communication systems employ continuous signals to encode information and communicate data in signal form that is near the original message. While often overshadowed by its digital counterpart, the use of this type of communication is important to numerous applications, including conventional broadcasting and sound transmission. These systems are waveform-modulation-based to convey information across the diversity of media, and they signify an amazing merger of engineering and physical principles. An understanding of the processes not only justifies the importance of analog methods in their eras, but also makes it possible for the technological progress arising from them to be comprehended. With the interconnectivity of the world increasing, the relevance of analog communication systems cannot be ignored. Despite the extensive dissemination of digital technology, analog systems continue to be applied to specific applications wherein continuous signal interpretation is necessary. A case in point is motion control systems wherein innovations have uncovered the limitations of conventional analog technology. Out-of-the-box thinking comes out of the complexity of tasks under programming, thus producing more advanced methodologies such as FPGA technology.

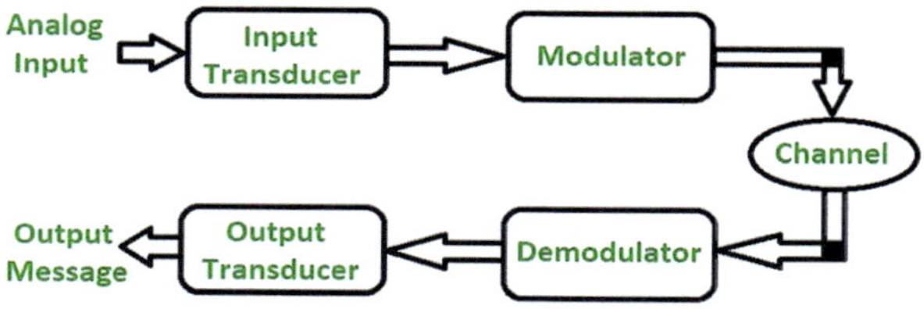

Figure 2.1. *Analog communication.*

Source: https://media.geeksforgeeks.org/wp-content/uploads/20240524132[illegible]49/Analog-Communication.webp.

The union of algorithmic precision in analog systems is evidence of the possibility of optimizing system performance, thereby presenting a pivotal breakthrough in hybrid systems that unite both analog and digital systems. The discovery not only highlights the advantages of analog communication but also the undeniable impact on modern technology (Koc et al., 2021). Secondly, the sophisticated components of analog communication extend beyond fundamental transmission procedures into specialized fields of utmost significance. To illustrate, studies on polarization faults in optics have revealed the critical role that analog methods play in cutting-edge applications like optical quantum communications. This blend of analog fundamentals and advanced optical elements illustrates how fundamental analog principles continue to steer the development of modern applications, in this case, for light polarization measurement and control. By synthesis of conceptual paradigms with real implementations in high-resolution measurement equipment, the use of analog technologies is kept in a dynamic setting. This adaptability and ongoing pertinence of analog communication emphasize its grounding in the development of complex systems that extend knowledge beyond limits of what is known (Kučera et al., 2009) (Figure 2.1).

Definition of Analog Communication

The phrase analog communication deals with the conveying of continuously variable signals that are employed to describe physical phenomena. Analog communication transmits information in the form of analog signals, and the signals are characterized by the fact that they carry an unlimited number of values. Unlike digital communication, where information is expressed in discrete values (0s and 1s), analog communication maintains the nuances of sound, light, and other waves. It is achieved with modulating methods such as amplitude modulation (AM) and frequency modulation (FM) to enable precise reproduction of original signals. The precursor to broadcasting and telephony, analog communication provided the platform for the technology used in early telecommunication systems. Knowledge of these concepts is necessary to further appreciate how communication technology evolved and how difficult it was to transition to digital systems. Analog communication systems utilize varied components to effectively transmit information along analog channels. System design begins with the observation of the communication process, specifying what each involved device requires. For instance, antennas are crucial by filtering input signals in frequency and space to obtain the optimum reception and transmission quality. More signal processing, including amplification and additional filtering, is through electronic circuits designed for enhancing signal integrity. Analog communication systems' success lies in their capacity to be adaptable under different transmission conditions, with reliability in connectivity across different environments. This adaptability is central in uses from radio broadcasts to traditional landline telephony, demarcating the flexibility of the systems in meeting communication needs (Gotthans et al., 2012). Appreciation of analog communication specifics is followed by an examination of its incorporation with modern standards such as Integrated Services Digital Network (ISDN). The continued relevance of analog theory in the current setting is illustrated using the example of satellite communication. Primary institutions are now engaged in the formation of enhanced standards and protocols that enable satellites to be integrated into larger communication networks. These efforts encompass technical performance specifications and implementation methods cutting across analog and digital systems. Such analyses reveal an era where mature analog technologies coexist with new digital technologies, suggesting the necessity to comprehend both paradigms for holistic design of communication systems (Barnes et al., 2025). Such interaction not only suggests the evolution of communication technologies but also guides upcoming developments in the field.

Historical Context and Development

The history of analog communication stretches back to the early part of the 20th century, where foundation breakthroughs paved the way for what would become a complex science. The achievement of transmitting sound over distance was a turning point and saw into existence concepts based on manipulating electrical signals to represent sound waves. This period saw dominant figures such as Alexander Graham Bell and Guglielmo Marconi whose innovations not only hastened communication technologies but also set the foundation for social changes. A historical look at their breakthroughs unfolds a story in which technological innovation was closely entwined with culture and economics, shaping the way communities engaged with new media. As these advances gained momentum, they began to influence public opinion as much as they were affecting the uptake of communications technology. Simultaneously, the advent of more sophisticated analog techniques, like amplitude and frequency modulation, further advanced the roles of communications systems. The mid-20th century, with its rapid technological progress, saw the advent of broadcasting media that completely changed the mode of information transmission over geographical distances. With the advent of radio and television as fame platforms for mass communication, they not only transformed entertainment and consumption of news but also defined the role of analog communication within the values and norms of society. The shift was technical, yes, but it brought new patterns of thought regarding identity, culture, and connection, showing how the historical context molded the trajectory of analog communication technologies. The interplay between technological advancement and cultural change goes on to define how these media are both perceived and consumed. However, the development of analog communication has also been seen with reference to its limitations, which have opened debate on whether it can be preserved and remain relevant in a world increasingly dominated by digital technology. Critics argue that even as analog systems were the focal point in staging the communication stage, they unknowingly got pushed to the periphery with the emergence of digital technologies. Analog being equated as being equivalent to obsolescence or second rate limits its value in the contemporary contexts. One should, however, realize that the historical value of analog communications goes beyond the measures of technology; it reflects broader cultural forces and effects on society. As practitioners and entrepreneurs move through this intricate terrain, understanding historical contexts can guide future development and encourage deeper connection with the traditions of analog communication and its continued impact on contemporary technology and society (Johnston et al., 2006; Wallace et al., 2001).

Importance in Modern Communication Systems

Existing communications systems are more and more reliant on analog communications methods, even in areas where digital technologies prevail. The effectiveness of such systems is partly due to their ability to transport continuous signals that reproduce information accurately. Analog systems are more effective in instances where bandwidth is limited or where signal integrity is an important factor to consider. For instance, analog signals are particularly valuable in telecommunication and broadcasting in the offering of subtleties in visual and audio material, enhancing the experience of the user. As there is more advancement in communication, the implementation of analog techniques within contemporary paradigms is not only welcome but unavoidable. Analog techniques ensure a richer, original form of information transmission that accommodates extensive applications from radio frequencies to high-fidelity audio systems. One of the key characteristics of analog communication is that it can cope with the problem of electromagnetic compatibility (EMC). Since the future looks towards wireless communications, it is a must that most communication systems co-exist without interfering with each other. Current EMC standards need drastic revisions, as conventional compliance methods are not enough to address the complexity of modern systems. Studies show that employing novel statistical techniques, for example, the amplitude probability distribution detector, can enhance the comprehension of degradation caused by electromagnetic interferences on digital communications (Azpúrua et al., 2019). This shift in perspective reflects the need for a new paradigm that optimizes compliance and performance, thereby reasserting the relevance of analog communication in maintaining efficient and trustworthy means of communication in the face of increasing rates of wireless technology. Furthermore, the use of analog techniques in high-frequency and broadband circuit design serves to indicate the long-lasting relevance of such techniques in the communication systems of today. Researchers are developing new circuit design approaches to address system integration requirements for nanometer CMOS technology, a critical point of analog performance in digital contexts (Nauta et al., 2007). Considering inherent limitations, such as noise and distortion, engineers are finding solutions that enhance wireless, wire-line, and optical communications systems' functionality. Ongoing innovation in advanced analog building blocks not only optimizes performance but also facilitates higher integration of circuits, leading to smaller, more efficient communication systems. This trend indicates that even with progress in digital technology, analog communication is still a critical component of contemporary systems, which ensures enhanced signal fidelity as well as higher operational dependability (Figure 2.2).

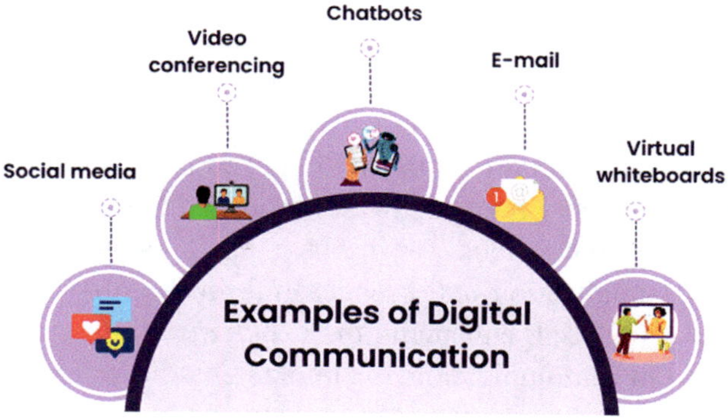

Figure 2.2. *Importance in modern communication techniques systems.*

Source: https://www.theknowledgeacademy.com/_files/images/What_are_the_Examples_of_Digital_Communication.png.

2.1. PRINCIPLES OF ANALOG COMMUNICATION

The principles of analog communications depend upon signal modulation to effectively transmit information through various media. Modulation is actually the modification of the properties of a carrier wave—specifically its amplitude, frequency, or phase—according to the input signal. Such coded data is still vulnerable to interference, necessitating sophisticated techniques of noise reduction and fidelity guarantee. Interestingly, Amplitude Modulation (AM) and Frequency Modulation (FM) are two fundamental techniques in this area, both possessing their own advantages and uses. AM is easier but more vulnerable to noise, whereas FM is better in sound quality and interference immunity. An understanding of these modulation techniques is vital in the design and execution of dependable analog communication systems. In contemporary practice, the fundamental principles of analog communication take on increased significance through their integration with emerging technologies. For instance, progress in recent years in tapping wind energy has put the use of analog modulation to regulate power output economically into focus. Single-phase DC-AC converter designs for developing work are a good example of extrapolation of these concepts into practical applications, where the objective is to control output voltage in response to varying operating conditions. Application of Artificial Neural Networks to the control of these converters indicates the convergence of traditional analog ideas with modern computational techniques, a demonstration of ongoing evolution in this field. The applications of these developments set the requirement to establish new technologies on tried principles of analog communication for optimizing performance and dependability in real applications (Buswig et al., 2011). Moreover, the evolution to more complex computing topologies demands revisiting communication methods inherent to analog platforms. New nanoscale memory technology is reshaping the landscape by challenging traditional CMOS architectures, i.e., scalability and integration density. They emphasize the need for effective on-chip communication networks, as they have to accommodate the high computational loads of modern analog and digital applications. Neuromorphic communication network engineering calls for a unique style of engineering that is different from conventional styles of engineering, with the optimism of transcending the limitations that are intrinsic to analog communication setups. All of these advances underscore the significance of a proper appreciation of analog communication fundamentals since they are the center focus in determining the direction of future smart computing and efficient signal processing (Manohar et al., 2018) (Figure 2.3).

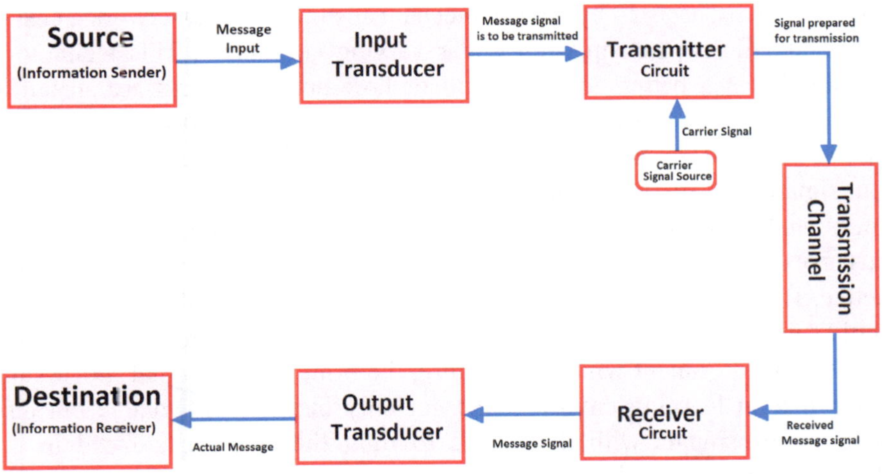

Figure 2.3. *Principles of analog communication.*

Source: https://blogger.googleusercontent.com/img/a/AVvXsEgI3E7TnNki8sXZ8Y-iqFX6xunlpAS7njPEzI1CIqK1wIJvqwG5s1ZKDkRVnyZAEYCDltVmpPDuGN-S7QzSPS8wbj-VKIysRGwSIC4fg14PFZoBzdW22N72XM3aQr-auH4GR2D7nDl-dKEFwqK0Rvf4D5h7_KStMt-NGg3c3xrW6A3WK23c5UcgRlW997=s16000.

2.1.1. Basic Concepts of Signal Transmission

In analog communication, transmission of signals forms the foundation of effective transmission of information. Signal transmission involves encoding data into an appropriate form and passing it through a physical channel, for example, air or cables. Signals can be altered using modulation techniques so that they can travel via various mediums in spite of interference. For instance, frequency modulation (FM) and amplitude modulation (AM) are typical practices which manipulate signal characteristics to expand their reach and reliability. The selection of a proper transmission medium—is in the form of coaxial cable, fiber optic cables, or radio signals—also is instrumental in improving the performance and quality of the signal. These basic principles, thus, form the cornerstone for proceeding with the practice and study of analog communication technology. The effectiveness of signal transmission is heavily reliant on encoding and decoding technology. Success in this field often comes with sophisticated gear and theories that ensure communication despite noise and environmental interference. One especially important area of innovation is converter design, such as DC-AC converters, which are vital in signal conversion for various applications, such as wind turbines that convert mechanical energy into usable electric form. Sophisticated mathematical models for such systems, assisted by control mechanisms like Neural Networks, ensure stable and efficient transmission under varying conditions. This technology layer facilitates a better understanding of how analog signals are sustained while in transit, thereby solidifying their integrity to the time of delivery (Buswig et al., 2011). Education plays a pivotal role in transferring knowledge about signal transmission technologies and their applications in real scenarios. Professional courses provided to specialists in the telecommunications sector focus on experiential, practical learning that bridges the gap between theory and practice. By using advanced training equipment, it is possible for instructors to hold laboratory seminars that facilitate interactive learning, thus enabling members to experiment with, and acquire, the principles of signal processing. Such hands-on learning environments not only increase knowledge but also provide professionals with the means to fight the challenges faced in the field of digital communications. The fusion of theory and practice underlies a broad comprehension of the fundamental principles of signal transmission that are foundational to future innovation and technology growth in analog communication (Albert et al., 1998).

2.1.2. Types of Analog Signals

The concept of defining analog signals is at the core of studying their applications within analog communication. Analog signals essentially consist of two main types: continuous-time and discrete-time signals. Continuous-time signals are defined for every instance of time, and a continuous flow of information exists. Audio waveforms and video signals are some examples, in which all the original information is retained when they are recorded. On the other hand, discrete-time signals have values at particular instants that determine how analog signals can be allowed to behave in a converted system. While quantized, these signals maintain the character of analog communication and may convincingly model real-world phenomena by means of techniques like sampling and quantization and thus serve as critical variables in modern telecommunications. In addition, some analog signal types such as sinusoidal waves and complex signals play a crucial role in communication systems. Sinusoidal waves are the foundation of periodic signals' representation that have a single frequency and incorporate harmonics through Fourier analysis.

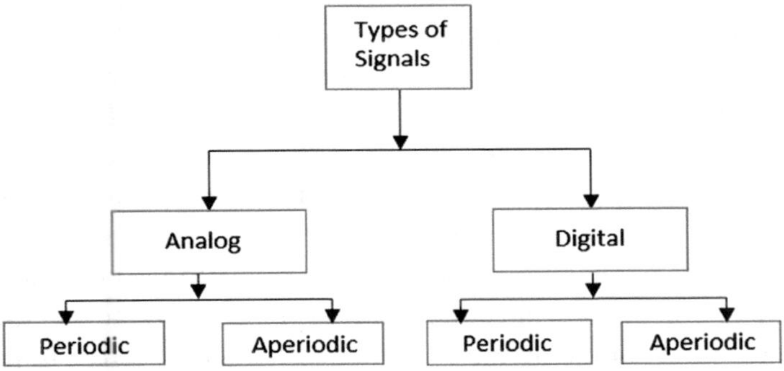

Figure 2.4. *Types of analog signals.*

Source: https://www.tutorialspoint.com/analog_communication/images/types_of_signals.jpg.

Complex signals are developed by combining more than one sinusoidal component and possess a larger bandwidth with higher capacity to transmit data in complex communication networks. These signals enable processes of modulation, important for inscribing information onto carrier waves to be transmitted effectively. The interaction among these various kinds of signals in modulation processes emphasizes understanding their properties to achieve the optimal transmission quality and minimize distortion, hence effective communication on vast networks. Lastly, practical uses of all types of analog signals vary into many arenas, particularly in sensor technology, audio processing, and telecommunication networks. For instance, in audio communication, analog signals capture sound waves with fidelity and therefore offer clarity and continuity in voice transmission. Similarly, in sensor networks, continuous-time analog signals are employed to monitor various environmental parameters like temperature and pressure, thus enabling improved real-time data collection and analysis for decision-making. The evolution of sophisticated control systems also reflects how different types of analog signals are integrated to manage complex data, leading to automation and machine learning technologies. As the world of analog communication continues to evolve, awareness of types of analog signals is necessary for designing robust and impactful communication solutions, bridging conventional methods with contemporary digital technologies (Piccinini et al., 2009; Sattouf et al., 2015) (Figure 2.4).

2.1.3. Modulation Techniques

The use of modulation methods is a core part of analog communication as they have a central role to play in determining how data is conveyed by various media. Modulation is the method by which one alters a carrier signals' parameter—amplitude, frequency, or phase—to modulate data for effective transmission. The method not only allows compatibility with various transmission channels but also enhances the power and effectiveness of the communication system. For instance, amplitude modulation (AM) and frequency modulation (FM) are prevalent in broadcasting since they enable the signal to be immune to noise and interference, hence keeping the quality of the received audio optimal. Ultimately, the mode of modulation used plays a significant role in the signal's integrity and transmission range, being an elementary component of effective analog communication. The advent of high-tech technologies has introduced novel innovations in modulation techniques, particularly in Software-Defined Radio technology. The SDR concept allows for adaptive radio systems that can accommodate a broad range of communication standards and bands via reprogramming via software to ensure more versatility in the rapidly evolving technical landscape. This reconfigurability makes it easy to include different modulation schemes without requiring extensive hardware modifications, thus making communication between devices simpler and enhancing user experience, especially in the case of mobile telephony. While multi-service system demand increases, research into SDR architectures emphasizes adaptive modulation techniques where dynamic adaptation can be performed according to channel conditions, network load, and specific user requirements as a fundamental step towards optimizing analog communication performance (Hoeksema et al., 2001). A corresponding significant field of research is the enhancement in modulation methods in optical communication systems, where the enhancements are essential in addressing fiber impairments. Coherent detection-based detection systems have transformed the landscape by being capable of practically supporting more sophisticated multi-level modulation formats, which increase spectral efficiency since they compensate for linear distortions present in optical fibers. As the optical systems continue to trend towards high data rates (e.g., 400 Gbps/1 Tbps), the issue of fiber nonlinearity becomes increasingly significant, significantly degrading transmission performance. Several nonlinear compensation methods have been studied in recent literature, suggesting that intelligent modulation techniques in conjunction with DSP algorithms are necessary. By adopting these sophisticated techniques, data fidelity in transmitted data is ensured, and more bandwidth services that are essential in modern-day communication networks (Amari et al., 2017). Thus, the creation of modulation techniques remains central to problem-solving and making analog communication systems function better.

2.2. COMPONENTS OF ANALOG COMMUNICATION SYSTEMS

In analog communication systems, there are several main elements that facilitate the transmission of information in various media. Foremost among these systems is the transmitter, which forms the original signal of information into a compatible form for transmission. The transmitter can incorporate a modulators, which alters the characteristics of a carrier wave—amplitude, frequency, or phase—facilitating effective signal transmission through different channels. This modulation process is crucial for ensuring the signal can be distinguished from noise when it is in transmission. With advancing technology, designing these components also gets better, enhancing their reliability and efficiency in conveying clear communication over varying distances. The transmitting medium is another vital component in analog communication systems that allows transmission of the modulated signal from the transmitter to the receiver. Various media like coaxial cables, optical fibers, and free space are used based on some applications and environmental conditions. Every medium has some advantages and disadvantages, which influence the overall performance of the communication system. For example, coaxial cables offer immunity to electromagnetic interference, while optical fibers offer high-bandwidth transmission over large distances with minimal attenuation. Therefore, it is essential to understand the properties and limitations of different transmission media in creating effective analog communication systems that can operate under different conditions. Finally, the receiver becomes the focal point of recreating the original signal from the received modulated wave. Its primary function is to demodulate the received signal, basically extracting the information within it. This demodulation operation needs great accuracy in an attempt to avoid noise and interference from spoiling the reconstructed signal's quality. Analog receivers generally consist of amplifiers and filters to enhance the quality of the signal, with much care being taken in minimizing distortion while maximizing readability of the output. Consequently, with systems moving to more advanced digital replacements that balance their analog precursors' weaknesses, innovative methods are an imperative to maintain the effectiveness of traditional analog communication systems (Guerrero et al., 2000; Messer et al., 2025).

2.2.1. Transmitters and Their Functions

Transmitters are the most important devices in the realm of analog communication that convert information into signal form that can be sent through various media. The primary responsibility of a transmitter is to modulate the input information, be it audio or data, onto a carrier wave. The modulation can take various shapes, some of which include amplitude modulation (AM) and frequency modulation (FM), both with their own features. By changing the properties of the carrier wave, transmitters ensure the information is transferred effectively and efficiently. Further, the design and complexity of the transmitter can play a significant role in the quality and reliability of the communication and call for the knowledge of the basic principles and the technologies involved in their operation. Evolving transmitter technology has been a significant factor in enhancing communication capabilities, particularly in high-speed application. For instance, the development of Dispersion Code Multiple Access (DCMA) exemplifies the new paradigms being employed in today's communications. According to the DCMA study, a number of mathematical and experimental models have been employed to optimize the efficiency of analog transmission systems. Chebyshev dispersion coding is an instance of employing high-level coding techniques to achieve maximum bandwidth and overall performance of communication systems. This innovation shows the ongoing revolution of transmitter technology, showing its capacity to cope with the growing demands for reliable and high-speed channels of analog communications, as exemplified in theoretical and experimental setups (Caloz et al., 2017). In addition to the traditional roles, transmitters are also being used in sophisticated systems, such as telemetry applications. Telemetry systems are designed for the transmission and broadcasting of data over long distances that often include position location and monitoring responsibilities. Transmitters are employed in these systems to deliver critical data from sensors to receivers in an attempt to acquire real time data and process it. System reliability is a major concern, especially in the fields of medicine, the military, and environmental monitoring. The accuracy of modern transmitters in maintaining correspondence in various conditions also brings out the importance of ensuring the accuracy and effectiveness in the transmission of data. It is due to this reason that the role of transmitters is more than just the production of signals, involving a wide array of functionalities that are inevitable in modern analog communication systems (Hilton et al., 1969) (Figure 2.5).

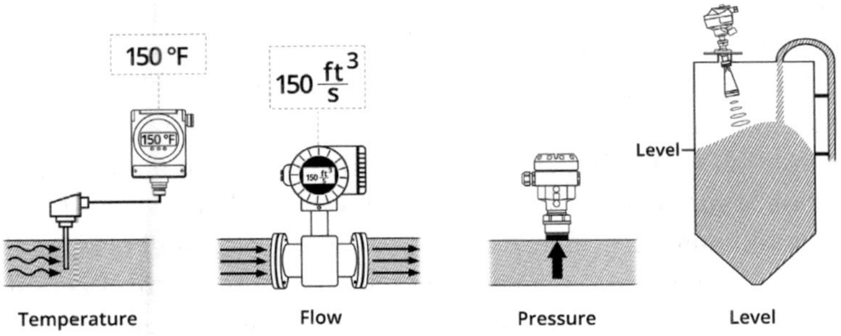

Figure 2.5. *Transmitters and their functions.*

Source: https://cdsentec.com/wp-content/uploads/2022/07/Variables-measured-by-the-transmitter-1024x453.jpg.

2.2.2. Receivers and Signal Processing

The design of analog communication system receivers is essentially not separable from signal processing principles. Receivers play a central role in reception on various media, their performance defining received information quality and purity. Analog receivers typically employ a chain of devices such as mixers, amplifiers, and filters to process incoming signals. All these constituents contribute to minimizing noise and distortion, which are the most critical considerations in signal integrity preservation. Effective signal processing makes possible the reception of unambiguously correct communications to facilitate an enormous variety of applications, from broadcasting to voice communication. As there is constantly increasing input towards yet more universal and effective communication systems, the importance of sophisticated techniques for signal generation and conditioning in receivers cannot be overstated. Recent advancements in digital signal processing (DSP) have revolutionized the analogue receiver world. Previously, most of the signal processing was done using analog components; however, the integration of DSP has dramatically enhanced performance and flexibility. Techniques such as filtering, demodulation, and error correction are now successfully done with DSP, not just making the system more dependable but also reducing the overall system cost (Harris, 1997). This shift to DSP emphasizes the need for an integrated synergy between analog and digital paths, ultimately leading to more sophisticated and responsive communication systems. In particular, the antenna interface — where the analog signals are initially detected — is a critical point of leverage for both DSP and traditional designs to get the most out of capability in achieving improved communication performance. The evolution to incorporate physical-layer security techniques in communication systems is necessary to address contemporary challenges such as data security and privacy. As noted in recent literature, concepts such as physical-layer service integration (PHY-SI) represent a significant step in managing multiple types of services simultaneously, resulting in improved efficiency (Chen et al., 2018). By combining heterogeneous service needs—i.e., multicast and secure transmissions—into a single collective communication stream, PHY-SI offers new solutions to ensure secure and reliable communications. In addition to maximizing bandwidth utilization, this approach enhances service performance via tailored transmission methods. As the research progresses toward developing new technology in this area, the integration of new DSP methods with PHY-SI principles could potentially provide revolutionary advancements to receiver engineering and signal processing that will enable the creation of the next generation of secure, efficient analog communication systems.

2.2.3. Transmission Mediums

In analog communication, the choice of transmission media greatly affects signal fidelity and integrity. Various media such as copper wires, fiber cables and wireless channels possess varying strengths and limitations that impact data transmission. Copper wires, used in telephony applications, are vulnerable to electromagnetic interference, which erodes signal quality, especially over long distances. Conversely, fiber optics transmit data by light, providing more bandwidth and immunity to noise. The innovation in wireless technologies also extended the applications of transmission mediums further, though their introduction is accompanied by the problem of environmental effects on signal strength. In order to maximize systems in efficient analog signal transmission, engineers and communications professionals must understand these differences. Lastly, the impact of emerging technologies cannot be overstated in regards to transmission mediums of analog communications. For instance, advancements in impulse radio ultra wideband signals have the potential to ease problems in self-organizing ad hoc and mobile ad hoc networks (MANETs) through offering improved capabilities in multiplex access. The technology signifies a shift towards highly versatile and efficient communication systems with the capacity to manage increasing demands for data capacity. Therefore, it is imperative to understand how new signaling technologies can be made part of existing mediums so as to redefine the communication environment. These mediums evolve to enhance performance but are a pivotal step towards becoming an information society in addressing some access and coverage area issues within viable applications (Bunin et al., 2014). Finally, the relationship between transmission media and analog communication portrays a complex context that is constantly evolving. Since each medium has unique characteristics and potential, understanding its influence on the quality and productivity of communication becomes critical. The development of innovative solutions, like impulse radio ultra wideband signals, reflects an industry's willingness to accommodate challenges inherent in analog transmission. As communication technologies advance, stakeholders need to recognize the significance of selecting the right medium, particularly under an increasingly changing information economy. Being able to design effective communication networks largely relies on the symbiotic integration of medium choice and emerging technologies, further underscoring the need for ongoing research and strategic innovation in this core area (N/A, 2002) (Figure 2.6).

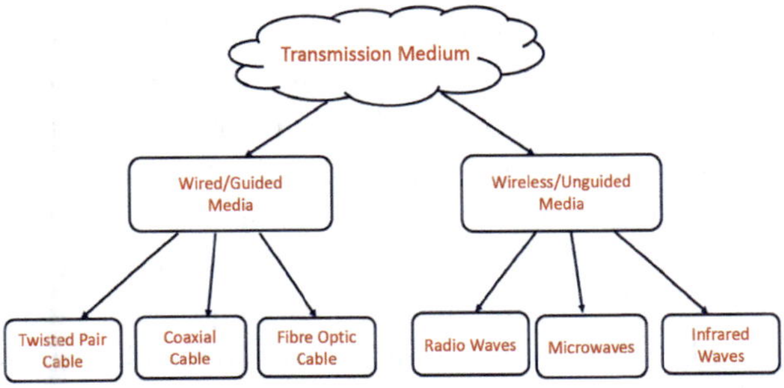

Figure 2.6. *Transmission mediums.*

Source: https://bcastudyguide.com/wp-content/uploads/2022/06/types-of-transmission-media-1024x463-1.webp.

2.3. TYPES OF ANALOG COMMUNICATION

Analog communication has a variety of forms and methods that seek to transmit information by varying the continuous signal. Amplitude modulation or AM is one of the primary forms involving variation of the carrier wave strength with respect to the information signal. It is commonly employed for broadcasting by radio and offers multiplexing of several transmissions in a single frequency band through varying various signals. Despite its common application, AM is susceptible to noise and interference, which can cause degradation of the signal. This presents a requirement for extremely robust communication systems that can be compensated for by these faults, such as those investigated in advanced research into how adaptive methods can provide assurance of performance stability when used in harsh environments (Acosta et al., 2025). Another popular form of analog communication is frequency modulation (FM), and it is quite distinct from AM in the sense that FM modulates the carrier wave's frequency rather than its amplitude. FM is particularly advantageous due to its high noise resistance, and therefore FM is the preferred method of high-fidelity sound transmission. As a result, FM is widely applied in television broadcasting as well as radio broadcasting with higher quality sound and less distortion. The mechanism involves encoding the information signals into higher frequencies, which helps in establishing a more stable communication channel. Considering the evolution of signal processing technology, reliable FM systems are usually used in contemporary communications infrastructure, supporting both analog and digital users simultaneously (Popok et al., 2014). Phase modulation (PM) is another significant type of analog communication, where the carrier signal's phase is varied in accordance with the information signal. It is possible to transmit digital data over analog channels by this technique and is the foundation for several sophisticated telecommunications applications, including satellite communication and data networking. PM is appropriate for use in high bandwidth and data transfer applications, serving a vital function in maintaining system integrity under varying conditions. Subsequent integration of phase modulation with other digital modulation forms has rendered it more useful, leading to novel applications such as adaptive phased array antenna systems, which facilitate efficient communication through real-time adjustment of environmental parameters (Acosta et al., 2025).

2.3.1. Amplitude Modulation (AM)

AM is a simple technique of analog communication, which exploits variations in the amplitude of a high-frequency carrier wave for communication. It is a technique founded on the superimposition of a modulation signal with the information to be communicated onto a high-frequency carrier wave. The result is a modified signal that may propagate over long distances. AM is particularly renowned for its simplicity to implement and immunity to noise and has been a long-standing choice for a variety of uses for broadcasting, most significantly AM radio. The inherent ability of AM to preserve signal integrity at longer distances makes it essential in analog communication systems, most significantly in systems where digital options can have issues with interference and degradation. Though it has its plus points, AM modulation is not without weaknesses, such as bandwidth and spectral efficiency weaknesses. The greatest weakness is the need for a larger bandwidth than most other methods of modulation, such as the Frequency Modulation (FM) which automatically leads to wastage of available frequency spectra. Additionally, the AM signals are more susceptible to distortion by noise, particularly in the amplitude factor, which undermines signal quality overall. Frequency synthesizer technologies, forming the cornerstone of a significant number of AM transmission systems, are required in performance optimization as well as erasing some of its limits.

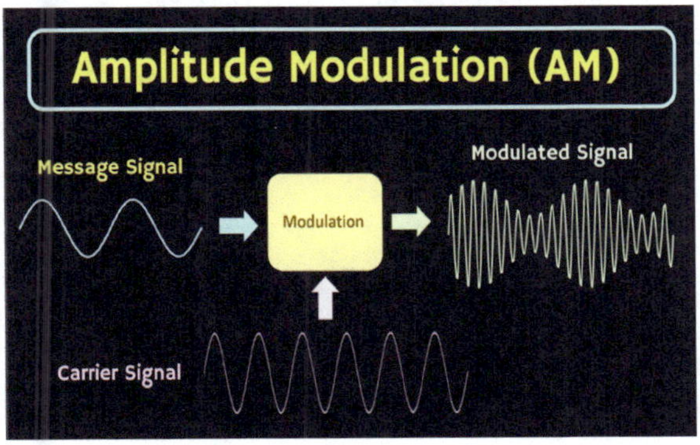

Figure 2.7. *Amplitude modulation (AM).*

Source: https://i.ytimg.com/vi/Yryi1bbmUjU/maxresdefault.jpg.

As has been noted in recent RF communication product innovation, high-end frequency synthesizers can enhance AM transmission efficiency and thus quality and reliability in analog communication (Nauta et al., 2002). With advancing technology in communication, the role of AM remains relevant, particularly in specialized applications when analog techniques are required. Though digital communication techniques have been in dominance for the transmission of general data, AM finds its niche in specialized uses such as air communication and emergency broadcasts where reliability under poor conditions is of most importance. Distinguished, however, are developments in complementary technologies, such as digital platforms for optical imaging, emphasizing the interdisciplinarity of innovation possibilities to stimulate traditional analog methods. With new analog implementations, driven by advances in digital technology, AM can be made even more accurate, ensuring its persistent relevance in niche analog communication applications (Torjesen et al., 2016). The history of AM thus demonstrates the intricate interdependence of communication technologies, blending conventional methods with cutting-edge innovations (Figure 2.7).

2.3.2. Frequency Modulation (FM)

FM is a core technique in the universe of analog communication, predominantly renowned for its robust ability to convey information via radio waves with improved fidelity. Unlike AM, whose signal strength directly indicates the information that is being transmitted, FM retains data within changes of frequency. This method minimizes interference and noise, thereby improving the audio signals. The power of FM renders it the preferred mode of high-fidelity music and two-way radio broadcasting. Through the modulation of a carrier wave based on the input signal, FM enjoys numerous benefits, including greater resilience to amplitude variation caused by electrical and atmospheric noises, which characterize most analog communication systems.

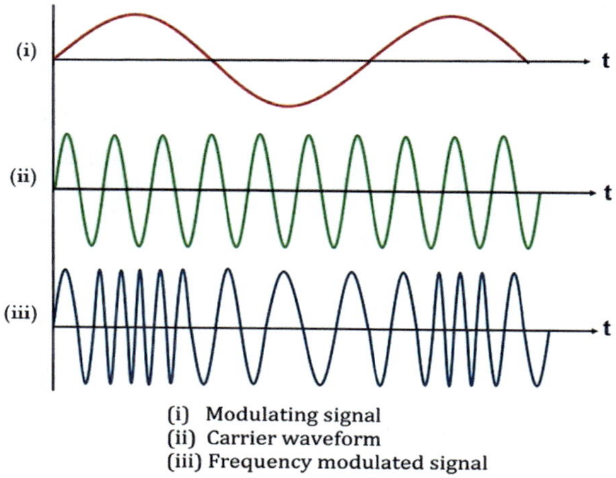

Figure 2.8. *Frequency modulation (FM).*

Source: https://electronicscoach.com/wp-content/uploads/2018/10/Waveform-for-frequency-modulation.jpg.

The mathematical foundation of FM revolves around the modulation index which defines the level of frequency deviation in relation to the frequency of modulating signals. The modulation index directly influences the bandwidth of the FM signal; the greater the level of frequency deviation, the broader the signals' bandwidth and consequently demands careful planning in allocating spectrums. According to recent studies, the interference-to-noise ratio (INR) significantly influences the performance of FM systems against numerous overlapping signals. Essentially, the INR measures permit the assessment of the resistance of FM to interference so that designers can tailor performance in any context, such as automotive radar systems, where accuracy and clarity are parameters of immense value under poor conditions (Alland et al., 2019). The fine-grained examination of these parameters facilitates the designing and implementing of FM systems such that they perform well in high-density spectral environments. New innovative advancements in the FM technology continue to unfold, and it is utilized in various fields, e.g., the Internet of Things (IoT). For instance, new technologies like PLoRa indicate the versatility of frequency modulation in offering far-distance communication for battery-less devices. There, ambient signals are recycled and relayed, demonstrating how FM is able to transcend traditional limits through the ability of devices to speak back to one another without direct battery supply (Peng et al., 2018). This sort of application not only demonstrates the adaptability of frequency modulation in modern contexts but also demonstrates its ability to enable more and more devices with efficient and renewable means of communication. By the same token, FM remains an integral part of the ever-evolving innovation of analog communications technology, ever-changing to fit into a more globalized world (Figure 2.8).

2.3.3. Phase Modulation (PM)

PM is a significant technique in analog communication renowned for its immunity to noise. Unlike amplitude modulation, which varies the carrier wave's amplitude as a function of the information signal, PM varies the instantaneous phase of the carrier according to the modulating signal. This type of modulation offers wider bandwidth utilization and more ability in transmitting data. In applications such as broadcasting and two-way radio communication, PM strength is converted into improved signal quality, allowing for less distortion and clearer sound, particularly when electromagnetic interference is high. Substantial theoretical concepts underpin the technical use of Phase Modulation, especially through the usage of understanding such as phase deviation and modulation index. The modulation index, being the phase deviation over the modulating signal frequency, is a key factor in determining signal fidelity. With proper control of this index, communications engineers can manage issues of bandwidth constraints issues to ensure compatibility with various transmission systems. Besides, advances in integrated circuit technology allow the development of sophisticated phase-locked loops (PLLs), which are key to the generation and demodulation of phase-modulated signals. This technology opens up new avenues for future enhancement of the efficiency and reliability of analog communication systems and thus PM is a cornerstone of modern communication infrastructure (Nauta et al., 2002). The theoretical significance of Phase Modulation is achieved through its applications in various applications, including telecommunications and signal processing. For instance, in wind power systems, PM techniques have been employed in power conversion system regulation to achieve stable output voltage under varying operating conditions. These applications show not only the versatility of PM but also its potential to complement other technologies, such as Artificial Neural Networks for next-generation control methods (Buswig et al., 2011). With continuing growth in communications demands, PM methods continue to be refined, possibly bringing in the next analog communications innovation wave. This union of sophisticated methods of control and robust modulation processes strengthens PM as a worthy commodity, emphasizing its viability in the midst of advancing technological settings and wireless communication in the coming years (Figure 2.9).

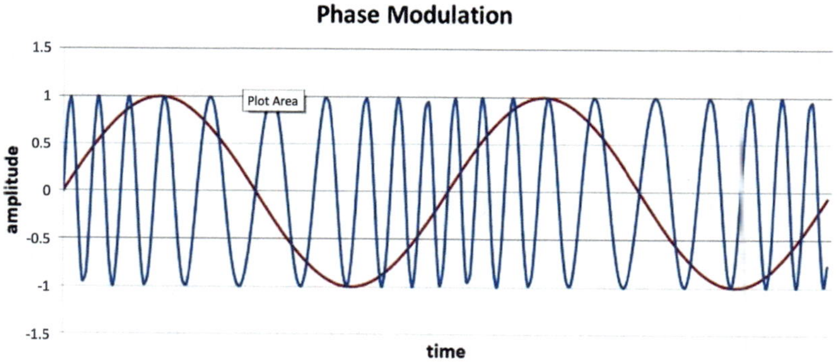

Figure 2.9. *Phase modulation (PM).*

Source: https://www.allaboutcircuits.com/uploads/articles/RFT_ch4_pg4_PMn4.jpg.

2.4. ADVANTAGES OF ANALOG COMMUNICATION

In the controversy surrounding communication technologies, the advantages of analog communication remain significant, particularly in terms of its simplicity and efficiency. Analog systems naturally deal with continuous signals, which facilitates them to communicate the expressiveness of sound and visual information more fluidly than digital systems. This characteristic results in a richer and more natural experience for consumers, particularly for uses such as audio broadcasting and initial video transmission. Besides, analog communication systems consume less sophisticated technology, hence making them simple to design and implement. Consequently, the systems can be easily installed in areas where resources are limited, making them very helpful in rural and developing communities where the infrastructure cannot support the more sophisticated requirements of digital systems. Another major advantage of analog communication is that it can withstand and be reliable under some circumstances. Analog signals are less susceptible to problems of latency and can maintain a consistent quality over varying distances and environments. For example, while digital signals may require rigid error correction techniques to maintain clarity, analog transmission would fail progressively in quality rather than suddenly cutting out. This smooth degradation makes it simpler to diagnose and repair. With the emergence of technology, the principles of analog communication still provide the impetus for emerging technologies such as Software-Defined Radio (SDR), where efforts are made to realize flexible, multi-service systems that can cover multiple standards without sacrificing the inherent benefits of analog systems (Hoeksema et al., 2001). Hence, the underlying principles of analog communication remain pertinent and relevant in the presence of emerging more sophisticated technologies. Finally, analog communication's cost-effectiveness is a crucial factor behind its enduring existence in the market. The low resource needs and uncomplicated circuitry required in analog communication systems translate to lower production and operational costs. Its cost-effectiveness is particularly helpful for small organizations and third-world nations, where fiscal constraints might limit access to more advanced digital infrastructure. Besides, analog equipment repair and maintenance are less likely to need specialized training and skills, thus lowering the costs of long-term operation. With the telecommunications landscape still changing, analog communications cost-effectiveness promises to make it a central aspect, particularly where high-fidelity transmission is not absolutely essential. While evaluating the strengths of analog communication, it is clear that even in a rapidly digitizing world, its strengths remain valid and are still called for (Dutta et al., 2016).

2.4.1. Simplicity and Cost-Effectiveness

Simplicity is a key problem in the design and implementation of numerous various systems in analog communication. The use of simple designs not only enhances user convenience but also simplifies maintenance and trouble-shooting. Simple analog communication systems, for instance, tend to be smaller compared to their digital counterparts because they require fewer components, which means a lighter and more portable system. This simplicity approach has a tendency to provide more resilience to failures because there are fewer parts and, consequently, less opportunity for breakdown. Furthermore, the relative simplicity of analog systems enables a simpler understanding of signal processing, which is advantageous to engineers and technicians. Ultimately, the focus on simplicity with analog communication not only increases the utility of these systems but also makes them solutions for the long term in cases where the complexity of an entity could otherwise be restrictive. Cost-effectiveness is an elementary feature that is also underlining the appeal of analog communications systems. Taking a benefit from less complex components and design, such systems usually incur lower production and operating costs than more advanced digital technologies. As an instance, using inexpensive, low-resolution circuit designs can significantly minimize expenditure while still delivering adequate performance in most situations. This is particularly the case in resource-poor environments where economic limitations govern technology adoption. Besides, investment saving occurs after the initial costs; simpler systems normally have lower maintenance costs in the long term. As suggested, traditional approaches in control systems for equipment such as Erbium-Doped Fiber Amplifiers (EDFAs) normally suffer from issues of cost and complexity but can benefit from new innovations such as memristor-based control systems that possess simplicity and increased cost-effectiveness (Zhao et al., 2024). The financial advantages of analog communication thus come to underscore its relevance in industry and education. The intersection of convenience and affordability in analog communication networks not only has ramifications for efficiency of function but further promotes broader accessibility. Lowering both cost and structural complexity, the systems afford greater access to technological innovation in communication for more users—small businesses to underdeveloped communities. For example, utilization of cost-effective tools, such as the MATLAB toolbox for synthesis and simulation of analog-to-digital converters, exemplifies how design simplification can translate into fantastic leaps in performance without a single cut in budgetary needs (Restituto et al., 2005). Even more, this openness encourages an open space where innovation shall prosper, and more development and competitiveness in the communications technology sector will ensue. Thus, as the world becomes in greater need of effective communication, the factors of simplicity and cost-effectiveness of analog systems position them not only as viable options but as necessary instruments for future growth.

2.4.2. Continuous Signal Representation

Continuously representative signal is a simple concept in analog communication that directs the techniques employed in signal receiving and sending. Generally, continuously representative signals have the ability to express information in the form of continuous change with time without discontinuity. The aforesaid continuous change is indicative of the communicated physical phenomena, such as voices, music, or any continuously varying data. The transition from discrete to continuous structures highlights the requirement of techniques ensuring signal integrity as they traverse through varied mediums. This type of representation allows for the consideration of varied mathematical models, which capture the complexity of actual signals, an essential aspect for communication systems to function effectively. In addition, technology advancements have imposed another necessity for high-fidelity techniques for representing continuous signals, particularly for filters and dynamic range optimization. For instance, research highlights the synthesis procedure geared towards continuous-time fully differential Gm-C filters that are crucial in analog signal processing. The synthesis in this case utilizes an extended state-space system model whereby noise and distortion performance can be tested in filters. Such innovation not only enhances the accuracy of signal purity but also transmission efficiency, as illustrated through the means presented for optimizing weakly nonlinear Gm-C filters under power constraints. Effective as they are, these approaches meet the gap between theoretical analysis and practical usage, enabling effective application in modern communication systems (Restituto et al., 2008). Continuous signal representation comes particularly into play in the area of high-frequency communication. With the emergence of millimeter-wave communication more specifically derived from the 60 GHz band, escalating bandwidth demands in analog communication systems become evident. This trend of increased frequencies is observed in the IEEE 802.11ad standard, which provides for communication by way of high-rate analog-to-digital converters. These types of technologies necessitate employing innovative techniques in signal reconstruction and channel estimation, particularly with regards to single carrier modulation. By exploiting sparsity within channel impulse responses, researchers are able to achieve efficient low-rate sampling techniques that maintain signal fidelity while optimizing usage of resources. These developments point to the pivotal importance of stable signal representation in ensuring effective, high-fidelity communication in a more digital, high-frequency scenario (Eldar et al., 2017).

2.4.3. Compatibility with Existing Technologies

Compatibility with the existing technologies is essential in the decision-making process of advancing analog communications. As systems evolve, the new technologies will have to be compatible with the infrastructure that exists so as to ensure performance and reliability. Compatibility is not just a matter of technology but also one of strategy, which guarantees long-term investment in communication systems. In analog communication, where signal quality directly correlates with system design, compatibility with legacy systems diminishes the risks of obsolescence. Further, having robust interfaces allows for more convenient upgrade to new methods with continued benefit from old technology. Hence, analog communication design principles shall have to turn their attention to backward compatibility as a foundation of technological advancement so that newer and older systems may coexist in harmony. Electromagnetic compatibility (EMC) is an important framework for evaluating the compatibility of analog communication technology with existing systems.

Figure 2.10. *Compatibility with existing technologies in analog communication.*

Source: https://www.mutare.com/wp-content/uploads/2020/07/Mutare_Work-From-Home_Advancements-in-Communication-Technologies.png.

As outlined in available literature, for instance, the importance of the verification and validation processes, EMC is vital in ensuring that new equipment does not generate interference that compromises the functioning of existing apparatuses. A prime example is the electromagnetic environments of interference that are encountered, for example, 400 Hz power and radio frequency fields that require careful design processes within set standards. By laying down protective measures like grounding, bonding, and shielding, designers are in a position to create advanced communication systems that not only meet existing operational needs but also harmonize with prevailing technologies. It is this end-to-end strategy for EMC that ensures all elements function within permissible limits, ultimately enhancing the dependability and performance of analog communication systems, as Clarke et al. (2025) note.

In the coming years, the rapid evolution of wireless technology is threatening and offering opportunities for compatibility of analog communication systems. While it promises more features, immediate revision of the existing standards for electromagnetic compatibility is necessary to guarantee interconnectivity of the systems. Evidence suggests that combining innovative statistical approaches to evaluate electromagnetic interference provides more accurate measures of compliance than past practices provide, and this can be essential to ensuring that communications can be effectively carried out among various platforms. As Azpúrua et al. (2019) explains, solutions to these essential issues involve harmonizing the international standardization organizations in order to redefine compliance indicators. By grasping these complexities, those responsible for analog communication can better prepare for the future while safeguarding legacy systems from obsolescence, maintaining them in good working and effective order in this very electronic age (Figure 2.10).

2.5. DISADVANTAGES OF ANALOG COMMUNICATION

The inefficiency of analog communications systems is glaringly seen in the susceptibility of signals to degradation. Digital signals, which maintain their integrity in a broad spectrum of transmission environments, will never be the same as analog signals, which get distorted by noise and interference. Such weakening tends to result in decreased quality of information being transmitted, leading to loss of fidelity in the output. For instance, in wind tunnel testing, where precision is critical, minor variations in signal quality can lead to inaccurate readings. Signal integrity is therefore shown by the development of sophisticated control mechanisms such as described in Buswig et al. (2011) with the aim of stabilizing outputs in high-order systems. Therefore, the inherent susceptibility of analog signals to undermine their reliability and efficacy in modern communication systems highlights a major limitation. Further, the narrow bandwidth of analog communication is a major limitation compared to its digital counterpart. Analog systems are inherently of limited bandwidth, therefore constraining the amount of information to be transmitted simultaneously. This limitation is most evident in high-usage environments, which are prevalent to the telecommunications and data transmission environments of today. For example, in several data stream applications, the inability of analog systems to handle bandwidth makes it lead to congestion, delay, and ultimately a degraded communication experience. The scaling limitation of analog signals often produces compromises on reliability and quality, necessitating more advanced technologies. Like in actual implementations, for example, in analog-to-digital conversion systems in Corwin et al. (1977), the future in improved communication must steer away from traditional analog systems. Lastly, the maintenance and management of analog systems require some level of skill that is costly and time-consuming. Analog circuit design and the requirement for constant calibration place an added burden on organizations that are dependent on these systems. This skill can transcend fields, from electrical engineering to applied physics, creating a skilled man-power shortage that can hamper working efficiency. Where precise control and measurement are critical, such as in the development of converters for alternate energy sources, risks caused by inadequate maintenance or operating expertise are multiplied. The reliance on sophisticated technical abilities, as well as the relentless revolutions in automated digital systems, substantiates the claim that analog communication is increasingly seen as something archaic and less useful in an accelerated technological environment.

2.5.1. Susceptibility to Noise and Interference

Analog communication systems are especially susceptible to interference and noise, which can greatly mar the quality of the received signal. Noise, generally called unwanted electrical signals, may be induced by other causes like environmental factors and electronic equipment in the system. Even though analog systems attempt to transmit continuous signals, noise may taint the signals and lead to receiving and interpreting errors. This issue is particularly critical in applications such as aviation, where failure to adequately counteract electromagnetic interference (EMI) has implications for the performance of avionics. The sophisticated design considerations encompassing grounding, shielding, and noise isolation become important to ensuring the intelligibility and dependability of signals in such risky environments, thereby highlighting the importance of adopting an integral approach towards EMC considerations in analog communication systems (Clarke et al., 2025). Other than interference from the environment, systemic noise can also result from other equipment using the same frequency band. This is particularly common in urban areas, where there are many communication devices that coexist in close proximity and may inadvertently interfere with each other's signals. Analog systems, by their nature, do not have the robust error-correction mechanism that is found in digital ones and are thus more vulnerable to interference. The consequences of this vulnerability are not merely those of inefficiency; it can result in drastic breakdowns in operation and degraded efficiency in critical communications. To counter, designers and engineers are resorting more and more to noise-immune design practices and filtering techniques to build protection into systems against interference, thereby trying to achieve the highest signal fidelity and transmission reliability in a highly congested electromagnetic environment (Ichiki et al., 2007). Furthermore, the inherent randomness of noise makes it difficult to analyze and predict the performance of analog communication systems. Stochastic processes such as synaptic noise tend to corrupt the desired information channel, and hence advanced methods for understanding and modeling such effects are required. Techniques such as self-consistent signal-to-noise analysis (SCSNA) offer useful frameworks to study the impact of noise on the retrieval and processing of signals in analog systems. These analytical methods aim at quantifying the interdependency among signal integrity, levels of noise, and resulting communication effectiveness. Such methodologies assist engineers to predict more precisely and combat the impact of noise, ensuring improvements in the hardness and reliability of analog communication systems compared to internal and external noise (Ichiki et al., 2007) (Figure 2.11).

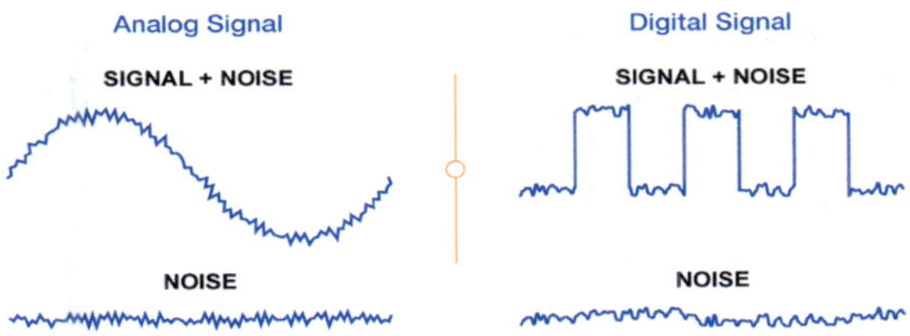

Figure 2.11. *Susceptibility to noise and interference.*

Source: https://external-content.duckduckgo.com/iu/?u=https%3A%2F%2Fwww.predig.com%2Fsites%2Fdefault%2Ffiles%2Fimages%2FIndicator%2F0216%2Fsig_noise_figure_1.jpg&f=1&nofb=1&ipt=52398d14371e2830636086ad43957ea7890487479df046185a927973bae11f23.

2.5.2. Limited Bandwidth

Limited bandwidth is a fundamental issue in analog communication that influences the efficacy and effectiveness of signal transmission. Bandwidth is the range of frequencies with which a communication channel can effectively transmit and influences the data rate as well as the quality of transmitted signals significantly. In environments where bandwidth is limited, like in dense electromagnetic environments, distortion and interference of signals take place, resulting in compromised perceived quality of communication. It is important to comprehend the implications of limited bandwidth in order to come up with strategies for maximizing signal transmission. For instance, multi-perspective sampling techniques can address the challenges provided by bandwidth constraints, which are central to the development of reliable and effective communication systems (Eldar et al., 2011). Another significant field impacted by limited bandwidth is the need for advanced coding techniques to extract as much information as possible from the given frequency range. Without encoding, critical information will be lost because of insufficient bandwidth. This can be exacerbated in analog-to-digital conversion processes, for which Shannon-Nyquist theorem dictates the minimum sampling rate necessary for consistent preservation of signal integrity. Breakthroughs such as sub-Nyquist sampling methods have been investigated by scientists to mitigate the factor of bandwidth constraint, where significant information is not lost (Eldar et al., 2011). It is a sign of the trend towards sophisticated methods that couple theoretical with practical applications in analog communication systems for utilizing restricted bandwidths more effectively. Furthermore, use of multiple description coding schemes has turned out to be one of the promising solutions for coping with restricted bandwidth in analog communication. By utilizing multiple descriptions, data can be encoded on multiple channels, and transmission is made more reliable against potential losses due to bandwidth constraints. This method gives the flexibility of receiving signals by receivers in order to receive different quality descriptions depending on the bandwidth available. Techniques optimized through non-convex optimization, such as deterministic annealing, have shown potential in achieving the best encoding and decoding mappings under this paradigm and outperforming traditional techniques. As research continues on the topic, coding technique development will likely be an important determinant to unlock the bottlenecks introduced by bandwidth constraints, paving the way for future analog communications innovation.

2.5.3. Degradation Over Long Distances

In analog communication, degradation over long distances is a significant impediment to signal quality and integrity. When analog signals are transmitted over large distances, they undergo a variety of impairments that include attenuation, distortion, and noise, all of which degrade the quality of information being conveyed. One of the significant issues in such degradation is the natural attenuation of electromagnetic waves by factors such as atmospheric absorption and scattering. This leads to signal attenuation that wearies the signals, weakening them and possibly failing to deliver the initial information at the receiving end. To combat such issues is crucial for guaranteeing proper means of communication, particularly in cases where distance affects the reliability and clarity of the signal received. Furthermore, the impact of distance on analog communication is reinforced by environmental factors outside of the system that introduce additional noise and interference. For instance atmospheric conditions such as rain, fog, and humidity have significant effects on the transmission of signals, causing additional distortion and compromising information integrity. In addition, as signals propagate through space, they are susceptible to electromagnetic interference from various sources such as industrial machinery and other transmission lines. Knowledge of these challenges is crucial for engineers and scientists aiming to create reliable communication systems. Systemic measures must be adopted to examine and counter the effects of degradation over long distances to maximize signal quality and minimize the occurrence of errors during transmission. Advances in recent technology, including the invention of reconfigurable intelligent surfaces (RIS), provide new solutions to resist degradation in analog communication systems. RIS can significantly enhance wireless communication by regulating electromagnetic wave propagation to create perfect transmission conditions. By adjusting the surface properties to facilitate improved channel realization, RIS allows the reconfiguration of unfavorable wireless environments into manageable systems, thus overcoming certain classical limits of distance-dependent degradation. The ability to code the surfaces heralds a new frontier in signal fidelity control over long distances, as highlighted in recent researches on the potential and uses of RIS technologies. These innovations are paramount not only to take analog communication forward but also to reshape wireless networks' futures as they evolve to respond to increasing demands for service quality against distance challenges (ElMossallamy et al., 2020).

2.6. APPLICATIONS OF ANALOG COMMUNICATION

The applications of analog communication are widespread and are still an important one in numerous areas. The most well-known application is the field of broadcasting. Radio and television broadcasts widely employ analog signals in conveying information highly efficiently. With this method, audio and video information is conveyed in real-time, connecting broadcasters and audiences instantly. Even though there has been a shift to digital broadcasting, most areas still rely on analog systems because they are easy and inexpensive. Furthermore, in less developed areas, analog communication systems are still crucial for emergency broadcasts because they offer vital information during crises. The continuity of the utility of analog technology in such places proves its indispensable role in the communication arena, reaffirming the need for standards that guarantee proper signal transmission and quality (Azpúrua et al., 2019). Analog communication, apart from broadcast, has significant uses across various industrial processes, particularly in machine automation and control. There are many industrial operations that use analog signals to monitor and control machinery, for instance, temperature, pressure, and flow within the production environment. Analog signals provide instantaneous feedback that allows operators to make key decisions to maximize efficiency and safety. Analog systems are particularly useful in terms of real-time response, allowing for on-the-spot adjustments without delay often attributed to digital processing. The use of analog communication technology in industries not only enhances the reliability of operation but also makes system design improvement easier, while simultaneously continuing to be critical to legacy systems that are very effective (Adesuyi et al., 2014). Healthcare is another field where analog communication becomes extremely important, particularly in medical instrumentation. The majority of diagnostic equipment, such as electrocardiograms and blood pressure monitors, utilize analog signals to give accurate physiological measurements. The equipment convert biophysical signals into readable analog waveforms for healthcare practitioners. The immediate data from these machines is essential for timely decision-making in patient care, making analog communication an essential aspect in the clinical setup. Additionally, the simplicity of analog systems in medical devices ensures they are simple to use and navigate through, particularly in resource-limited environments. With the evolution of technology, there should be an equilibrium between analog and digital means of communication to ensure successful healthcare delivery while conforming to the certain needs of both systems (Azpúrua et al., 2019).

2.6.1. Radio Broadcasting

The radio broadcasting domain is a fundamental aspect of analog communication that offers a means to transmit diverse information across large distances without the need for physical links. The intrinsic nature of radio waves allows for the transmission of sound signals, effectively reducing geographic distances and facilitating a globalized world. This process predominantly utilizes amplitude modulation (AM) and frequency modulation (FM) techniques to stamp sound waves on carrier waves. Radio broadcasting, in the past, has not only been a source of entertainment and news gathering but also an emergency communication lifeline for catastrophe. While media coverage is continued to be redefined by technological advancements, radio broadcasting remains a strong relevance, adapting into new forms and merging with digital media so as to continue presence in public discussion. One of the most important issues confronting radio broadcasting within analog communication is the issue of electromagnetic compatibility (EMC). The establishment of effective EMC standards is important to ensure that radio broadcasts do not conflict with one another and this is ever more imperative as the demand for wireless communication increases. New research emphasizes the necessity of a revision of existing practices of compliance on EMC towards the use of new statistical methods that have a better capability of addressing the complexities of electromagnetic interference in broadcast networks. The requirement for such a shift is due to the requirement to adjust test methodologies and standards of compliance to further enhance the resilience of analog broadcasting against increasing congestion of the radio frequency spectrum. The resolution of these problems by facing them squarely will ensure the future of radio broadcasting in sustaining quality and dependability in analog communications with evolving technology (Azpúrua et al., 2019). Apart from technical challenges, the radio broadcast environment is also affected by policy and practice factors in the shape of public service broadcasting paradigms. Comparative studies indicate that strategies employed in the U.S. to support PSB would benefit the European counterparts aiming to handle contemporary broadcasting challenges. This becomes particularly vital considering the fact that both continents hope to balance public interest and business sustainability. The American model is concerned with the necessity to reform public broadcasting to suit the needs of today's audiences without compromising commitments to core public values such as accessibility and programming diversity. Through study of these practices, stakeholders can learn to enhance radio broadcasting's relevance and efficacy and ultimately to a better-informed citizenry and more robust democratic discourse (Huntsberger et al., 2010) (Figure 2.12).

Figure 2.12. *Radio broadcasting.*

Source: https://achievedstrategies.com/wp-content/uploads/2019/12/Radio-Broadcasting3.jpg.

2.6.2. Television Transmission

The legacy of television broadcasting has significantly shaped the landscape of analog communication, providing a medium of delivering entertainment and information in equal measure. Initially, television signals were transmitted over analog waves using amplitude modulation (AM) for sound and frequency modulation (FM) for images. With this two-way modulation, rich audio-visual content could be effectively broadcast, creating a combined viewing experience. The analog transmission model, though pioneering in its day was plagued by degradation in signals and bandwidth limitations. With growing requirements for quality and additional channels, the limitations of traditional analog systems were realized, ultimately leading to the development of digital communication technologies that could offer more clarity and efficiency in content delivery (Abbott et al., 1985). Despite such advancements, analog television transmission remains a significant area of study within the broader framework of communication technologies. At its core, analog transmission relies upon continuous modulation of signal qualities that directly relate to the physical waveforms of audio and video content.

Figure 2.13. *Television transmission.*

Source: https://static1.howtogeekimages.com/wordpress/wp-content/uploads/2023/09/oldstatickytv1.jpg.

The method might be less effective than digital methods, but it still maintains significant implications concerning the historical development of media communication. Moreover, principal characteristics of analog transmission are discussed in terms of source parameters of digitized analog data. By reviewing the parameters against different data compression techniques, researchers can account for how analog television has adapted to rising demands for quality and speed without sacrificing its inherent value in communication (Ingels et al., 2025). The intrinsic role of bandwidth in television transmission cannot be overstated, as it is a limiter and an enabler of the type of content that can be broadcast efficiently. Analog television systems normally operate on designated frequencies, which set the number of channels dedicated to broadcasting. Such limiting often results in compromises in content quality since different channels share the same bandwidth. Apart from that, bandwidth compression techniques are imperative in enabling the possibility of broadcasting an even larger number of video signals simultaneously. Understanding such techniques illuminates one on matters pertaining to analog-to-digital switchovers' dynamics, reflecting on a conceptualism of applicability in the television transmission methodology in the context of analog communication (Ingels et al., 2025) (Figure 2.13).

2.6.3. Telephony

Telephony, the foundation of analog communication, transformed communication between individuals and societies. The telephone's discovery initiated fresh paradigms of personal and business communication, allowing for instantaneous voice delivery via existing networks over long distances. Analog signals commanded this technology, with audio waves converted into electrical currents, transmitted via wires to be converted back into audible speech upon arrival. The simplicity of analog telephony made it readily accessible to all, and it facilitated the boom in global communication. In addition, with the advent of technology, telephony evolved into more advanced modes like trunk lines and switchboards, which enhanced the efficiency of communication and facilitated the evolution of the modern telecommunication system in use today. Ultimately, the impact of early telephony remains significant due to the fact that it created the precedent for all subsequent innovation in communication technologies. The impact of telephony is not merely in the communications mechanics but also has serious socio-economic implications. Telephony has the ability to induce economic growth and enhance people's livelihood in developing and rural countries. Mobile telephony, for example, has had mass coverage in developing countries where fixed landline infrastructure may not have been economically feasible. Since the arrival of mobile technology, communities are able to reach services, enhance opportunities in trade, and achieve easier communication over long distances. This has been profoundly explored using the Gompertz model of diffusion of technology, which portrays in a clear manner the intricacies of how emerging technologies infiltrate markets, especially in areas with differing economic abilities. Telephony dependence as a tool to social change underscores its role in the evolving texture of society, with the need for sustained investment in communication technologies (Rouvinen et al., 2025). The movement from analog to digital communication has garnered considerable discussion in the context of the information society. Although analog telephony was robust, digital technologies seek to correct the defects of the earlier systems. For example, digital services offer higher data transmission rates, clearer reception, and more reliability compared to their analog counterparts. But the gains of digitalization are heterogeneously distributed, particularly between developed and less-developed countries. As has been emphasized in ICT infrastructure debate, investment in network infrastructures, regulatory frameworks, and public readiness is central for any nation that aims to leverage information technologies for the good of society. Thus, while analog telephony established basic communication processes in place, the future trend as far as telephony and its contribution toward an information society is yet to be dependent on all-encompassing infrastructure development and access to technological advancements for everyone (Kordha et al., 2025) (Figure 2.14).

Figure 2.14. *Telephony, as a cornerstone of analog communication the telephone*

Source: https://thumbs.dreamstime.com/b/vintage-phones-smartphones-first-telephone-old-cellphone-modern-smartphone-set-call-communication-invention-retro-mobile-273166814.jpg.

SUMMARY

In summary, exploration of analog communication reveals its ongoing applicability in a rapidly digitizing age. While digital communication is everywhere, analog methods are still a crucial part of many applications, particularly broadcasting and audio. The natural properties of analog signals, such as smooth waveforms, enable a more delicate conveying of information that others believe digital systems cannot accomplish. Notably, the analog medium also generally is more resistant to certain types of distortions, thereby contributing further to its enduring relevance. As theorists strive to adopt a multidimensional approach to understanding information processing and computation, it is essential to appreciate the unique contribution of analog communication for a comprehensive understanding of the history of media channels and transmission modes (Piccinini et al., 2009). In addition, the implications of analog communication extend past technological systems into cultural and social interactions. The personal and communal connections through analog mediums, such as radio and vinyl records, give rise to a sense of shared experience that digital communication is sometimes lacking. This cultural relevance is particularly poignant in talks of nostalgia and the revival of old technology, to show how analogue communication still shapes identity formation and solidarity in communities. As we gaze in the direction of the innovation that has been experienced in technology and the need for digital technology, it is noteworthy to reflect on what values are embedded in analogue activity. Comprehension of past functions of analog systems inspires greater responsiveness to their impact on contemporary communication patterns and evokes a more vociferous celebration of media development (Sattouf et al., 2015). Finally, the green future of communication must consider the essential insights gained from analog systems. At a time in society when everyone is more eager to minimize digital footprints and adopt green ways, the adaptability of analog communication is very important. While digging into how analog devices can coexist with and support digital devices, new paths for innovation might be brought about that would lead one towards a hybrid way of communication where both are valued. Thus, the conclusion drawn from analog communication studies identifies not only its historical value but also its potential in informing future communication approaches. Embracing the advantages of analog communication can drive creative solutions to address current challenges, confirming the importance of maintaining a wide range of tools in a continuously changing media landscape (Piccinini et al., 2009).

The most important aspect of analog communication is the explanation of traditional modes of information transmission with the assistance of continuous signals. This involves knowing the modes of modulation, which have a vital role in putting information within carrier waves. Analog systems primarily rely on amplitude modulation (AM) and frequency modulation (FM), which have varied advantages depending on the requirements of the medium for transmission. The quality and purity of the obtained signals can be significantly decided by channel states and interference, which are inherent characteristics of the environment of communication. The rich background of evolution in the field is brought out in applications today, as analog communication, despite the invention of digital systems, remains with its applications in broadcasting, audio transmission, and other fields where continuous signal representation is useful. Such a background also allows for an exhaustive analysis of the technological advancements that continue to redefine the face of analog communications. In the backdrop of digital communication breakthroughs, it is necessary to highlight the role of tried and trusted analog approaches in current practice. The never-ending flow of ideas between analog and digital technologies necessitates a symbiotic relationship wherein one learns from the other. More significantly, sampling theory and signal processing subtleties' studies emphasize how underlying analog principles still manifest in modern digital schemes. As a point of illustration, the Shannon-Nyquist theorem itself remains a cornerstone in the study of information transfer capacity through sampling, tying theoretical underpinnings to practical realities. Moreover, the trend towards incorporating such theoretical insights into hardware design indicates how analog communication ideas remain applicable today. Lastly, being able to identify these intersections helps to present a clearer picture of both analog and digital modalities within current communication systems. As the topic proceeds to dig deeper into electromagnetic compatibility (EMC) and wireless communication, the need for new standards becomes mandatory. Current approaches, which focus primarily on verifying compliance with specified quasi-peak levels, can perhaps not fully convey the complexities introduced by current wireless environments. New statistical techniques, as have been developed in recent literature, foresee better means for estimating electromagnetic interferences effects on communication networks. A re-examination of EMC testing and conformity models to reflect increasing complexity in wireless applications appears necessary. Emphasis on such innovations not only adds to the present discourse in analog communication but also the essential requirement for flexibility in a steady performance amidst rapidly evolving technology. Resolving these issues makes the road to more robust and efficient communication methods in an ever more globalized world available.

REFERENCES

1. Abbott, G. L. (1985). Video-based information systems in academic library media centers. https://core.ac.uk/download/4816744.pdf (accessed on 12th September 2025).
2. Acosta, R. J., Terry, J. D., & Zaman, A. J. (2025). System overview on electromagnetic compensation for reflector antenna surface distortion. https://core.ac.uk/download/pdf/42806114.pdf (accessed on 12th September 2025).
3. Adesuyi, R. S. (2014). The future of mobile industry. https://core.ac.uk/download/19447812.pdf (accessed on 12th September 2025).
4. Akyol, E., Mehmetoglu, M. S., & Rose, K. (2015). Analog multiple descriptions: A zero-delay source-channel coding approach. http://arxiv.org/abs/1510.00450 (accessed on 12th September 2025).
5. Alland, S., Stark, W. E., Ali, M., & Hegde, M. V. (2019). Interference in automotive radar systems: Characteristics, mitigation techniques, and current and future research. IEEE Signal Processing Magazine, 36, 45–59. https://doi.org/10.1109/msp.2019.2908214 (accessed on 12th September 2025).
6. Amari, A., Ciblat, P., Dobre, O. A., et al. (2017). A survey on fiber nonlinearity compensation for 400 Gbps and beyond optical communication systems. http://arxiv.org/abs/1708.06313 (accessed on 12th September 2025).
7. Asmar, S. W., Dyal, P., Lau, E. L., Fimmel, R. O., et al. (2005). The study of the Pioneer anomaly: New data and objectives for new investigation. http://arxiv.org/abs/gr-qc/0512121 (accessed on 12th September 2025).
8. Azpúrua Auyanet, M. A., Pous Solà, M., & Silva Martínez, F. (2019). Improved electromagnetic compatibility standards for the interconnected wireless world. https://core.ac.uk/download/288625020.pdf (accessed on 12th September 2025).
9. Barnes, F., Byerly, R., Codding, G., et al. (2025). NASA and the challenge of ISDN: The role of satellites in an ISDN world. https://core.ac.uk/download/pdf/42832623.pdf (accessed on 12th September 2025).
10. Bertran Albertí, E., Montoro López, G., & Tarrés Ruiz, F. (1998). An experimental course on digital communications. https://core.ac.uk/download/81582988.pdf (accessed on 12th September 2025).
11. Bunin, S. G. (2014). The self-organizing radio networks and ultra-wide band signals. https://core.ac.uk/download/47231512.pdf (accessed on 12th September 2025).

12. Buswig, Y., & Mohammed, Y. (2011). Development of a DC-AC power conditioner for wind generator by using neural network. https://core.ac.uk/download/12007862.pdf (accessed on 12th September 2025).

13. Caloz, C., Gupta, S., & Zou, L. (2017). Real-time dispersion code multiple access (DCMA) for high-speed wireless communications. http://arxiv.org/abs/1703.10516 (accessed on 12th September 2025).

14. Chen, Z., Fang, J., Li, S., et al. (2018). Physical layer service integration in 5G: Potentials and challenges. http://arxiv.org/abs/1707.01064 (accessed on 12th September 2025).

15. Clarke, C. A., & Larsen, W. E. (2025). Aircraft electromagnetic compatibility. https://core.ac.uk/download/pdf/42836482.pdf (accessed on 12th September 2025).

16. Corwin, R. R., Creel, T. R., Jones, R. A., et al. (1977). Apparatus for determining thermophysical properties of test specimens. https://core.ac.uk/download/pdf/10327258.pdf (accessed on 12th September 2025).

17. Delgado Restituto, M., Fernández Bootello, J. F., & Rodríguez Vázquez, Á. B. (2008). Matrix methods for the dynamic range optimization of continuous-time Gm-C filters. https://core.ac.uk/download/28656477.pdf (accessed on 12th September 2025).

18. Delgado Restituto, M., Rodríguez Vázquez, Á. B., de la Rosa Utrera, J. M., Ruiz Amaya, et al. (2005). Simulation-based high-level synthesis of Nyquist-rate data converters using MATLAB/SIMULINK. https://core.ac.uk/download/286565261.pdf (accessed on 12th September 2025).

19. Dutta, S., Ford, R., Mezzavilla, M., et al. (2016). Frame structure design and analysis for millimeter wave cellular systems. http://arxiv.org/abs/1512.05691 (accessed on 12th September 2025).

20. Eldar, Y. C., & Mishali, M. (2011). Sub-Nyquist sampling: Bridging theory and practice. http://arxiv.org/abs/1106.4514 (accessed on 12th September 2025).

21. Eldar, Y. C., & Mishra, K. V. (2017). Sub-Nyquist channel estimation over IEEE 802.11ad link. http://arxiv.org/abs/1705.03943 (accessed on 12th September 2025).

22. ElMossallamy, M. A., Han, Z., Li, G. Y., et al. (2020). Reconfigurable intelligent surfaces for wireless communications: Principles, challenges and opportunities. http://arxiv.org/abs/2005.00938 (accessed on 12th September 2025).

23. Gotthans, T., Hrubos, Z., Kadlec, P., et al. (2012). Communication subsystems for emerging wireless technologies. https://core.ac.uk/download/30311907.pdf (accessed on 12th September 2025).

24. Guerrero, A. F., Miranda, J., & Partal, P. (2000). Implementation of digital modulator using digital multiplier for wireless applications in Verilog and Cadence. https://core.ac.uk/download/pdf/82008140.pdf (accessed on 12th September 2025).

25. Harris, F. (1997). Digital signal processing in radio receivers and transmitters. https://core.ac.uk/download/357380486.pdf (accessed on 12th September 2025).

26. Hilton, G. E., Hollenbaugh, R. C., Laughlin, C. R., et al. (1969). Position location and data collection system and method. https://core.ac.uk/download/pdf/10289782.pdf (accessed on 12th September 2025).

27. Hoeksema, F. W., Schiphorst, R., & Slump, C. H. (2001). The front end of software-defined radio: Possibilities and challenges. https://core.ac.uk/download/pdf/11457590.pdf (accessed on 12th September 2025).

28. Huntsberger, M., & Stavitsky, A. G. (2010). With the support of listeners like you: Lessons from U.S. public radio. https://core.ac.uk/download/143958380.pdf (accessed on 12th September 2025).

29. Ichiki, A., Amari, S., Anderson, C. R., Benzi, R., Choi, M. Y., Frank, T. D., Gardiner, C. W., et al. (2007). Thouless-Anderson-Palmer equation for analog neural network with temporally fluctuating white synaptic noise. http://arxiv.org/abs/0706.2048 (accessed on 12th September 2025).

30. Ingels, F. M., & Shahsavari, M. M. (2025). A study of data coding technology developments in the 1980–1985-time frame, volume 2. https://core.ac.uk/download/pdf/42871670.pdf (accessed on 12th September 2025).

31. Johnston, S. F. (2006). Absorbing new subjects: Holography as an analog of photography. https://core.ac.uk/download/223236201.pdf (accessed on 12th September 2025).

32. Koc, O., Koç, O., Kunt, E. D., et al. (2011). A field programmable gate array based modular motion control platform. https://core.ac.uk/download/pdf/11742747.pdf (accessed on 12th September 2025).

33. Kordha, E., Gorica, K., & Ahmetaj, L. (2025). Managing IT infrastructure for information society development: The Albanian case. https://core.ac.uk/download/pdf/6386857.pdf (accessed on 12th September 2025).

34. Kučera, P. (2009). Polarization imperfections of light in interferometry. https://core.ac.uk/download/30310924.pdf (accessed on 12th September 2025).

35. Manohar, R., & Moradi, S. (2018). The impact of on-chip communication on memory technologies for neuromorphic systems. http://arxiv.org/abs/1809.06016 (accessed on 12th September 2025).

36. Messer, D. D., & Park, S. (2025). A simple implementation of the Viterbi algorithm on the Motorola DSP56001. https://core.ac.uk/download/pdf/42812745.pdf (accessed on 12th September 2025).

37. N/A. (2002). The American Assembly: Art, technology, and intellectual property. https://core.ac.uk/download/71349265.pdf (accessed on 12th September 2025).

38. Nauta, B. (2007). Analog IC design at the University of Twente. https://core.ac.uk/download/pdf/11468379.pdf (accessed on 12th September 2025).

39. Nauta, B., & Vaucher, C. S. (2002). Architectures for RF frequency synthesizers. https://core.ac.uk/download/pdf/11471090.pdf (accessed on 12th September 2025).

40. Peng, Y., Shangguan, L., Hu, Y., Qian, Y., Lin, X., Chen, X., Fang D., et al. (2018). PLoRa. https://doi.org/10.1145/3230543.3230567 (accessed on 12th September 2025).

41. Piccinini, G., & Scarantino, A. (2009). Computation vs. information processing: Why their difference matters to cognitive science. https://core.ac.uk/download/pdf/87347.pdf (accessed on 12th September 2025).

42. Popok, A., Trushakov, D., Попок, А. А., et al. (2014). Застосування цифрової автоматичної телефонної станції SI2000 для побудови міських та сільських коммунікаційних мереж зв'язку.

43. Rouvinen, P. (2025). Diffusion of digital mobile telephony: Are developing countries different? https://core.ac.uk/download/pdf/648388.pdf (accessed on 12th September 2025).

44. Sattouf, M. (2015). Data acquisition and control system of hydroelectric power plant using internet techniques. https://core.ac.uk/download/30311149.pdf (accessed on 12th September 2025).

45. Torjesen, A. A. (2016). An ultra-fast digital diffuse optical spectroscopic imaging system for neoadjuvant chemotherapy monitoring. https://open.bu.edu/bitstream/2144/19470/1/Torjesen_bu_0017N_12323.pdf (accessed on 12th September 2025).

46. Wallace, D., & Wallace, R. (2001). Predicting health impacts of the World Trade Center disaster: 1. Halogenated hydrocarbons, symptom syndromes, secondary victimization, and the burdens of history. https://core.ac.uk/download/pdf/84906.pdf (accessed on 12th September 2025).

47. Zhao, Z. (2024). Memristor based gain-varying PI control for erbium-doped fiber amplifiers (EDFAs). https://core.ac.uk/download/644632735.pdf (accessed on 12th September 2025).

CHAPTER 3

RANDOM PROCESSES AND NOISE COMMUNICATION SYSTEMS

Contents

INTRODUCTION

Random process theory constitutes a key role in understanding communication systems, particularly signal integrity and noise-induced interference issues. Random signals' occurrence constitutes a main source of complexity in the majority of engineering applications, which has fueled efforts towards the formulation of robust signal-processing methods. The character of communication systems is built on the ability to distinguish useful signals from other kinds of noise that could destroy intended information. Effective signal processing techniques, as proposed in adjacent literature, enable engineers to enhance quality and performance in communications systems, video and audio. In their focus on stochastic processes, introductory forms not only describe the interaction of these signals in frequency and time domain but also provide a platform for comprehending the strategies used to achieve optimal system performance in interference (Etten et al., 2006). Analysis of noise in communication systems extends beyond the mere categorization, including the dynamic response of systems to random interference.

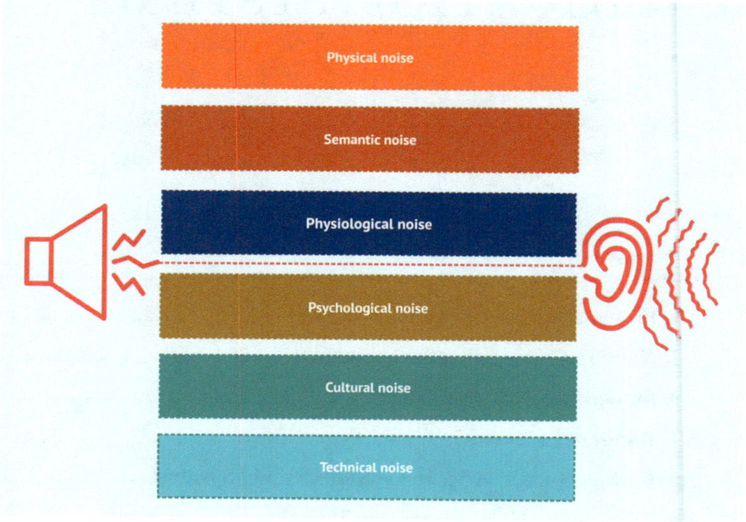

Figure 3.1. *Random processes and noise communication systems.*

Source: https://fourweekmba.com/wp-content/uploads/2022/10/noise-in-communication.png.

The same interference, or more commonly referred to as noise, has the potential to cause devastating effects on performance expectations, particularly in distributed-memory systems where there are intricate delay factors involved. Information regarding the interaction of noise and communication mechanisms is crucial as it bears testimony to the quality of signal transmission in actual operating environments. Synthetic benchmarks research highlights the complexity of modeling performance metrics against the occurrence of random noise. From the analysis of propagation delays and their correlations to communication attributes, researchers will be able to determine ways of minimizing the influence of noise to maintain system dependability and effectiveness (Afzal et al., 2019). Here, a preliminary examination of random processes and noise in communications systems is employed to establish a theoretical foundation for further investigation. Familiarity with the basic principles of stochastic processes allows identification of techniques applied to optimize signal detection and filtering and maximize overall system performance. Signal processing techniques and noise characteristics are of vitally essential importance in designing communications systems with the capability of performing well under uncertain conditions. As interactions of this sort become more and more sophisticated, a firm grasp of fundamental principles becomes important to engineers and scientists in seeking alternative solutions to the problems of the day for signal integrity and noise control (Afzal et al., 2019; Etten et al., 2006) (Figure 3.1).

Definition of Random Processes

The random process theory plays a crucial role in explaining the behavior of systems under uncertainties, particularly in communication system systems. A stochastic process or a random process is defined as a collection of random variables indexed based on time or space. All samples of a random process can have varying results, defining the randomness inbuilt concerning different phenomena. Random processes are used in the area of noise communication systems to model the noise experienced in signal transmission, which may affect performance metrics such as signal-to-noise ratio and capacity. Stochastic characterization becomes more significant with this application, particularly in the examination of non-stationary systems where the behavior of the random variables could change over time, complicating analysis but requiring it (Yüksel et al., 2012). In noise communication systems, channel characteristics and random process interaction are major. Specifically, in performance estimation of such systems, stochastic stability is the notion which is defined as the characteristic that ensures that some of the system's parameters will be bounded in time even when there is noise present. This concept finds extensive use in areas such as control theory, where system stability against random perturbations is of utmost importance. For example, conditions for guaranteeing stochastic stabilizability of linear systems in the presence of Gaussian noise are obtained from the understanding of the underlying random processes. These factors lead to methods of ensuring asymptotic mean stationarity so that engineers can design more robust communication systems with a capability to accommodate the inevitable fluctuations encountered in practical usage (Agrawal et al., 2015). Furthermore, relay network advancements best capture the significance of random processes for enhancing communication rates within noisy conditions. Since such networks have a tendency to rely on multiple nodes for carrying signals, the use of relay methods taking correlated noise into account becomes a necessity. Random processes allow one to take into account the uncertainty at each node and how these compound as signals pass along various routes. It is indicated by research that setting lower bounds on capacities in such systems is aided by the knowledge of the underlying random processes and leads to better coding strategies that achieve maximum throughput. These innovations demonstrate how an explicit grasp of the definition and implications of random processes can lead to creative solutions particular to the challenges posed by noisy communication channels, thus propelling improvements in wireless and multi-hop communication systems (Yüksel et al., 2012; Agrawal et al., 2015).

Importance of Noise in Communication Systems

The role of noise in communication systems is not just interference; it affects the performance and reliability of these systems. Noise introduces uncertainty, tampering with the integrity of the signal and making data transmission produce errors. Yet through proper use, this randomness can actually aid in enhancing methods of signal processing. For instance, the entropy of binary symmetric Hidden Markov Process illustrates how different parameters of noise can be mapped into successful models for real-world applications to facilitate efficient communication under uncertainty in noise. From the analysis of noise in mathematical modeling terms, engineers can design useful solutions that minimize the adverse effect of noise so as to produce superior overall system performance without compromising data integrity. These analytical techniques provide a gateway to more sophisticated communication protocols that make use of noise rather than just combating it, highlighting the intelligence and importance of studying noise in communication systems (Schliep et al., 2005). Additionally, the importance of noise encompasses its impact on systems' control processes. In a vast majority of applications, like the operation of nuclear reactors, noise has along with it challenges that must be addressed to provide stability for operation. Power corruption by network-induced random delays and sensor noise demonstrates how noise complicates feedback mechanisms. With the application of sophisticated control techniques such as fuzzy proportional integral derivative (PID) controllers, system dynamics can be controlled in response to noise. These controllers use genetic algorithms to optimize performance, which is a positive approach to handling noise. Therefore, noise control enhances robustness in systems such that communication systems ought to be able to adapt and react dynamically to changes in operating conditions at high-efficiency levels (Das et al., 2013). Thus, the intricacy of noise in communication systems relies on its inherent randomness as well as its analytical power. The research into noise leads to tremendous advances in telecommunications and control systems, proving its dual character of a problem and an opportunity. By learning about noise and its influences thoroughly, scientists and engineers are able to utilize the properties of noise to design communication systems stronger against the interfering impact of many noise agents. The progress in the future will likely continue to further enhance the techniques of coping with and utilizing noise, so that communication systems become capable of handling the challenges posed by continuously increasing complex environments. Therefore, the study of noise is a fundamental area of study within the overall context of random processes and communication systems and, in the end, defines the direction of future technology and information exchange.

3.1. FUNDAMENTALS OF RANDOM PROCESSES

Random processes theory forms the foundation of comprehension of noise in communication systems. Random processes, defined as sets of random variables indexed by time or space, enable engineers to model and analyze all kinds of signals that necessarily encounter random fluctuations. These are critical for real-world applications since they constitute the noise that can interfere with the effectiveness of signal transfer. For instance, analog, discrete-time, and bandpass signals mimic the complexity of engineering scenarios whereby isolation of meaningful signals from accompanying unwanted interference is essential. In line with literature, signal processing innovations give engineers an avenue to suppress noise, thereby maximizing system functionality and reliability in diverse environments, thereby making the concepts of random processes essential in effective communication system design (Etten et al., 2006). It is important to understand the relationship between random processes and noise, particularly for heterogeneous cellular networks, a dominant model for modern telecommunications. These networks incorporate multiple layers of heterogeneous power, data rate, and density BSs. Stochasticity in user connections coupled with being guided by the strongest available BS demands an extensive study of signal-to-interference-plus-noise ratio (SINR) to provide coverage. The phenomenon of Rayleigh fading also complicates this scenario, as environmental conditions heavily influence signal reception. As explained in recent studies, achieving probability expressions for coverage allows understanding of network performance under varying conditions and, as a result, better understanding of the behavior of random processes in complicated systems (Harpreet et al., 2011). In addition, communication system optimality hinges on the application of effective filtering methods in accordance with random process principles. Matched filtering and correlation receivers play a critical role in ensuring signal integrity is at its best, particularly in an environment replete with interference. These methods benefit from mathematical underpinnings in stochastic processes to facilitate detection and recovery of desired signals amidst noise. As engineers innovate and deal with the complexities of modern communication networks, a correct grasp of random processes becomes a crucial ability. Through the integration of theoretical underpinnings and applications, one can design robust systems that can survive the stresses generated by noise and random variations, ultimately leading to the development of better communication networks (Etten et al., 2006) (Figure 3.2).

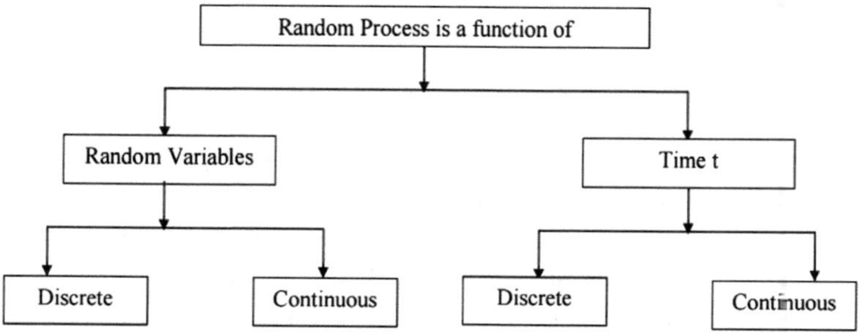

Figure 3.2. *Random processes is foundational.*

Source: https://img.brainkart.com/imagebk7/8AeglyW.jpg.

3.1.1. Characteristics of Random Processes

Random processes are central to dynamic modeling in communication systems with uncertainty and variability. At the center of these processes is the concept of randomness, which is characterized by unreliability in the timing and occurrence of events. In communication systems, unreliability can be caused by an assortment of factors such as external environmental factors, systems overload, and inherent delays. These random moves can generate fluctuations that create interference with the expected function of a communication system, calling for the analysis and modeling of the characteristics of these random processes in an efficient way. Analyses aid in the formation of communication systems that can combat the undesired impact of noise, therefore delivering reliable data transfer under varied conditions. Random processes have one notable characteristic they will have periods of idleness, which are inevitable in distributed message systems. Idle times arise when a process anticipates data from another and tend to cause deterioration in performance over time. It is useful to learn how these periods of idleness propagate in debugging and message-passing program optimization. With techniques that measure such idle periods, programmers can detect patterns where delays have a bearing on inter-process communication. The distribution of idle periods will resemble waves, giving rise to the work of designing mechanisms to curb the impact of such delays on system performance. This information is particularly valuable for applications where precise timing and synchronization are needed, with even minimal amounts of disturbances having a significant cumulative effect on overall performance (Gioiosa et al., 2017). Furthermore, the inter-relation of idle time with other kinds of noise disturbance is a challenging task in the modeling of performance. Random disturbance, also referred to as noise, has the ability to greatly disrupt the predictability of operations in distributed-memory applications. Inspecting these interruptions reveals their impact on propagation delay and how they can increase or decrease the impact of natural idle times. Through systematic research with artificial benchmarks simulating real applications, it is apparent that the type of these random processes can utilize a system's functioning effectiveness in its favor or against it. Besides, the interfacing of noise with delay propagation underscores the need for optimizing communication methods to preemptively counter probable performance bottlenecks (Afzal et al., 2019). These results form part of our wider understanding regarding how random processes affect communication system functionality and resilience.

3.1.2. Types of Random Processes

In considering the types of random processes, it becomes extremely crucial to differentiate between discrete and continuous-time processes. Discrete random processes operate on a collection of countable results, with events occurring at well-determined intervals. Such a characteristic is particularly significant in digital communication systems, where signal samples are measured at well-determined time intervals. Conversely, random processes in continuous time can accommodate an uncountable number of results and, as such, can precipitate more complex modeling scenarios involving different mathematical tools. This difference not only specifies the communication system theory but also teaches the practice applications, for example, how to design the noise filtering systems that must accept the specific characteristics of the random processes employed. One of the most popular kinds of random process is the stationary process, where statistical properties do not vary with time. Stationarity facilitates analysis, allowing very powerful statistical techniques to be applied on the basis of these constant features. For instance, with noise communication systems involving a stationary process, such as white Gaussian noise, engineers are able to apply known performance metrics without having to keep adjusting for variations. As technology continues to evolve, for instance, in neural networks, understanding stationary processes can assist in modeling spiking and rate-based communication dynamics among neurons. In practice, such understanding is a starting point that enables the enhancement of the application of communication strategies to maintain efficiency and reliability in the face of randomness, as evidenced in frameworks established in recent research (Bolten et al., 2016). Another significant distinction between random processes is between ergodic and non-ergodic processes. Ergodic processes possess the property that ensemble averages are equal to time averages and they are extremely useful in analyzing over long durations of time. This property is useful in communication systems because it allows engineers to predict performance outcomes using finite observation data. On the other hand, non-ergodic processes are challenging because their statistical properties tend to change with time, requiring a more dynamic system design paradigm. In wireless networks with impulsive noise, for example, engineers must devise robust algorithms that adapt under changing circumstances rather than anticipate consistent behavior. Thus, the insight gained through understanding the nuances of ergodic and non-ergodic processes is the basis for planning effective handling of communication in unstructured contexts, as illustrated in channel parameter estimation studies in impulsive noise scenarios (Meng et al., 2013) (Figure 3.3).

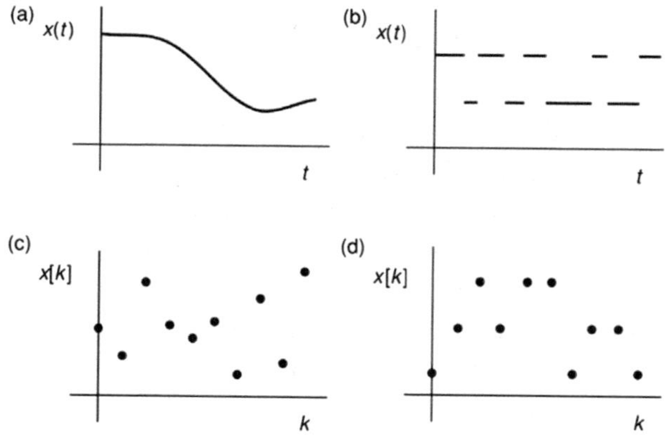

Figure 3.3. *Types of random processes.*

Source: https://www.oreilly.com/api/v2/epubs/9781118393956/files/OEBPS/images/nc06f006.jpg.

3.1.3. Mathematical Representation of Random Processes

Mathematical formulation of random processes is the basis in understanding their behavior in various systems, particularly in communications and signal processing. Random processes are defined as collections of random variables parameterized with time as the index, which is the foundation in the investigation of signals with noise. The complexity of both the noises and signals requires a robust framework for description, which is largely achieved through the use of stochastic processes. Algorithms such as Gaussian processes develop a mathematical basis for filtering random signals and applying filtering techniques. Engineers utilize mathematical models in order to filter out desired signals from interference so that more effective communication systems can be developed. This mathematical basis enhances our ability to interpret and further the stability of systems in opposition to perturbing noise, ensuring more secure information in communication systems, as highlighted by Etten et al. (2006). Mathematical representation of signal and noise properties also allows for the development of a variety of analytical methods. For instance, the Horizontal Visibility Graph (HVG) approach supplies insight into the underlying structures of time series data, distinguishing deterministic and stochastic components effectively. This approach presents a new insight into the interpretation of random processes, emphasizing the necessity of node degree distributions to characterize both chaotic and random dynamics. By doing this, the significance of proper mathematical modeling is achieved, not only because it helps in the classification of such processes, but also because it enhances the understanding of their behavior in real systems. This injection of sophisticated mathematical theories with practical signals showcases a key conceptual link between theory and application, critical for engineers working in noise communication systems, as found in Carpi et al. (2014). Finally, effective filtering methods play an important role in minimizing the impact of unwanted signals on communication systems. Mathematical modeling allows the use of such methods to predict and manage noise interference so that the quality and reliability of transmitted information is enhanced. Methods like matched filtering capitalize on the statistical behavior of random processes for maximizing signal detection under conditions of noise. These methods stress the requirement for mathematical foundation so that engineers would be able to implement their strategies based on the characteristics of the noise and wanted signals. Thus, the interaction between applied signal processing and mathematics points to the requirement of cautious mathematical abstraction in developing efficient communication systems that can efficiently function regardless of random noise, as discussed in Etten et al. (2006).

3.2. NOISE IN COMMUNICATION SYSTEMS

Knowledge of communication noise is significant for the best design and operation of distributed systems. Communication noise could be due to either environmental or system structure reasons and leads to wild and random disturbances in data transmission. They affect the predictability that conventional analytic models usually rely on and make it more difficult to model performance in distributed-memory programs. Where such inactivity periods exist—like for instance when waiting on data—such noise components travel and can significantly impact system-wide performance. These inactivity periods introduce latencies that, aside from its influence on a process can propagate through the entire communication infrastructure, demonstrating why system designers must be aware of how noise could impact their applications and, by proxy, the users' perception in a networking world.

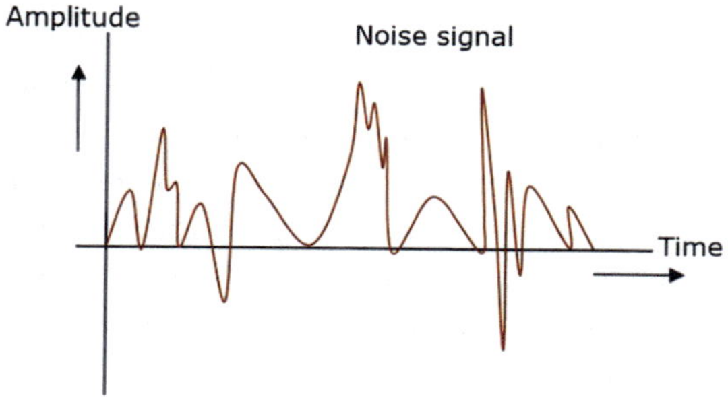

Figure 3.4. *Noise in communication systems.*

Source: https://www.tutorialspoint.com/principles_of_communication/images/noise.jpg.

The propagation of idle time, or noise, is a prime example of issues that occur in communication systems. An experiment to study the effect of idle waves revealed that such events can be transmitted from one process to another as waves in a system. By employing a means of tracking idle periods in Message Passing Interface (MPI) codes, researchers were capable of illustrating that idle time was actually synchronizable for multiple processes, especially where performance variability is present. It is through such findings that application programmers learn how to understand the dynamics of transmission of idle periods as being imperative to compensating for performance degradation. Techniques of detection and resolution of such problems can actually enhance system efficiency and reliability in communication networks as well as make an impact on future evolution. Most importantly, noise in communication systems actually significantly affects significant dimensions of performance measures, particularly with enhanced delays with building-influences. These random interrupts build complexity induced by distributed-memory applications, so it is important to examine their effect on performance extensively. Studies aimed at synthetic benchmarks have highlighted mechanisms of noise influence on delay propagation. For instance, outcomes from research of such systems indicate how various amounts of noise influence the speed and rate of decay of idle times. Through the discrimination between causes such as application execution characteristics and communication properties, understanding may be gained on potential mitigation. As exemplified in recent studies, fine-scale noise intensities can enhance a system's immunity to idle wave propagation, paving the way towards robust communication paradigms in upcoming applications (Figure 3.4).

3.2.1. Definition and Types of Noise

Noise is a natural phenomenon in communication systems, typically defined as any type of unwanted interference that disrupts the reception and transmission of signals. Interference comes in various forms, each with unique features that influence the reliability of communication. Noise can typically be categorized as thermal, shot, and flicker noise, each due to a different physical process. Thermal noise, for instance, is due to the random motion of the charge carriers in conductors and is a function of temperature with a Gaussian distribution. This is larger at higher temperatures, hence the increased noise at high-temperature levels. Understanding these types is necessary in order to develop systems that are able to handle or mitigate their effects, especially in situations that demand high fidelity, i.e., telecommunications and data transmission. Aside from thermal noise, impulsive noise is another significant class, where there are bursty pulses of short duration and high amplitude interference. It is particularly prevalent in environments like wireless sensor networks, where the bursts are due to electrical interference or signal distortion. In engineering usage, for instance, initialization of channel iterative decoders like low-density parity-check codes, channel parameter knowledge and estimation are paramount in order to counteract impulsive noise. The state-of-the-art has come up with estimators that can be utilized under the stringent environment, with promising development being documented for several modulations like MPSK and MQAM. Characterization of impulsive noise and the use of strong countermeasures are the primary advancements towards enhancing the reliability of communication systems, as evidenced in works that explore the empirical characteristic function to estimate parameters precisely (Meng et al., 2013). Further, the theoretical framework including the types of noise is extended to implications for the simulation and modeling of communication networks. Various modeling approaches, including biologically inspired spiking neuron models and rate-based units, provide a comprehensive description of how several forms of noise influence communication dynamics. Significantly, spiking network modeling conveys the notion of how instantaneous and delayed interactions are engaged in modulating noise. With the progress of integration of several models, it gives way to rich comprehension of rate dynamics and corresponding noise found within communication systems. Researchers suggest adaptive frameworks for investigating interaction between noise in these models so that simulation accuracy for a communication network can be enhanced along with operating reliability (Bolten et al., 2016). By examining various instances ranging from neural field models to random networks, the place of noise continues to unwrap, necessitating an overall strategy in the fight against its adverse impact (Figure 3.5).

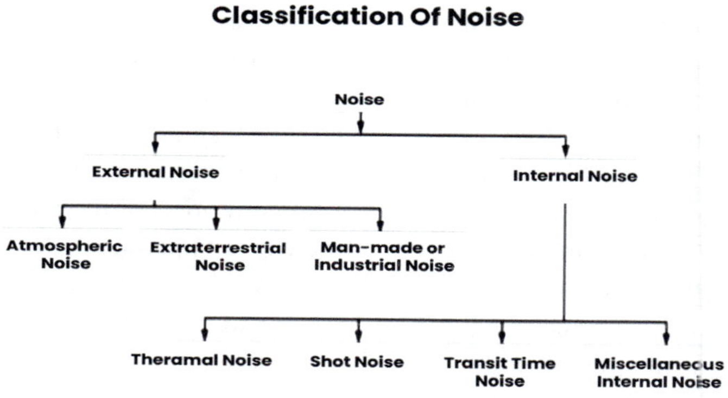

Figure 3.5. *Types of noise.*

Source: https://symbl.ai/wp-content/uploads/2021/01/understanding-speech-to-noise-ratio-3.jpg.

3.2.2. Sources of Noise in Communication Systems

In communication systems, it is necessary to understand the causes of noise to provide signals with a higher integrity and system performance. Noise may be caused by various sources like thermal fluctuations, electromagnetic interference, and quantization error. Thermal noise, for instance, is induced by random electron flow through electronic devices and infects all electronic equipment. This type of noise is temperature-sensitive and continues to increase with rising temperatures, becoming a challenge for systems operating in hot conditions. Additionally, electromagnetic interference can also be generated externally, through radiofrequency signals or neighboring electrical apparatus, which can potentially disrupt the communication process. The net impact of these noise sources can lead to signal degradation and distortion of received signals, necessitating robust error correction techniques to ensure reliable communication. In quantization noise in digital communication, noise is introduced when digitizing, further contributing to inaccuracies that can complicate it even more to transmit the signal. When digital signals are converted from analog, they are quantized and rounded off to discrete values.

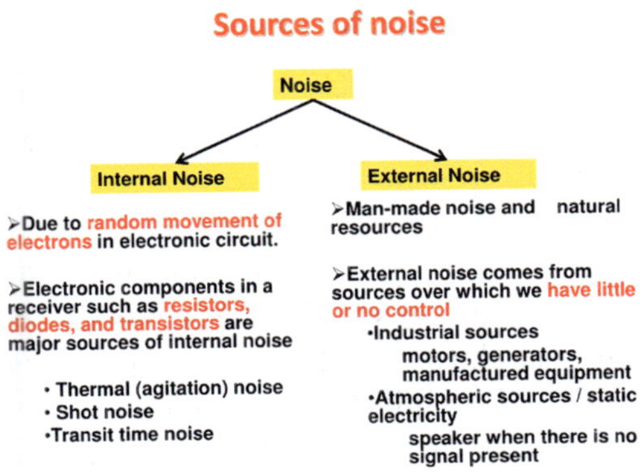

Figure 3.6. *Sources of noise in communication systems.*

Source: https://image2.slideserve.com/4706848/sources-of-noise-l.jpg.

During this process, some information necessarily is lost, giving rise to quantization noise. Secondly, randomness in quantized outputs can cause an adverse impact on the system performance overall. In an attempt to offset the detrimental effects of quantization noise, engineers usually resort to techniques such as dithering, through which a specified amount of noise is added to the signal to improve the accuracy of the digitized signal. Because noise in all its manifestations always devastates communication systems, combating the sources thereof becomes necessary for achieving ultimate operating efficiency and signal purity. Lastly, the impacts of sources of noise are not only on direct conduction of information but also on system design and stability. For instance, the use of stochastic processes in modeling and controlling noise can lead to better communication techniques. It has been noted by some research that under what conditions it is feasible to make communication systems stable in the case of unbounded noise and stochastic stabilizability is a critical element to maintain signal quality. The challenge lies in creating systems that can counteract the effect of noise through feedback loops and advanced error correction techniques. Moreover, an understanding of phenomena such as idle period propagation in parallel computing systems illustrates how noise propagates through communication channels to impair system performance in general (Gioiosa et al., 2017). Lastly, the general knowledge of noise sources helps in the direction of developing strong communication systems that are capable of sustaining high-performance operation in noisy environments (Yüksel et al., 2012) (Figure 3.6).

3.2.3. Impact of Noise on Signal Integrity

Noise impact on signal integrity remains a critical concern in communications systems, particularly as data rates and throughput requirements continue to increase. Noise will induce many forms of interference, such as jitter and crosstalk, that inherently compromise signal integrity. Jitter, which is represented by inconsistency in the timing of signals, can potentially severely degrade error rates and synchronization procedures in high-rate communications. Particularly, as argued in Buckwalter et al. (2006), data-dependent jitter can limit bit-error rate (BER) performance and complicate the operation of critical data recovery systems. In order to develop strategies for alleviating these adverse effects, it is crucial to understand how noise impacts signal integrity. Also, different signaling techniques will influence the function of noise in impacting signal integrity, particularly with future-generation integrated circuit designs. For instance, three-dimensional (3D) integrated circuits' through silicon via (TSV) interconnects illustrate that signaling mode selection determines performance outcomes. Voltage-mode (VM) signaling could struggle with high data rates due to fundamental noise issues, whereas current-mode (CM) signaling should be less prone to noise, as stated by Grange et al. (2010). Through the utilization of field-solver extracted parasitics in simulations, scientists are able to identify how various signaling techniques affect key performance characteristics. This debate emphasizes the need for selecting appropriate signaling methods to counter noise and ensure optimal signal integrity in modern communication systems. To counter the issues induced by noise, equalization techniques and circuit topologies play a significant role in ensuring signal integrity. As highlighted in research documented in Buckwalter et al. (2006), deterministic jitter equalizer usage can lead to significant improvements in clock recovery, effectively reducing jitter and expanding timing margins. The optimization of such circuits in terms of jitter control enables communication systems to give better BER performance, which is essential for high-speed operations. Ongoing innovation in equalization methods testifies to the interactive convergence of noise control and circuit design, demanding continuous innovation in the area.

Through constant research and development activities, the impact of noise on signal integrity can be mitigated, enabling communications systems to meet ever-growing requirements for information transfer.

3.3. STATISTICAL ANALYSIS OF NOISE

Statistical analysis of noise is one of the essential components in understanding the efficiency of random processes in communication systems. Noise as an unstructured interference can drastically affect signal transmission integrity by causing increased error rates and degradation of data fidelity. Researchers have tried to measure the impact of these effects of noise through theories that explore delay propagation and the impact on distributed systems. By the examination of traces collected from benchmarks, it is apparent that noise will desecrate regularity assumptions that are customarily called on in analytic models. The challenge is to capture the way disruptions take place and impact system performance, particularly in scenarios of long delays that can ring through a network. This complex interplay demands a rigorous statistical environment to inform mitigation strategies as well as maximize communication reliability in noisy situations, thus making it critical for experimental and theoretical explorations of this field (Afzal et al., 2019). Statistical analysis of the behavior of vehicles is another enlightening perspective on measuring noise. With the help of several public datasets, researchers have used Information Theory to study how noise affects the speed of cars under different traffic conditions. Bandt-Pompe method along with the Complexity-Entropy plane serves as a major tool to identify various traffic regimes from which it has been observed that car speeds experience different levels of correlation and randomness under the influence of environmental noise. For instance, during congestion, vehicle speeds are equivalent to a random velocity profile, which is described by a power spectrum that has minimal correlation. In free-flowing traffic, though, a more structured, correlated velocity pattern emerges. These findings confirm the relevance of applying statistical frameworks not only to describe observations of noise but also to inform traffic management policy, demonstrating the broader utility of statistical noise analysis for classical communication systems (Almeida et al., 2015). In random processes, statistical analysis of noise yields profound understanding of system parameter interaction. Noise can create non-linear behaviors to cause complex dynamics to baffle conventional communications system models. By the defining of noise as a normal aspect of system functioning, scientists can introduce controls to minimize its effects. Understanding the degradation of delays under different levels of noise, and the collision of delays in a noisy environment, is central to improving system robustness. Moreover, the methodologies developed to cope with noise can bring significant gains to parallel application efficiency within distributed memory systems. Such research not only facilitates theoretical insight but also has a practical significance, making a window of opportunity for innovation in communication technologies that are immune to random interference (Afzal et al., 2019) (Figure 3.7).

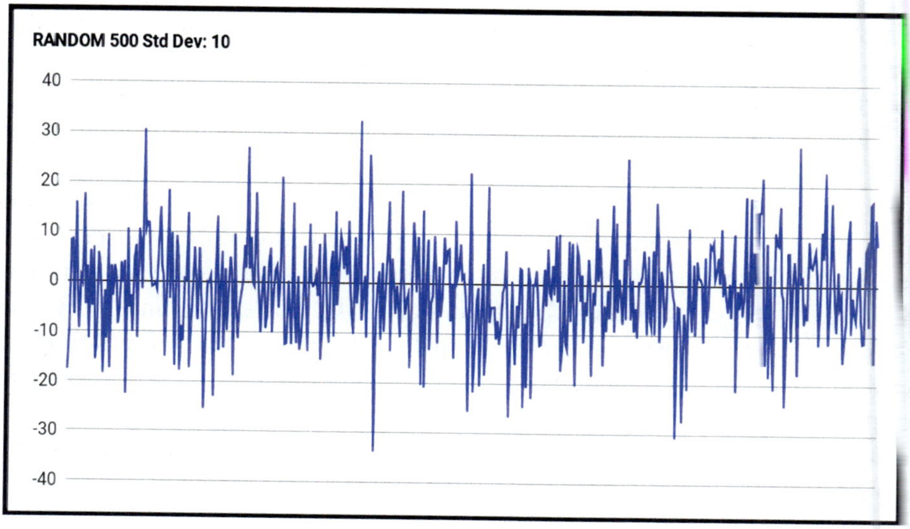

Figure 3.7. *Statistical analysis of noise.*

Source: https://www.allaboutcircuits.com/uploads/articles/statisticalNatureofNoisewithCalculations_fig01.png.

3.3.1. Probability Distributions in Noise Analysis

The analysis of noise in communication systems demands a good understanding of probability distributions, as these provide the fundamental framework for dealing with and measuring randomness. Especially significant, in this context, is the role of Gaussian noise in the testing of the performance of various channels of communication. In the majority of applications, noise may be approximated as a Gaussian process due to the central limit theorem that dictates that the sum of numerous independent random variables has a tendency of conforming to a normal distribution. This aspect becomes crucial in ascertaining signal integrity and system reliability under practical environments so that engineers can utilize probability distributions in predicting the probability of error rates and loss of capacity. Understanding of these distributions is central to the construction of robust communication systems that can function irrespective of the amount of interference that is present. In exploring the capacity of noisy communication systems, one cannot rule out the significance of phase noise and its probabilistic effects. Phase noise, most often caused by oscillator instability, presents unique challenges to signal processing and transmission. Research indicates that different configurations, such as Single-Input Multiple-Output (SIMO) and Multiple-Output Single-Input (MISO), have different characteristics when it comes to high-Signal-to-Noise Ratio (SNR) capacities in such noise. Research indicates that while high-SNR capacity enhancement is largely scenario-independent, the second-order phase-noise term is necessarily scenario-dependent. Specifically, greater diversity gains are attained in certain configurations, which impact overall system performance (Durisi et al., 2015). Hence, precise compensation of probabilistic features of phase noises is very important in design and optimization of modern communication systems. Additionally, noise analysis in communication systems often becomes entangled with concepts of stochastic stability, particularly for non-stationary linear systems. Stabilization is an issue of concern as noise may affect system operation seriously, jeopardizing signal fidelity. As research has shown, establishing necessary and sufficient conditions for stochastic stabilizability is critical, particularly in channels influenced by Gaussian noise. These conditions assert that for effective stabilization, the channel capacity must exceed certain thresholds derived from the systems dynamics. The range of these stochastic stability conditions is also extended to performance metrics such as asymptotic mean stationarity and sample path ergodicity upon which noisy system long-term dependability is based (Yüksel et al., 2012). Operating within the realm of probability distributions in the context of this argument not only further establishes the theoretical foundation of noise dynamics but also enforces practical avenues for engineering resilient communication systems.

3.3.2. Mean and Variance of Noise Signals

In random process and noise communication system research, one needs to determine the mean and variance of noise signals to effectively process and analyze signals. Mean, or expected value, is a quantity used in statistics to comprehend the average level of noise present in a communication system. It plays a crucial role in establishing the baseline performance of various communication methods. When measuring mean levels of noise, it is necessary to consider factors of interference and distortion that have significant effects on the integrity of a signal. By computing the mean, engineers can establish thresholds of optimal system performance, ensuring signals' strength against the adverse effects of noise in a given setting. Variance quantifies the level of deviation or dispersion of noise signals from the mean. Increased variance implies more inconsistency, which might result in turbulent interferences in signal clarity. This theory is directly applicable to systems based on predictability and stability, e.g., cognitive radio networks. Engineers are able to design detection algorithms to minimize high-variance noise by analyzing variance, thereby improving system performance. For example, the generalized detector (GD) has been shown to improve spectrum sensing effectiveness by effectively canceling variances with noise signals and better recovering information from input data streams than traditional detectors such as the energy detector (ED) and generalized likelihood ratio test (GLRT) detector (Shbat et al., 2015). Also, variance and mean have a significant role to play in filtering mechanisms used for noise control. New techniques like finite impulse response (FIR) and autoregressive moving average (ARMA) graph filters are efficient in dealing with randomness of noise signals in communication networks. These filters operate on random time-varying graph signals, and the great advantage they have is that they optimize the output based on statistical characteristics of the noise. Importantly, it is realized that even though such filters behave similarly to their deterministic counterparts, knowledge of statistical characteristics of noise can yield efficient techniques for cancellation and processing complexity reduction (Isufi et al., 2017). By taking advantage of both the mean and variance, communication systems can be built not only to survive but thrive in conditions of uncertainty and variation (Figure 3.8).

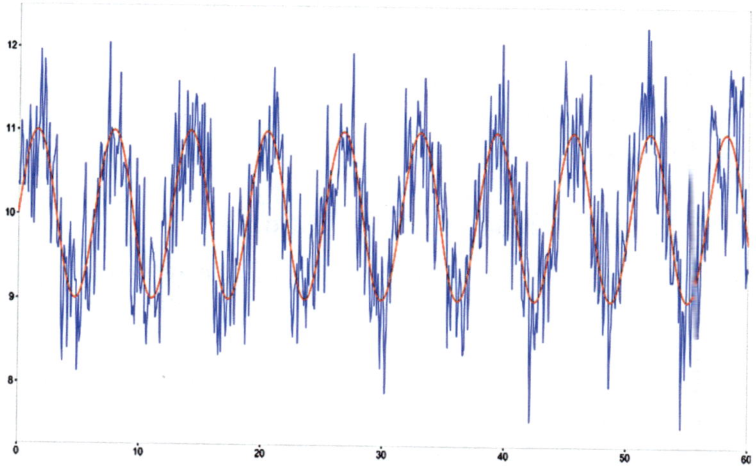

Figure 3.8. *Mean and variance of noise signals is crucial for effective signal processing.*

Source: https://assets-global.website-files.com/5babb9f91ab233ff5f53ce10/5c4a_2 39563a357d8c2b8455_Signal-processing.jpeg.

3.3.3. Autocorrelation and Power Spectral Density

The concepts of autocorrelation and power spectral density are central to the understanding of random processes in communication systems. Autocorrelation illuminates the persistence of a signal over time, showing how a signal autorelates with itself over different instances of time. It is a measure that is central to the study of signal characteristics, particularly where there is noise. The structure of the autocorrelation function can also detect periodicities and other more subtle patterns within a random process, thus enabling forecasting of future signal behavior. For communication systems having noise, understanding these correlations helps the assessment of the signal-to-noise ratio, which will measure the efficiency and dependability of data transmission. The power spectral density complements the analysis provided by autocorrelation through the provision of a frequency-domain representation of the signal. PSD is the method by which power of a signal is dispersed among various frequency components so that dominant frequencies and the nature of noise in a system can be identified. With the use of the Wiener-Khinchin theorem, it is clear that the relationship between autocorrelation and PSD is such that the Fourier transform of the autocorrelation function provides the power spectral density directly. This association is an important one for communication engineers in that it enables the quantification of interference effects and noise on propagating signals and, consequently, the design of filtered systems that can optimize signal reception as well as efficiency of signal processing in real-world applications. The interlinking of autocorrelation and PSD analysis is of even higher significance in problem-solving involving network interference and environmental interference in wireless communications. For instance, in multipath fading conditions, the time and frequency properties of the signal can be severely affected, resulting in a degradation of communication quality. The framework established in Moe et al. (2007) emphasizes the importance of these statistical examinations by detailing how network interference affects the capacity in the link. Through the utilization of autocorrelation and PSD measurements, engineers can gain insight into the collective noise performance that eventually informs the creation of stable communication systems that can endure the unpredictable nature of signal transmission within complex environments. Through these processes, the insight derived also informs the establishment of spectral controls that are critical in minimizing interference within dense communications networks, exemplifying their essential role in modern communications.

3.4. RANDOM PROCESSES IN SIGNAL PROCESSING

Random process analysis is the foundation of studying how signals are affected by noise and other stochastic variables. Signal processing is essentially the analysis and manipulation of signals that are deterministic and random in nature. Random processes with their probabilistic nature are significant to the modeling of a variety of real-world phenomena, particularly in communication systems where noise pervades everywhere. Understanding such processes enables engineers to design systems with the ability to differentiate between desired signals and interference effectively, which is at the heart of modern communication technology. Using techniques adopted from stochastic analysis, engineers are able to use matched filtering, description of signal space, and correlation receivers to enhance the purity of messages transmitted, demonstrating the applicability of these theoretical principles in a solution to the noise-induced problems in practice (Etten et al., 2006). Detection of digital signals occupies a key position in designing communication systems to optimum performance. With technology coming of age, the application of digital techniques ensures greater precision in isolating the signal of interest from interfering background noise. Various algorithms, some of which are founded upon optimal filtering, resolve the problems occasioned by random processes. As an example the characteristics of shot noise and Poisson processes gain more significance in analyzing the capability and limitations of various detection mechanisms. By realizing these aspects, engineers are able to design improved robust systems that perform well in reaction to the inherent variability in communication channels. The introduction of these advanced concepts highlights the significant relationship between practice and theory in developing noise-immune systems (Etten et al., 2006). Cognitive radio systems represent a milestone in the application of random processes in signal processing. Based on generalized detectors, they enhance spectrum sensing capability when compared to current detection methods. This enhancement enables better utilization of the radio frequency spectrum, which is becoming increasingly congested with the growth in wireless devices. Sophisticated techniques consider receive antenna array element correlation and noise differences, enabling better signal evaluation and decision-making. Simulation analysis has revealed the enhanced performance of generalized detectors over other detection architectures, illustrating the benefit that the integration of sophisticated spectro-temporal analysis provides to cognitive radio engineering. Implementation and formulation of such strategies highlight ongoing developments in the technology of signal processing and how random processes facilitate the overcoming of issues posed by noise in communications.

3.4.1. Role of Random Processes in Modulation Techniques

In modern communication systems, modulation techniques play a central role in transmitting information over noisy channels. Stochastic processes underlie determining these modulation techniques in terms of performance and reliability. Utilizing randomness, engineers can devise modulation schemes that resist various types of noise, including Gaussian noise and interchannel interference. For instance, with the help of stochastic model concept, communication signal properties can be optimized for the purpose of minimizing ill effects caused by random noise. This aligns with the ideas of random processes, wherein random uncertainty in the environment is utilized in order to build fault-tolerant modulation techniques that enhance signal fidelity and overall system efficiency.

Figure 3.9. *Role of random processes in modulation techniques.*

Source: https://www.bitsathy.ac.in/wp-content/uploads/1-1-1.png.

One fascinating application of random processes to modulation techniques is in optical wireless communications (OWC). The unique difficulties of OWC, such as light control over various parameters, call for innovative modulation structures based on the theory of random processes. Specifically, randomness has the capability to manage signal-dependent optical channel properties that complicate modulation and demodulation operations. As shown in recent developments, the use of deep learning (DL) technologies offers a potential solution to addressing these complexities using algorithms that are capable of recognizing patterns within the noise and shaping the transmission based on it. Such developments harness the fundamental role of random processes in designing adaptive modulation techniques that optimize the performance of OWC systems, greatly contributing to the advancement of wireless communications (Lee et al., 2018). And yet another significant area of randomness within modulation techniques is that of countering interchannel nonlinear interference (NLI), and more so within high-speed fiber optics. The randomness of the noise is bringing about uncertainty with respect to channel capacity and spectral efficiency, which results in severely degraded transmission quality. Recent research on mitigation strategies of NLI has pointed out the benefits of employing phase and polarization noise compensation combined with subcarrier multiplexing. Importantly, the interaction between these techniques allows for improved performance even when it is designed to operate with higher-order modulation formats. By examining the coherence properties of NLI and by taking advantage of advanced computational models, researchers can analyze the impacts of various factors such as amplifier spacing and dispersion compensation on overall system performance. Hence, the coupling of random processes in these different modulation techniques turns out to be instrumental in augmenting the efficiency and resilience of contemporary communication systems (Agrell et al., 2018) (Figure 3.9).

3.4.2. Filtering Techniques for Noise Reduction

Effective filtering techniques are of prime importance in addressing noise reduction in communications systems, where signal integrity during transmission is of the utmost importance. Various methods, such as linear and non-linear filters, try to eliminate the impact of noise on the received signal. Linear filters, such as the moving average filter or Wiener filter, operate by minimizing noise without altering the signal fidelity. These filters achieve this by averaging a predetermined number of input samples, thereby smoothing out random noise-induced fluctuations. On the other hand, non-linear filters such as median or adaptive filters perform well in impulsive noise environments, serving as a strong alternative when conventional linear techniques fail. With the selection of the appropriate filtering technique that aligns with the specific noise characteristics of the signal, engineers are able to optimize the clarity and reliability of communication, thus the system performance. Statistical models are often employed by the design and application of filtering techniques to model the noise in the communication system. Advanced methods have subsequently emerged, utilizing feedback controls and statistical inference to provide optimized performance in filters. For example, quantum control principles-based techniques illustrate the potential of employing rigorous modeling for noise reduction in continuous monitoring environments. The blending of statistical investigation with real-time noise management enables a responsive and nimble filtering process necessary to maintain data integrity in dynamic environments. The ability to adaptively re-tune filter parameters based on real-time estimates of the levels of noise is a key step forward in the technology, emphasizing the need for sophisticated systems with the capability to address the challenges of noise in multiple channels of communication (Mabuchi et al., 2005). Most importantly, actual implementation of the filtering of noise has revolutionized data processing within digital communication systems. Most notably, optical diagnostics integration of noise data has encouraged the development of advanced filtering subroutines that enhance digital data sensing purity. Such advancements affect not only enhanced data communication but also better data resistance to corruption by noise interference. The continued development of these filtering techniques is significant in many applications, from communications to digital storage, in order to maintain the information readable and reliable. With the change in noise patterns due to improving technology, continued research and development in the area are significant, demonstrating the ongoing relationship between random processes and the requirement for effective filtering strategies (Palgen, 2025).

3.4.3. Applications of Random Processes in Digital Communication

Random processes are essential in digital communication systems to improve the reliability and efficiency of data transmission. These processes, specifically, help model various types of noise and interference that can contaminate communication channels. In order to combat these problems, modern systems often use stochastic models to predict and resist the effects of noise. For instance, the best transmission policies can be achieved through techniques such as dynamic programming, which help to minimize the long-run average cost of both the transmission energy and the mean estimation errors. Such methods demonstrate the significant role played by random processes in ensuring reliable communication, particularly in randomly varying environments and uncertain parameters, as evident in research on smart sensor technologies (Dey et al., 2014). Moreover, the advancements in digital signal processing technologies have enabled more effective applications of random processes within communication systems. The uses of high-performance hardware, i.e., Field Programmable Gate Arrays (FPGAs), facilitate more capability in handling large amounts of data while dealing with the intrinsic challenges introduced by noise. As digital communication systems such as radio telescopes increase in size and complexity, the need for sophisticated correlator architectures becomes apparent. Such architectures exploit random process theories to optimize phase and correlation of signals in a large number of antennas. Utilizing Ethernet switches and open-source libraries, designers can create adaptable systems that support the computational necessities for astronomical procedures of high-sensitivity imaging and prove the practical applications of random processes in digital communication systems (Parsons et al., 2009). Second, random processes enable the generation of new communication protocols required for digital transmission. Randomization techniques can be used to harden protocols against interference and degradation of signals, a common vulnerability in today's networks. For example, by modeling packet drop events and fluctuating feedback systems, researchers can create systems that control transmission strategies adaptively based on network conditions in real time. These policies not only optimize performance but also provide for efficient use of resources in communication networks. The interface between random processes, noise, and adaptive communications policy is a salient illustration of a high step toward obtaining more reliable digital communication frameworks, and the importance of these concepts in modern technology is accentuated.

3.5. PERFORMANCE METRICS IN COMMUNICATION SYSTEMS

Performance metrics in communication networks are important to quantify the efficacy and reliability of information transmission through various channels particularly under conditions of interference and noise. The metrics provide quantitative measures that help engineers and researchers assess how effectively a communication network operates under diverse conditions. For example, important performance metrics are bit error rate (BER), signal-to-noise ratio (SNR), and capacity. These measurements' relationship gets more advanced when considering random phenomena due to network interference and environmental sources. The understanding of these measurements is key to the process of optimizing system design and ensuring reliable communication, especially for wireless communications where the signal propagation is influenced by many fading effects and noise characteristics (Moe et al., 2007). The capacity of a communications link, which is also partially dependent on both interference and noise, is a simple performance measure. Recently, the notion of spectral outage probability (SOP) has been developed as a novel definition of cumulative radio-frequency emissions in wireless systems. Analyzing the spatial distribution of interferers, modeled as a Poisson process, engineers can gain insights into the links capacity for different noise and interference environments. This observation is in favor of developing spectral policies and enhancing network planning, especially in covert military use where quality communication must be maintained even in hostile environments. Capacity measurements therefore enable system development that can adapt to varying situations while optimizing the efficiency of communication and minimizing error rates (Moe et al., 2007).

In addition to conventional performance measures, emerging technologies such as the Internet of Things (IoT) necessitate innovative paradigms for analyzing the performance of communication systems. Precisely, new WP-BackCom paradigm provides for the simultaneous power transmission and device-to-device communication, with which numerous low-power IoT devices can effectively communicate. Its performance metrics are stochastic-geometry based, considering network coverage probability and transmission capacity as a function of various parameters like density of power beacons and backscatter characteristics. Under the condition of high density of devices, adapting these parameters to ensure consistent connections irrespective of interference and noise is critical. Therefore, compatibility of performance metrics with next-generation technologies is crucial for future communication systems (Figure 3.10).

Figure 3.10. *Performance metrics in communication systems.*

Source: https://fastercapital.co/i/Performance-metrics--Beyond-Words--The-Role-of-Performance-Metrics-in-Contract-Negotiation--Using-Performance-Metrics-for-Effective-Communication.webp.

3.5.1. Signal-to-Noise Ratio (SNR)

Theory of SNR plays a key role in measuring the reliability and performance of a communication system, particularly those with random process and noise. SNR quantifies the ratio of power of a signal to ambient noise power, which is an important measure for determining the fidelity and clarity of information to be transmitted. Increased SNR translates to a purer signal, which is required to successfully mitigate the adverse impact of noise during transmission. In communication system operating environments, such as in cognitive radio networks, and recognizing and optimizing SNR can greatly improve system performance. Employing advanced detection techniques, like the generalized detector (GD), could considerably enhance SNR by optimizing spectrum sensing performance and error rates under noisy environments, according to the findings presented in the research on cognitive radio systems in Shbat et al. (2015). Antenna system design is also a significant factor determining SNR in communication systems, especially phase noise being taken into account. In cases such as Single-Input Multiple-Output (SIMO) and Multiple-Output Single-Input (MISO) systems, the impact of phase noise may vary significantly based on whether independent local oscillators (SLO) or a common local oscillator (CLO) is applied. Research reveals that the initial SNR capacity remains the same in various antenna configurations; however, the phase-noise number—a second-order term essential for diversity gains characterization—exhibits scenario-dependent variations.

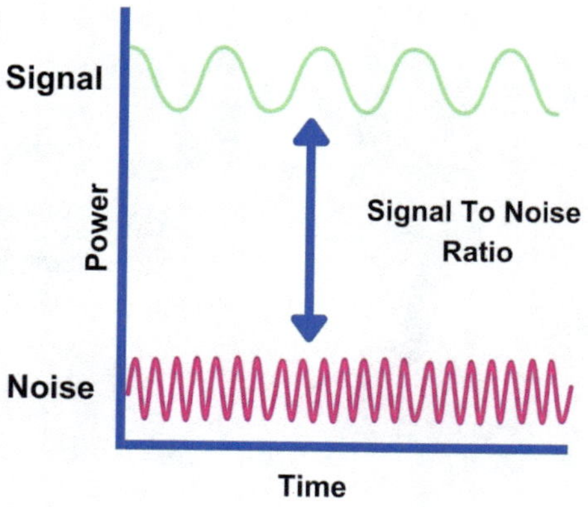

Figure 3.11. *Signal-to-noise ratio (SNR).*

Source: https://routenote.com/blog/wp-content/uploads/2022/08/Signal-to-noise-ratio-RouteNote-Blog-1.jpg.

This subtle observation on SNR dynamics emphasizes the need for adjusting antenna configuration and signal processing techniques so as to optimize communication fidelity, particularly at high-noise conditions, such as in the phase-noise channel analyses in Durisi et al. (2015). Essentially, the study of SNR in random processes and communication systems based on noise is part of the evolution of communication technology that grows more robust. With wireless communication becoming more omnipresent, the ability to distinguish signal from noise in fluctuating and hostile environments is of key importance. Technologies that enhance detectors, such as the GD, are employed to highlight the ongoing progress in developing methods for enhancing SNR. Consequently, greater insight into the determinants of SNR, from environmental considerations through system design aspects to noise dynamics, will drive ongoing research. With more sophisticated systems, a complete comprehension of such factors will be required not only to optimize performance but to push the boundaries of what can be done with communication systems operating in the presence of random processes and noise (Figure 3.11).

3.5.2. Bit Error Rate (BER)

BER is a key performance metric in communication networks, which quantifies the proportion of received bits that have errors out of the transmitted bits. BER in random processes and noise is a function of several parameters such as transmission media, interference, and encodings. The processes that affect changing the BER must be known if one needs to optimize communication systems as the demand for high-rate data transfer is always increasing. Because these systems are meant to operate in stochastic environments with random noise, the estimation of BER is put at the fore in making dependable models that reflect real-world performance. Proper estimation of BER allows engineers not to take risks of data loss with dependability in contexts ranging from telecommunications to satellite communications. A proper analysis of BER entails examination of different types of noise and the impact of such noise on the quality of the signal. For instance, intersymbol interference (ISI) has a tendency to occur when the bit duration is close to the channel's bandwidth constraints. In optical transmission, in the case of avalanche photodiodes (APDs), for instance, ISI can produce extreme performance degradation as the transmission rate increases. The randomness of the avalanche buildup time in APDs captures the challenge in exactly calculating BER in such environments. Recent studies highlight that traditional Gaussian approximations prove to be insufficient in the presence of ISI noise, hence requiring more sophisticated techniques of analysis such as the large-deviation theory (LDT). Proper analysis of BER in such an environment proves valuable and helps direct the development of more reliable communication systems that are able to withstand greater rates of data while accepting reduced error (Das et al., 2009). In addition to noise properties inherent within, external factors such as radio frequency interference (RFI) also make a large contribution towards the Bit Error Rate. Addition of advanced simulation software, for example, LinCsim, has assisted in modeling RFI impacts on satellite communications channels, particularly in channels like the Shuttle/TDRSS channel. With the aid of revised channel models, engineers are able to evaluate how RFI degrades the reliability of communication signals, hence influencing overall BER. This is a very important aspect when designing around complexities of contemporary communication structures, where several interferences come together to affect the clarity of the transmitted information.

By enhancing simulation methods and overcoming hardware-related changes, professionals can better anticipate BER-related issues, ultimately leading to effective measures in reducing errors in future communication systems (An et al., 2025) (Figure 3.12).

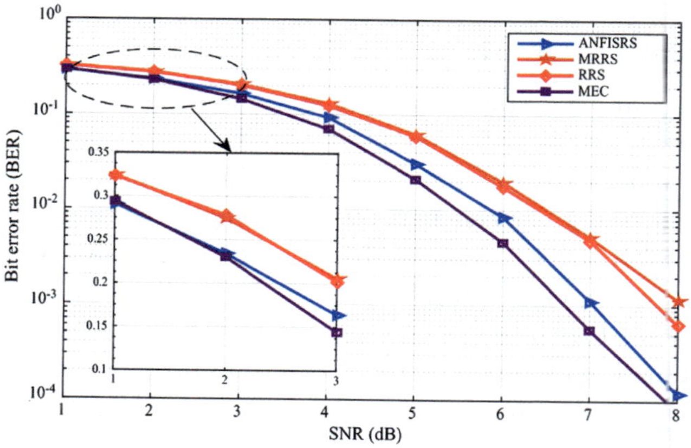

Figure 3.12. *Bit error rate (BER) serves.*

Source: https://www.researchgate.net/publication/343566530/figure/fig2/AS-952220137672718@1606422566533/Bit-Error-Rate-BER-performance-corresponding-to-Signal-to-Noise-Ratio-SNR.png.

3.5.3. Capacity of Communication Channels

Channel capacity is the fundamental idea in understanding how information can be transmitted effectively in various systems, most notably those corrupted by noise and random processes. Channel capacity as a concept in terms of information theory is the maximum rate at which data can be successfully transferred through a communication channel. This capacity limit is typically reduced by the presence of noise, which corrupts the received signals and introduces errors. The relationship between the noise and the channel parameters has direct impact on the performance of the communication systems, highlighting the need for studying these factors in devising effective communication systems. Comparing various classes of channels, one would obviously find that bandwidth, power limitation, and modulation methods employed play significant roles in determining the overall capacity for transferring information. For covert channels, the term "capacity" assumes an even more complicated meaning. Covert channels are designed to present inconspicuous means of communication that do not necessarily need to be visible to anyone viewing from the outside. As recognized in previous research, traditional interference measures are not able to delineate the unique patterns of information transmission that result in the detection of covert channels. With the study of iterated multivalued interference, we are further able to determine the limitations of traditional capacity measures, particularly with regard to the situation where covert channels have less than one bit capacity per use. The subtlety of interaction between controlled information flow and external interference hints at the subtlety of complexities of challenges arising whenever there is an attempt to quantify the capacity of such covert means of communications, thus necessitating more sophisticated analyses in the capability to encompass their realities of operation (Hélouët et al., 2010). The increased call for ultra-reliable low latency communications (URLLC) also highlights the significance of channel capacity in modern communication systems. URLLC uses, which require strong latency and reliability commitments, require a shift of focus from mere throughput optimization to ensuring reliable end-to-end communication. Analysis of multiple-input, single-output (MISO) systems gives critical clues on performance in statistical delay-constrained systems. Closed-form descriptions of service processes and probabilistic delay bounds illustrate how channel capacity can be managed by utilizing diverse parameters, such as transmit power and antenna designs. These analyses not just add to our understanding of MISO channels but also set the most important guidelines for the design of communication systems capable of meeting the demanding specifications of URLLC applications (Arnau et al., 2017).

3.6. ADVANCES IN NOISE MITIGATION TECHNIQUES

Research into ways to reduce noise has come a long way, particularly in communication systems. Modern techniques now involve sophisticated algorithms that make use of stochastic modeling to reduce unwanted noise. Random processes play a key role in describing noise patterns in communication channels so that engineers can effectively design systems. For example, through the use of high-level signal processing techniques like adaptive filtering and echo cancellation, it is possible to eliminate noise from the signal, thereby making the signal more clear. These advances not only maximize data transmission quality but also extend the operating range of wireless communication systems. As a result, the use of such techniques becomes increasingly essential in maintaining the integrity of the signal in adverse environments exposed to multiple sources of disturbance. In addition to software methods, developments in hardware have also contributed significantly to noise mitigation methods. Antenna arrays have also been an effective weapon in improving communication system performance, especially under non-line-of-sight (NLOS) conditions. Antenna arrays can counter the effects of multipath fading—a pervasive form of noise in wireless channels—through the use of spatial diversity. According to recent studies, the integration of space-time information processing and these arrays can enable enhanced positioning and data transmission, with the potential to render the system much more resilient. The study further indicates that the use of geometrical constraints based on time difference of arrival measurements has the capability to considerably improve the accuracy of positioning systems in noisy environments, demonstrating how the physical deployment of communication infrastructures can provide pragmatic pathways towards noise-free signal transmission (Garzon et al., 2015). Moreover, the innovation in novel materials and technologies offers unprecedented opportunities for eliminating noise from communication systems. Material science advancements have enabled the development of advanced shielding techniques that can effectively minimize electromagnetic interference, a prime source of noise in the majority of communications systems. Such developments as metamaterials with their unusual wave manipulation characteristics enable the use of tailored noise reducing techniques. This is particularly useful where the complexity of urban life is added to due to noise pollution and signal clarity is essential. Furthermore, application of machine learning algorithms has been fruitful in predicting and adapting to fluctuating noise levels in real-time, achieving optimal communication performance. These advances promise that a multi-faceted approach, which combines algorithmic innovation and material advancements, holds the key to achieving effective noise cancellations in the evolving environment of communication technology (N/A, 2017).

3.6.1. Adaptive Filtering Methods

The application of adaptive filtering methods is at the core of enhancing communications systems subject to random noise. Fundamentally, adaptive filtering is the methodology of designing filters that counteract fluctuating signal characteristics and environmental conditions in order to minimize the impact of noise. With the aid of algorithms such as the Least Mean Squares (LMS) and Recursive Least Squares (RLS), interference in real-time communication channels is effectively combated by engineers. These algorithms continuously adapt filter coefficients based on received data, improving signal purity and system dependability as a whole. The significance of adaptive filtering is emphasized in several applications, including intelligent buildings on the basis of Intelligent Building (IB) technologies in which systems utilize algorithms to predict occupancy through CO_2 concentrations, temperature, and humidity levels with a remarkable improvement in prediction accuracy using adaptive approaches, thus demonstrating their suitability in practical situations (Bilík et al., 2019). The distributed recursive least-squares (D-RLS) algorithm is a sophisticated method in the field of adaptive filtering, particularly for wireless sensor networks with random noise. The method achieves optimal data collection efficiency while reducing the computation complexity significantly, thus finding application where there is a necessity for constant updates and real-time processing. By embracing a cooperative estimation approach, sensors perform local operations, improving their estimates via iterative processes as they exchange information with nearby nodes. This cooperative framework makes it possible to estimate slowly-varying nonstationary signals, which is important in communication systems under time-varying environments. Mean-square error (MSE) performance analysis portrays the robustness and accuracy of the D-RLS algorithm under noisy inter-sensor communication, hence assuring its applicability in enhancing noise-corrupted environments' robustness (Giannakis et al., 2011). Lastly, adaptive filtering methodologies form the core of the design of noise communication systems, as they efficiently manage the impairments caused by random processes. Dynamically adapting the parameters of filters based on the changes in signal conditions is critical in achieving optimal communication quality. By applying sophisticated algorithms such as LMS and D-RLS, researchers and practitioners are able to efficiently improve the accuracy of predictions and overall performance of communication systems. Besides, applications in the real world of such techniques can be identified through current applications, where adaptive filtering supports energy-saving automation and occupancy detection in smart buildings, contributing to improved end-user experience as well as a reduction in operation costs. As continuous technology evolution takes place, there is likely going to be growing need for effective adaptive filtering systems, further solidifying their presence in the future of noise communication systems.

3.6.2. Error Correction Codes (ECC)

In communication networks, error correction codes (ECC) are the most important tool in ensuring data passed via noisy channels remains secure. In the knowledge that interruption will be felt by the signal in transmission, these codes are precious in minimizing errors caused by noise from various sources. ECCs work by adding redundant information to the data being conveyed so that the recipient can detect and correct errors without necessitating the original data to be retransmitted. With the evolution in technology, conventional approaches to error correction, such as block and convolutional codes, continue to develop further toward even more sophisticated implementations to cover the intelligent needs of modern digital communication systems. Effective application of such codes will come a long way in making wireless communications extremely resilient, with information remaining secure and trustworthy even in adverse environments. Modern research has sought to revolutionize error correction techniques to achieve maximum performance, particularly in power-conscious applications like Orthogonal Frequency Division Multiplexing (OFDM) systems. A novel cross-layer scheme, for instance, was proposed that includes resolution-adaptive analog-to-digital converters (ADCs) alongside Fountain codes with the goal of reducing power consumption.

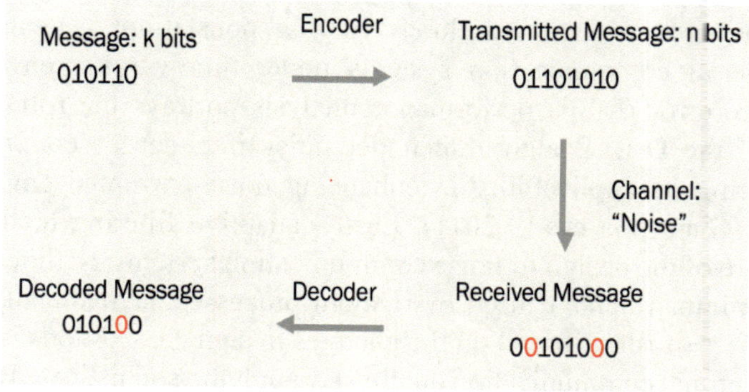

Figure 3.13. *Error correction codes.*

Source: https://lh3.googleusercontent.com/9y96GkfFf6FBCUKoufmsqUGfXt-Vm-DqAuCQI1IaOfy4DB-VmJWEvSwL9c1mj9QP9gVYq49haRBI96eNMx5qPpr-B-hhGuWvTv4wixNbLLuTuSnI3xv36xaVa64D3DvshMs_XwmT.

The essence of this approach is to selectively examine only packets transmitted over good sub-carriers and filter out packets transmitted through bad channels. Not only does this save energy in the ADC, which may constitute a significant portion of overall baseband energy consumption, but it also maintains communication efficient. This novel technique proves it possible to specialize error correction codes for the special demands of specific applications and deliver energy saving as well as communication reliability as described in Schiphorst et al. (2009). Besides, theoretical backgrounds of error correction codes are also normally revealed by means of models representing similarities between information theory and life systems. Recent debates have established the linkages between genetic information transmission and ECC concepts and enriched the two fields with improved understanding and complexity. Using elements of Shannons rate distortion theory to analyze error correction processes in gene networks, researchers suggest that the identical principles of robustness apply to biological and communicative operations. The idea, or frozen hypothesis, is that certain evolutionary innovations can optimize information transmission in living organisms, mirroring concepts in communications theory regarding redundancy and error tolerance. Such interfield connections provide evidence that error correction codes not only enhance digital communication but also comply with fundamental processes in nature, as treated in James et al. (2011) (Figure 3.13).

3.6.3. Recent Innovations in Communication Technology

The landscape of communication technologies has undergone drastic shifts in the recent years, with most of them being spurred by developments in algorithmic architecture and network design. One of these developments has been the construction of distributed recursive nonlinear least squares (NLS) estimators that facilitate parameter estimation of multi-agent systems. It offers a mechanism through which agents are able to estimate true parameters from noisy measurements, thus enhancing the consensus process. The constructed algorithm, termed Consensus-Innovation Weighted Nonlinear Least Squares (CIWNLS), ensures that local parameter estimates are always adjusted by processing real-time innovations and incorporating neighboring agents' consensus information. Key to the methodology is the weak connectivity promise, ensuring that all agents enjoy the finest convergence rates comparable to centralized estimators, thus revolutionizing the manner collaborative systems have been able to function under noisy environments (Kar et al., 2016). These advances therefore show the convergence of technology and mathematics towards promoting communication reliability. Recent advances also highlight the role of source coding in communications systems with zero-delay constraints. The functional zero-delay vector-valued Gaussian rate distortion function (RDF) shows efficiencies realized using state-of-the-art coding schemes applicable to Markov sources. With the help of interactive rate distortion models, scientists have derived lower bounds on causal performance to counter fundamental difficulties in achieving mean-squared error fidelity. These new realizations typically involve predictive coding architectures that optimize resource allocation and quantization techniques, which are instrumental to the effectiveness of transmission over noisy channels. For example, the use of feedback mechanisms and lattice quantization indicates a significant reduction in the difference between lower and upper bounds of operational performance (Charalambous et al., 2018). Thus, such coding innovations are crucial constituents in enhancing the robustness and effectiveness of communication systems. Interplay between innovations in coding methods and distributed algorithms is a critical contribution towards governing random processes within communication systems. These innovations, among other things, not only solve intrinsic noise problems but also enable successful data exchange between networks, thereby redefining the potential of modern communication platforms. With various structures of codes becoming available, their integration with schemes based on consensus estimation will result in stronger systems for operation under dynamic working environments. Moreover, as systems further evolve to meet the demands for real-time data exchange, additional breakthroughs in both applied and theoretical communication technology will assuredly be revealed through future studies. This combination of coding methods and parameter estimation algorithms ultimately ensures better fidelity and efficiency in signal processing, which contributes greatly to the future of intelligent communication networks (Figure 3.14).

Figure 3.14. *Recent innovations in communication technology.*

Source: https://i.ytimg.com/vi/36bF2zN4lFs/maxresdefault.jpg.

SUMMARY

In learning random processes and noise in communication systems, the conclusions of the analysis highlight the intricate interdependency of system performance and noise. Random disturbances create a multiplicity of complications for simplistic model-building approaches. The disturbances disrupt the deterministic pattern descriptions of simpler analytic models and therefore demand a more realistic understanding of system operation under varied conditions of noise. The findings show that the detection and quantification of noise effects are necessary towards developing effective communication strategies because simple predictive models will not capture the complexity introduced by random processes. Therefore, it is crucial to deal with the effects of noise in order to achieve maximum performance in different applications, particularly in distributed-memory systems, where latency can be dispersed far and wide and affect the overall system performance (Afzal et al.,

2019). Also, the research into advanced schemes for secure communication, such as those developed from the Kirchhoff-Law-Johnson-(similar)-Noise frameworks, further illustrates the potential of countering the adverse effects of noise while enhancing security. The advancement of these protocols illustrates how embracing noise can allow one to create innovative solutions, such that systems can surf stochastic behavior to function better under less-than-ideal conditions. By incorporating noise properties into the architecture of such protocols, scientists have been able to enhance security and performance, and considerable advances have been made in secure key exchange practical applications. Noise-security interaction highlights the necessity to find new methods to not only control noise, but also leverage it to drive the advancement of communication technologies (Kish et al., 2013). Lastly, the results of the research on noise communication systems and random processes prove the need for continued research in these areas. With communication systems becoming more and more complex and more susceptible to various forms of debilitating noise, the compounded effects of these interferences will have to remain under investigation. The possibility of predicting and controlling noise impact is of prime importance in the design of systems to come that will be capable of achieving high reliability and robustness under practical operating conditions. Through a greater understanding of the involved dynamics with random processes and their interaction with noise, researchers can unlock the doors to new approaches that have the potential to enhance communication fidelity and system performance. So, embracing the challenge brought about by noise is essential for the growth of both theoretical frameworks and practical applications in this critical field.

In examining the interaction of random processes with noise communication systems, it is imperative to properly understand network architectures. Particularly, the advent of wirelessly powered backscatter communication (WP-BackCom) networks is a case in point of innovation in this field. In facilitating device-to-device (D2D) communication through passive nodes, these networks are pivotal in overcoming the limitations incurred by traditional battery technologies on Internet-of-Things (IoT) devices. The addition of wireless power transfer enables these nodes to reflect and modulate incoming signals at minimal power loss, thereby significantly improving transmission efficiency. To describe WP-BackCom networks as a random Poisson cluster process is to highlight the strengths of power beacon deployment for coverage and capacity optimization because stochastic geometrical approaches facilitate strict performance assessments by variable parameters, ensuring that the vast potential of IoT could be realized in an exponentially scalable manner (Han et al., 2016). The entry of chaos and noise in communication systems further pushes the understanding of signal behavior under randomness. Recent work on chaotic nature exhibited by laser diodes introduces interesting opportunities for tapping into the nonlinearity in communication protocols. The study of chaos caused by nonlinear coupling in vertical-cavity surface-emitting lasers introduces new ideas on how randomness can be harnessed and not merely evaded. From the comparison with traditional noise behavior, researchers have unveiled the potential of chaos to be employed in high-speed optical communication, thereby contradicting the anticipated use of deterministic chaos as a destructive force. Bifurcation leading to attractors with behaviors close to Lorenz chaos opens up new avenues for investigation, particularly when connecting it to the merging of high-speed laser performance and controlled chaotic effects (Argyris et al., 2013). These advancements indicate the necessity of formulating newer methodologies in noise communication systems and random processes. With increasing levels of complexity in systems, newer models that combine random phenomena are necessary to simplify communication plans. The study of WP-BackCom networks and the new respect for chaotic dynamics in laser diodes is a broader movement that views randomness not as something to be avoided but as an essential element that can be harnessed in order to make systems more efficient and stable. Future research needs to focus on building such models and applications so noise can be used strategically instead of simply being seen as a destructive force. This paradigm shift emphasizes the need for interdisciplinary thought and cooperative work among disciplines in an attempt to capitalize on all advantages that randomness in communications technology has to offer (Han et al., 2016; Argyris et al., 2013).

REFERENCES

1. Afzal, A., Hager, G., & Wellein, G. (2019). Propagation and decay of injected one-off delays on clusters: A case study. http://arxiv.org/abs/1905.10603 (accessed on 12th September 2025).
2. Agrawal, M., Balakrishnan, V., & Love, D. (2015). Communicating over filter-and-forward relay networks with channel output feedback. http://arxiv.org/abs/1305.4403 (accessed on 12th September 2025).
3. Agrell, E., Camara, M. R., Forestieri, E., et al. (2018). Nonlinearity mitigation in WDM systems: Models, strategies, and achievable rates. https://core.ac.uk/download/199919811.pdf (accessed on 12th September 2025).
4. Almeida, E. S., Aquino, A. L. L., Cavalcante, T. S. G., et al. (2015). Characterization of vehicle behavior with information theory. https://core.ac.uk/download/603321520.pdf (accessed on 12th September 2025).
5. An, S. H., Choi, H. J., Lindsey, W. C., et al. (2025). Ku-band system design study and TDRSS interface analysis. https://core.ac.uk/download/pdf/42851216.pdf (accessed on 12th September 2025).
6. Argyris, A., Garfinkel, A., Provenzale, A., Scirè, A., Uchida, A., Wolf, A., Andronov, A. A., et al. (2013). Deterministic polarization chaos from a laser diode. http://arxiv.org/abs/1407.5810 (accessed on 12th September 2025).
7. Arnau, J., & Kountouris, M. (2017). Delay performance of MISO wireless communications. http://arxiv.org/abs/1707.08089 (accessed on 12th September 2025).
8. Bilík, P., Gorjani, O. M., & Vaňuš, J. (2019). Novel proposal for prediction of CO_2 course and occupancy recognition in intelligent buildings within IoT. https://core.ac.uk/download/322825115.pdf (accessed on 12th September 2025).
9. Bolten, M., Dahmen, D., Diesmann, M., et al. (2016). Integration of continuous-time dynamics in a spiking neural network simulator. http://arxiv.org/abs/1610.09990 (accessed on 12th September 2025).
10. Buckwalter, J. F., & Hajimiri, A. (2006). Analysis and equalization of data-dependent jitter. https://core.ac.uk/download/4871722.pdf (accessed on 12th September 2025).
11. Carpi, L. C., Frery, A. C., Gonçalves, B. A., et al. (2014). Distinguishing noise from chaos: Objective versus subjective criteria using horizontal visibility graph. https://core.ac.uk/download/159304072.pdf (accessed on 12th September 2025).

12. Charalambous, C. D., Ostergaard, J., & Stavrou, P. A. (2018). Zero-delay rate distortion via filtering for vector-valued Gaussian sources. https://arxiv.org/abs/1810.00298 (accessed on 12th September 2025).

13. Chon, Y. T., & Hutchinson, C. E. (2025). An analysis of the effect of a particular class of PFM on noise inputs. https://core.ac.uk/download/pdf/85243627.pdf (accessed on 12th September 2025).

14. Das, A. K., Hayat, M. M., & Sun, P. (2009). Bit error rates for ultrafast APD based optical receivers: Exact and large deviation based asymptotic approaches. https://core.ac.uk/download/286965452.pdf (accessed on 12th September 2025).

15. Das, S., Das, S., & Pan, I. (2013). Fractional order fuzzy control of nuclear reactor power with thermal-hydraulic effects in the presence of random network induced delay and sensor noise having long range dependence. http://arxiv.org/abs/1306.3679 (accessed on 12th September 2025).

16. Dey, S., Leong, A. S., Nourian, M., et al. (2014). An optimal transmission strategy for Kalman filtering over packet dropping links with imperfect acknowledgements. https://core.ac.uk/download/162214711.pdf (accessed on 12th September 2025).

17. Dhillon, H. S., Andrews, J. G., & Ganti, R. K. (2011). Modeling and analysis of K-tier downlink heterogeneous cellular networks. http://arxiv.org/abs/1103.2177 (accessed on 12th September 2025).

18. Durisi, G., Eriksson, T., & Khanzadi, M. R. (2015). Capacity of SIMO and MISO phase-noise channels with common/separate oscillators. http://arxiv.org/abs/1409.0561 (accessed on 12th September 2025).

19. Garzon, E., Jativa, E. R., & Vidal Manzano, J. (2016). Space-time diversity for NLOS mitigation in TDOA-based positioning systems. https://core.ac.uk/download/81581300.pdf (accessed on 12th September 2025).

20. Giannakis, G. B., & Mateos, G. (2011). Distributed recursive least-squares Stability and performance analysis. http://arxiv.org/abs/1109.4627 (accessed on 12th September 2025).

21. Gioiosa, R., Kestor, G., Laure, E., et al. (2017). Idle period propagation in message-passing applications. http://arxiv.org/abs/1704.08244 (accessed on 12th September 2025).

22. Glazebrook, J. F., & Wallace, R. (2011). 'The frozen accident' as an evolutionary adaptation: A rate distortion theory perspective on the dynamics and symmetries of genetic coding mechanisms. https://core.ac.uk/download/pdf/289861.pdf (accessed on 12th September 2025).

23. Grange, M., Pamunuwa, D. B., Tenhunen, H., et al. (2010). On signaling over through-silicon via (TSV) interconnects in 3-D integrated circuits.

https://core.ac.uk/download/pdf/68701.pdf (accessed on 12th September 2025).

24. Han, K., & Huang, K. (2016). Wirelessly powered backscatter communication networks: Modeling, coverage and capacity. https://core.ac.uk/download/95560745.pdf (accessed on 12th September 2025).
25. Hélouët, L., & Roumy, A. (2010). Covert channel detection using information theory. http://arxiv.org/abs/1102.5586 (accessed on 12th September 2025).
26. Isufi, E., Leus, G., Loukas, A., et al. (2017). Filtering random graph processes over random time-varying graphs. https://core.ac.uk/download/148032506.pdf (accessed on 12th September 2025).
27. Kar, S., Moura, J. M. F., Poor, H. V., et al. (2016). Distributed constrained recursive nonlinear least-squares estimation: Algorithms and asymptotic. http://arxiv.org/abs/1602.00382 (accessed on 12th September 2025).
28. Kish, L. B. (2013). Enhanced secure key exchange systems based on the Johnson-noise scheme. https://core.ac.uk/download/231873639.pdf (accessed on 12th September 2025).
29. Lee, H., Lee, I., Lee, S. H., et al. (2018). Deep learning framework for wireless systems: Applications to optical wireless communications. http://arxiv.org/abs/1812.05227 (accessed on 12th September 2025).
30. Mabuchi, H., Stockton, J. K., & van Handel, R. (2005). Modelling and feedback control design for quantum state preparation. https://core.ac.uk/download/4870471.pdf (accessed on 12th September 2025).
31. Meng, L. M., Wang, K., & Xu, Z. J. (2013). Channel parameters estimation algorithm based on the characteristic function under impulse noise environment. https://core.ac.uk/download/30312063.pdf (accessed on 12th September 2025).
32. N/A. (2017). SciTech News, 71(1). https://core.ac.uk/download/83049835.pdf (accessed on 12th September 2025).
33. Palgen, J. J. O. (2025). Applications of optical processing for improving ERTS data, volume 1. https://core.ac.uk/download/pdf/85234302.pdf (accessed on 12th September 2025).
34. Parsons, A., Siemion, A., Parsa, A., Blackman, R., Bradley, R., Werthimer, D., MacMahon, D., et al. (2009). A scalable correlator architecture based on modular FPGA hardware, reusable gate ware, and data packetization. http://arxiv.org/abs/0809.2266 (accessed on 12th September 2025).
35. Schiphorst, R., Shao, X., & Slump, C. H. (2009). An opportunistic error correction layer for OFDM systems. https://core.ac.uk/download/pdf/11468947.pdf (accessed on 12th September 2025).

36. Schliep, B., Derrida, D., MacKay, D. J. C., Fisher, D. S , Domany, E., Grinstein, G., Kanter, I., et al. (2005). The entropy of a binary hidden Markov process. http://arxiv.org/abs/cs/0507060 (accessed on 12th September 2025).
37. Shbat, M. S., & Tuzlukov, V. P. (2015). Generalized detector as a spectrum sensor in cognitive radio networks. https://core.ac.uk/download/30312305.pdf (accessed on 12th September 2025).
38. van Etten, W. (2006). Introduction to Random Signals and Noise. https:// core.ac.uk/download/pdf/11469291.pdf (accessed on 12th September 2025).
39. Win, M. Z., & Pinto, P. C. (2007). Communication in a Poisson field of interferers–Part II: Channel capacity and interference spectrum. http:// arxiv.org/abs/1001.4520 (accessed on 12th September 2025).
40. Yüksel, S. (2012). Characterization of information channels for asymptotic mean stationarity and stochastic stability of non-stationary/unstable linear systems. http://arxiv.org/abs/1201.5360 (accessed on 12th September 2025).

CHAPTER 4

SAMPLING AND PULSE MODULATION

Contents

INTRODUCTION

In communications engineering, pulse modulation and the theory of sampling represent fundamental concepts of signal processing and transmission. Sampling, which is the translation of a continuous-time signal into a discrete-time signal, is crucial for digitalizing analog information. Sampling makes it possible to process and store information in an efficient manner within modern electronic systems. Nyquist theorem provides an essential guideline of sampling, such that a signal should be sampled at least twice its highest frequency so as not to induce aliasing. There has to be proper sampling rates to maintain the integrity of the original signal since any inadequacy would produce distortions that impact the fidelity of the information sent. Pulse modulation techniques, such as pulse amplitude modulation (PAM), pulse width modulation (PWM), and pulse position modulation (PPM), also improve the efficiency of signal transmission.

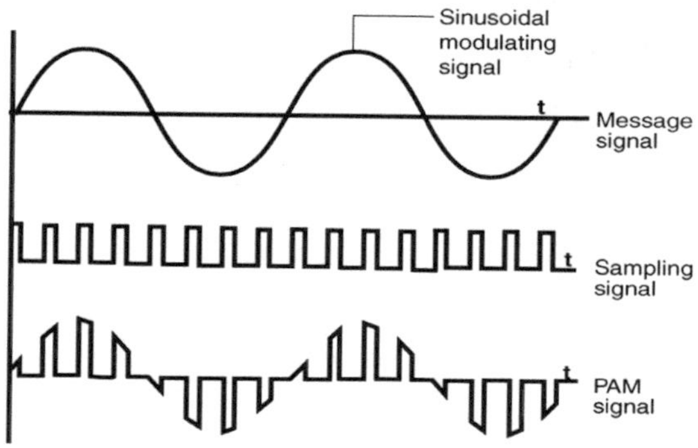

Figure 4.1. *Sampling and pulse modulation in pulse amplitude modulation (PAM).*

Source: https://cdn1.byjus.com/wp-content/uploads/2020/11/Pulse-Modulation-2.png.

The techniques provide a way of transmitting information through the form of discrete pulses, rendering them suitable for various communication environments, such as digital telecommunications and broadcasting. Pulse modulation is also optimal in surviving noise and interference due to its critical nature in most signal processing applications today. By translating analog signals into pulses, modulation facilitates more effective bandwidth use to offer robust data transfer across numerous media. The integration of sophisticated systems incorporating these pulse-based techniques also lends support to progress in remote sensing and digital signal processing (Campbell et al., 2011). The latest improvements in minimal-rate sampling schemes have introduced new techniques for processing highly complicated signals. These schemes aim sampling infinite streams of pulse shapes based on the rate of innovation principle, which has been measured in terms of degrees of freedom per unit time (Eldar et al., 2010). This signal processing innovation finds wide-ranging ramifications, particularly for applications involving multichannel architectures capable of sampling specially designed waveforms efficiently. Through the use of a bank of integrators coupled with modulation techniques, these systems are able to achieve the fine balance of low hardware without sacrificing noise robustness. In doing so, they not only enhance the recovery of pulse streams but also reduce vulnerability to component failures in sampling channels. Their creation represents the pinnacle of the ongoing advancements that sampling and pulse modulation engineering have undergone, marking their essential role in the future of communication technologies (Figure 4.1).

Definition of Sampling

Sampling, used for signal processing and telecommunications, refers to the discrete representation of continuous analog signals. It is a very crucial process allowing a signal to be transformed into a format that is easily processed, analyzed, or communicated in digital systems. Nyquist theorem focuses on sampling frequency in the belief that in order to record an analog signal sufficiently well without losing data, it must be sampled at least twice the highest frequency. The rule based on frequency holds for most engineering uses, addressing the precise balancing act between adequate resolution and avoiding aliasing, or distortion in reconstructing the original signal. As technology advances, sampling techniques continuously evolve, providing engineers with novel means of enhancing communication systems, particularly in pulse modulation engineering applications. The delicacy of sampling plays a crucial role because it directly influences the performance of pulse modulation systems. Pulse modulation processes, which transmit information in pulsed signals, rely heavily on efficient sampling techniques to safeguard data over long distances and through mediums. As an example, the effectiveness of a PAM system depends on the ability of the amplitude of the sample signal to be measured accurately and transmitted. The advancement of complex coding and modulation methods, such as those analyzed in present literature, has led to engineers investigating diversified sampling strategies capable of achieving the highest trade-off between bandwidth usage and signal integrity. The effects of low sampling rates are extensive; inadequate sampling methods may introduce excessive signal distortion and increased error rates, hence impacting the overall system efficiency, as can be evident in discussions on optimal spectral efficiency in long-haul optical links (Colavolpe et al., 2014). Additionally, the influence of sampling rates on pulse modulation cannot be overstated in modern communication systems. Optical communications advancement demands robust sampling schemes capable of managing efficiently non-linear impairments over transmission in long distances. High-order modulation formats, while adequate for spectral efficiency maximization, introduce complexity that is linked with requiring aggressive sampling schemes for effective recovery of signals.

Research into the optimization of low-order constellations with time-frequency packing demonstrated viable solutions to the above challenges, part of an overall trend towards maximizing efficiency within the wake of technological progress and the constraints of signal processing theory (Hartman et al., 2025). In building further upon the subject discipline, definition and utilization of sampling will continue to evolve, forming the basis in the development of the future of pulse modulation engineering (Figure 4.2).

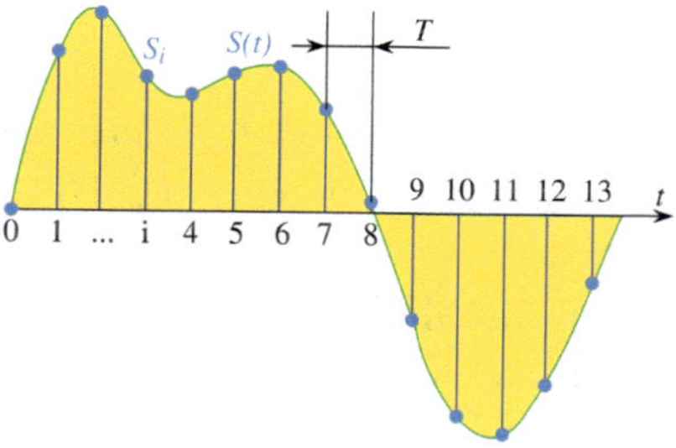

Figure 4.2. *Sampling, in the context of signal processing.*

Source: https://upload.wikimedia.org/wikipedia/commons/thumb/c/c3/Signal_Sampling.svg/1200px-Signal_Sampling.svg.png.

Overview of Pulse Modulation

Pulse modulation is an essential signal processing concept that offers a reliable means of information encoding for transmission via various media. Pulse modulation relies on the creation of discrete pulse signals with the actual analog information. Pulse width modulation (PWM), pulse position modulation (PPM), and pulse amplitude modulation (PAM) are a few types of pulse modulation. Both techniques have their individual merits and applications, which are mainly based on the nature of the signal to be communicated and the characteristics of the transmission medium. The utilization of pulse modulation offers advantages such as increased noise immunity and the ability to make efficient use of bandwidth, which are vital in modern communication systems, especially in digital communications and data processing. Advanced modulating schemes with enhanced performance and efficiency have appeared in recent practice. Nanophotonic devices integrated, for instance, have made outstanding advancements in this area using squeezed light generated by spontaneous four-wave mixing. These techniques generate quadrature squeezed vacuum and photon number difference squeezing, an interplay between quantum mechanics and traditional pulse modulation techniques with a wide range of applications in communication and quantum information processing. Modern technologies refine these conceptual notions to push the limits of spectral efficiency and power optimization, establishing a new paradigm within pulse modulation systems. These advances unveil new possibilities, resolving theoretical as much as practical issues in the field of communication engineering, particularly in maximizing utilization of channel resources in increasingly demanding environments (Collins et al., 2020). Moreover, the significance of flexible trellis modulation encompasses the revolutionary potential of present-day digital communication systems. Design of such architectures as the flexible trellis modem/codec (FTMC) transmitter not only integrates advanced modulation formats—BPSK, QPSK, and 16-QAM—into a compact device but also facilitates programmability in modulation depending on the transmission condition of interest. This feature of programmability is pivotal in maximizing performance across various operation environments and channel conditions. By applying digital baseband pulse-shaping and precompensation techniques, the FTMC can effectively eliminate power and spectral inefficiency issues to enhance overall system performance. Such technological advancements in pulse modulation are an indication of future advancement in the science of sampling and modulation engineering, with significant implications for researchers and engineers with tools capable of executing future communication and data processing (Budinger et al., 2025).

Importance in Communication Systems

The function of effective communication systems cannot be overstated in our more integrated global system of today. As the central methodology in these systems is the process of sampling, it plays a crucial role in the conversion of continuous signals into discrete representations to be processed digitally. The conversion is essential in applications ranging from telecommunication to computer networking, where data needs to be transmitted effectively and precisely. Proper sampling techniques ensure that the essence of the original signal is sustained, thus avoiding distortion that could occur during transmission. Secondly, technological innovations in sampling techniques ensure greater bandwidth efficiency and interference reduction, allowing for more data to be processed with less compromise on quality. Thus, sampling forms the foundation for developing robust communication systems capable of meeting the demands of contemporary society. Pulse modulation techniques also enhance the performance and reliability of communication systems by allowing effective transmission of sampled signals. This kind of modulation allows encoding information based on variations in pulse width, amplitude, or position, which itself aligns with digital signal processing frameworks. The versatility of pulse modulation techniques brings useful advantages to numerous applications, such as radar and wireless communications. Through the incorporation of pulse modulation in sampling systems, system designers can design systems with increased noise tolerance and increased adherence to the statistical properties of the communication channel. Progress in the technology, particularly via spread spectrum modulation, is very critical because they are making the systems more interference-resistant and better detection sensitivity, as outlined in the research encouraging such techniques in cases of complications (Messerschmitt et al., 2011). Progressions in detection technologies are also pivotal towards developing more effective communication systems. The application of soft noncoherent detection methods, employing a posteriori probabilities with no channel estimation, has exhibited encouraging sides in high-dimensional wireless communication. They offer improved detection performance, particularly in complex cases where signal conditions vary. Recent progress, including the belief propagation-based soft multiple-symbol differential detection scheme presented, illustrates the potential for large performance improvement in impulse radio systems. Through efficient performance of signal processing tasks like sampling and detection, such breakthroughs play a vital role in the reliability of communication systems, providing the foundation for iterative detection and decoding (Gao et al., 2016). Therefore, the integration of sampling, pulse modulation, and advanced detection techniques heightens the significance of such constructs in shaping dependable communication networks.

4.1. FUNDAMENTALS OF SAMPLING THEORY

The sampling theory principles form the basis of the understanding of how continuous signals are translated into discrete data points, which can be processed using digital means. Based on sampling theory, a continuous signal is best represented by its samples, provided that the sample rate is a minimum greater than double the highest frequency in the signal, as stated in the Nyquist-Shannon sampling theorem. This theorem is the basis of most pulse modulation engineering applications and, in a sense, bridges the gap between digital and analog systems. Because of this, the selection of an appropriate sampling rate is important; under-sampling can lead to aliasing, where different signals cannot be distinguished from one another. These ideas assume a significant significance in the current integrated systems, where signal manipulation can make or break processes like data encoding and transmission. In pulse modulation technology, the application of sampling theory facilitates signal integrity and system performance enhancement. Both PAM and PWM methods, for instance, rely on robust sampling methods to ensure that the original information is preserved when transmitting. These kinds of modulations convert analog information into digital forms where it becomes easily processable using digital signal processing algorithms. With the enhanced need for higher data rates and better communications, the need for accurate sampling in order to provide clarity and fidelity has increased further. Furthermore, new technologies based on integrated photonic circuits are beginning to unveil new horizons in sampling theory to enable yet newer sophisticated forms of data manipulation and transmission. Progress in related areas over recent years serves to emphasize sampling theory's ongoing relevance. For instance, in ultrafast electronic switching experiments, it has been demonstrated that novel materials possess sub-picosecond timescale threshold switching mechanisms. Here, this research demonstrates that with the evolution of digital systems, the theory of sampling evolves with it. Similarly, progress in quantum optics demonstrates how sampling can be employed towards advanced measurement techniques and further towards devices employing optical squeezing and photon number statistics. These findings indicate that sampling theory is not only a strict set of rules but an ongoing area of research that will play a crucial role in the future of pulse and sampling modulation engineering, as systems call for more sophisticated methods of signal processing and transmission, such as in Brongersma et al. (2016) and Collins et al. (2020) (Figure 4.3).

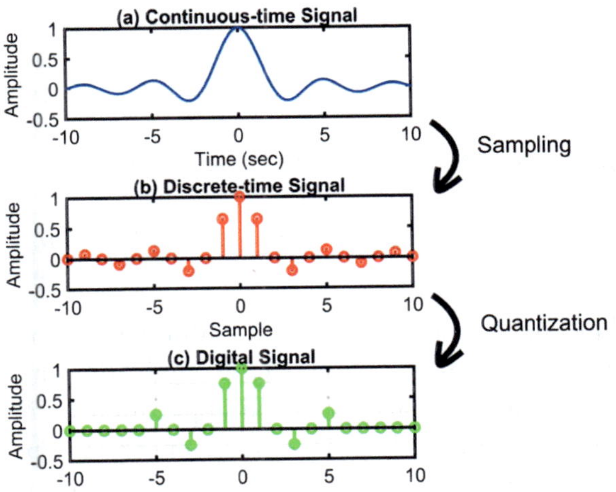

Figure 4.3. *Fundamentals of sampling theory a continuous signal.*

Source: https://lh7-us.googleusercontent.com/SoK1jS8civky5KtzARCs0D-F72ccG0KtGZMpPPSAATyaeYBIHkMFo4LmEIiFIDwBcgammZrLaE__EBOfLlbPvCwgSa4_0yGW6mx_afCLFHFGGbGM5Fpc9QMer293ORzisSeUpB-gaEz6hq47aJysiBfw.

4.1.1. Nyquist Theorem

The Nyquist Theorem is a key column in digital communication engineering and signal processing. Basically, the theorem states that to be able to reconstruct a continuous-time signal based on its samples with accuracy, one must sample at least twice as much as the highest frequency in the signal. This lowest sampling rate, or the Nyquist rate, is the one that prevents information loss when converting. The effect of this theorem is profound in that it not only facilitates the design of sampling plans but also the design of many digital systems. With modern applications pushing the science of signal processing to newer horizons, the understanding of the subtleties of the Nyquist Theorem is as crucial for engineers and researchers today to utilize digital systems effectively. Developing beyond the work of Nyquist, recent advances in sampling theory have begun exploring the potential of sub-Nyquist sampling.

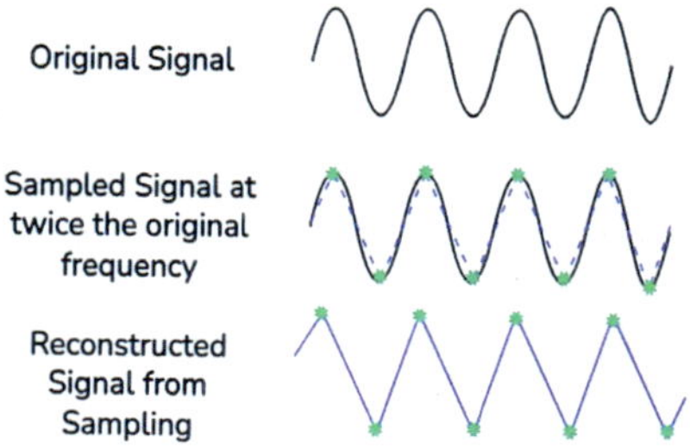

Figure 4.4. *Nyquist theorem of digital signal processing.*

Source: https://media.geeksforgeeks.org/wp-content/uploads/20240223131125/image-248.webp.

This technique allows for the recovery of information at rates sub-Nyquist and is therefore a useful instrument to apply in bandwidth-limited applications or with high-dimensional signals. With the application of concepts such as innovation rate, referring to the degrees of freedom per unit time for parametric signals, engineers have developed new sampling architectures for pulse streams. These advances, described in recent literature, emphasize the need for prefiltering input signals with carefully designed waveforms. These innovations have the potential to provide substantial reductions in hardware complexity as well as in the immunity of the sampling process to noise and other disturbances (Eldar et al., 2010). Further, research into sub-Nyquist methods has rekindled arguments as to whether the Nyquist Theorem can truly be applied in practical engineering practice. Sampling techniques map the trajectory of theoretical advancement developed since the beginning of the theorems, making theoretical knowledge formal with real application. Up-to-date studies endorse the utilization of complex algorithms that allow signal modeling in a union of subspaces, pushing the bounds of traditional sampling. By demonstration of the correspondence between the theoretical principles and the real systems, such investigations not only validate the applicability of the Nyquist Theorem but also support its use in the design of contemporary hardware. This, as demonstrated through contemporary literature, can potentially convert theoretical solutions into tangible applications for contemporary and future signal processing challenges (Eldar et al., 2011) (Figure 4.4).

4.1.2. Aliasing and Its Effects

The aliasing effect becomes an important concern in digital signal processing, particularly with the sampling of continuous-time signals to discrete streams. This occurs when a signal is sampled at a rate that is insufficient to capture its most frequent components and hence causes waveform distortion within its original presentation. The rigorous application of the Nyquist theorem dictates that, in a quest to prevent aliasing, the sampling frequency must be at least twice the highest frequency found in the signal. However, in practice, as explained in recent advancements, engineers commonly face challenges fulfilling this requirement due to bandwidth or resource-limited considerations for processing. Through the lack of utilizing correct sampling schemes, useful information is lost or misrepresented, effectively disturbing the integrity of data transfer and processing in systems like programmable rate digital satellite modems (Eldar et al., 2011). Moreover, the impact of aliasing extends beyond mere distortion; it has potential system-wide performance influence in systems that rely on precise signal reproduction. For instance, if aliasing occurs, high-frequency components of a signal will be undercover as low frequencies and produce misleading outputs. This can result in misinterpretation of data and reduce the effectiveness of algorithms meant to process such signals. In pulse modulation engineering, these errors could hinder effective communication, reduce audio and video systems' fidelity levels, or render data transmission protocols ineffective. The integration of contemporary theoretical methods, as offered in studies bringing together signal modeling and engineering sub-Nyquist methods, exemplifies a major path to minimize such issues and supports the need for ongoing study in minimizing aliasing effects in contemporary engineering (Eldar et al., 2011). Finally, aliasing fighting calls for the establishment of robust sampling systems, achievement of which is generally attained at the expense of balancing performance and complexity. Techniques such as anti-aliasing filters can be inserted into the signal path prior to sampling to limit the bandwidth of the input signal, ensuring that only of interest portions are recorded. Advanced digital signal processing techniques can also allow sampled data to be interpolated, which enhances signal representation and reduces the likelihood of aliasing. These techniques are instrumental to systems that are examined in engineering studies targeted towards the optimization of communication systems within the digital environment, whereby reliability under lower sampling rates corresponds with practical concerns of efficiency and cost in the real world (Bunya et al., 2025). Continued studies on these new techniques confirm the applicability of aliasing reduction in the context of evolving technological demands, making it a topical topic in sampling and pulse modulation engineering.

4.1.3. Types of Sampling Techniques

In signal processing, sampling processes are now needed to convert continuous signals into a discrete form compatible with digital processing. Among the various processes, uniform sampling and non-uniform sampling are two prominent methods. In uniform sampling with uniform time intervals between samples, the reconstruction of signals is simple due to regularity. In contrast, non-uniform sampling methods offer flexibility in sample acquisition by reacting to signal properties in real time and thus potentially are more efficient for complex waveforms. This versatility is best seen in pulse modulation engineering, where sampling methods must have the ability to accept various signal behaviors in a manner that will preserve integrity during modulation. Thus, choosing between these two types of sampling requires finding an equilibrium between both the efficiency in data capture and the preservation of the integrity of the rebuilt signal. One such advanced concept in sampling designs is the rate of innovation, which is the minimum rate of sampling required for extracting the features of pulse streams. Recent developments have presented multichannel structures that sample arbitrary pulse shapes effectively while preserving this rate of innovation. These developments enable pulse streams to be reconstructed from low-rate samples, something that uses techniques from spectral estimation for returning stable signals even at high innovation rates. Hardware simplification is available from the architecture, obviating problems of practical implementation often quoted for traditional sampling methods. By the use of a bank of integrators along with appropriately chosen waveforms for modulation, such new approaches are highly promising to enhance noise robustness and overall performance in uses of pulse modulation and mirror the progress of sampling techniques in engineering practices (Eldar et al., 2010). The pairing of these sampling techniques transcends their theoretical applications; practical applications have been constructed making use of everyday technology, such as personal computers and audio equipment, to create sophisticated systems. Applications that utilize pseudorandom noise (PN) codes in carrier modulation, for example, show how novel signal processing techniques can be applied towards object detection and approximate measurement. This balance of simplicity and complexity illustrates the pedagogical value of sampling techniques in learning concepts related to remote sensing and digital signal processing. Beyond enabling the understanding of theory models, it engages students in experimental hands-on work with low-cost but effective instruments. The research on sampling techniques, thus, not only encourages continued innovation in pulse modulation engineering but also supports pedagogic methods within the field (Campbell et al., 2011) (Figure 4.5).

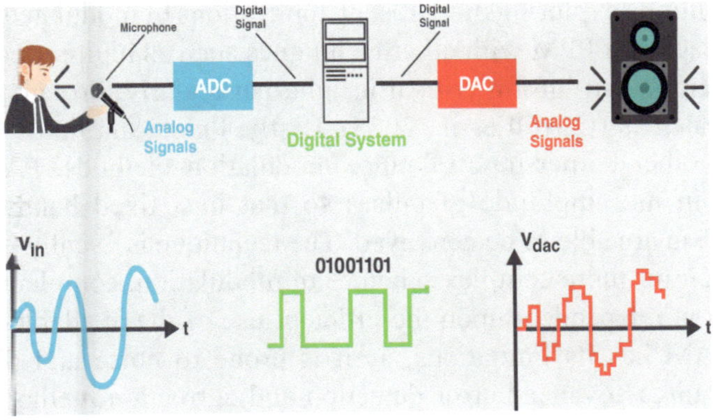

Figure 4.5. *Types of sampling techniques for digital processing.*

Source: https://www.allaboutcircuits.com/uploads/articles/An-Introduction-to-Digital-Signal-Processing-(3).png.

4.2. TYPES OF PULSE MODULATION

In the field of pulse modulation, varied techniques are employed to encode information in a pulse signal. One such is the Pulse Width Modulation (PWM), which is characterized by its effectiveness and ease of implementation. PWM encodes information by varying the pulse width based on the input signal amplitude. This process is prevalent in motor drive systems and transmission of signals due to its ability to utilize energy efficiently. Consequently, PWM systems are capable of high fidelity with minimal distortion, which suits the majority of real-time applications. The most significant virtue of PWM is its noise immunity since it can maintain signal integrity in long-distance transmission, particularly through the application of advanced modulation techniques like those in semiconductor optical amplifiers in mode-locking operations in distributed Bragg reflectors (Hou et al., 2018). And yet another significant category of pulse modulation is Pulse Position Modulation (PPM), which represents information dependent on the position of a pulse in a given time interval. PPM's new approach provides built-in noise immunity as the pulse shape itself contains less information than its location. This characteristic makes PPM particularly appropriate for low-bandwidth applications, where bandwidth use becomes preferred. In addition, PPM performs better in situations requiring precise timing, including optical communication and synchronization. These scenarios see the reduced complexity of PPM best equated with bandwidth-hungry schemes and thereby enhancing system reliability as a whole. As engineers continue exploring more recent applications of modulated signals, the relative efficacies of PPM with other techniques such as high-speed analog-to-digital converters can inform critical insight, particularly through redundancy encoding strategies (Carroll et al., 2025). Lastly, Pulse Amplitude Modulation (PAM) is another cornerstone of pulse modulation methods. PAM conveys information in the amplitude of pulses so that in a fixed bandwidth, more quantity of data are able to be conveyed. The technique is usually employed as the foundation for more complex schemes of modulation, complementing well the uses that are dependent upon the efficient use of the available bandwidth. However, PAM has its limitations, as it is prone to noise and distortion. It normally requires advanced error detection and correction methods to ensure data integrity in order to be effectively utilized. As the work goes on advancing in this field, the exploration of the synergy between PAM and other types of modulation can reveal new paths in communication technology, uniting concepts like those found in the recent advances of mode-locking and fiber-optic systems (Hou et al., 2018) (Figure 4.6).

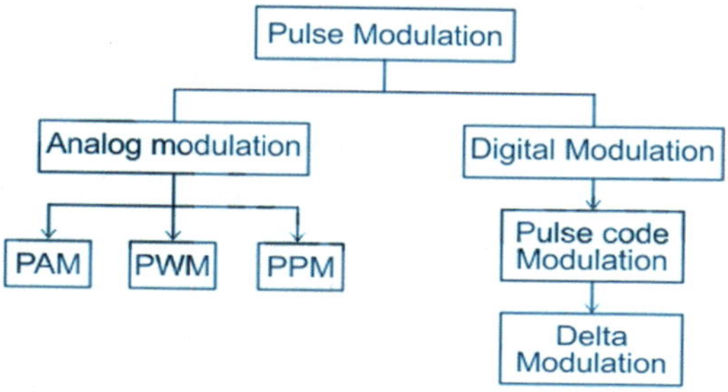

Figure 4.6. *Types of pulse modulation.*

Source: https://testbook.com/question-answer/consider-the-followingi-pulse-position-modulat--5fbd2389cb0fdbfbbd7f8cdc.

4.2.1. Pulse Amplitude Modulation (PAM)

In communication signal processing, PAM is a significant technique employed to convey information between communication media. PAM sends information encoded in the amplitude of individual pulses and facilitates effective signal modulation for effective usage of bandwidth. The basic idea behind PAM is sampling an analog signal in which levels of its amplitude are sampled at uniformly spaced intervals. This method is the foundation of most telecommunication devices, especially in cases where bandwidth is scarce. Digital communication has also brought into focus the role of PAMs, as it has a direct impact on the design of more complex modulation schemes, like Quadrature Amplitude Modulation (QAM), that has been extensively studied because of its ability to transmit at high-data-rate systems (Amaral et al., 2010). PAM has gained popularity due to its ease of use and effectiveness in real application, particularly when integrated with optical fiber transmission systems.

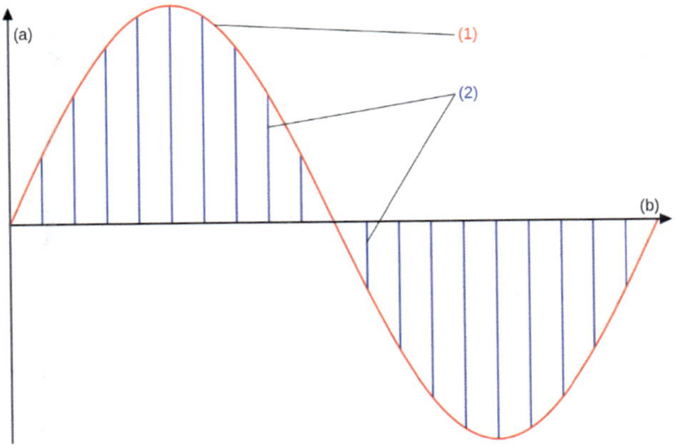

Figure 4.7. *Pulse amplitude modulation (PAM).*

Source: https://upload.wikimedia.org/wikipedia/commons/thumb/e/e2/PAM_neutral.svg/1200px-PAM_neutral.svg.png.

Modern technological development has led scientists to research PAMs capability for multi-gigabit data transfer, specifically short distance transmission by plastic optical fibers (POF). PAM-based methods such as intensity modulation with direct detection have been found to achieve tremendous data rates of up to 10 Gbit/s on large-core fibers and are hence suitable for residential and datacom. Such developments are particularly relevant in contemporary network design, as they cater to the growing need for enhanced capacity and faster transmission speed (Charbonnier et al., 2011). Therefore, PAM continues to be a major contributor to communication technology development. The PAM analysis also reflects its shortcomings, particularly with respect to susceptibility to noise and signal integrity during transmission. Heavy distortion in real-world applications due to variations in the transmission medium necessitates efficient error correction techniques and adaptive filtering schemes. Also, as the need for faster data rates becomes the top requirement, the modulation schemes of PAM must be modified to support compatibility with advanced communication systems. The ability to integrate PAM with more advanced modulations, as observed in the high-energy physics needs of CERN, goes towards demonstrating ongoing work on the development of higher modulation efficiency with reduced error rates (Amaral et al., 2010). This study not only identifies the strategic significance of PAMs in engineering but also paves the way for further innovations that can make future communication networks more reliable and perform better (Figure 4.7).

4.2.2. Pulse Width Modulation (PWM)

PWM is a critical technique in pulse modulation and sampling engineering, particularly its application in power electronics. PWM encompasses the modulation of the width of the pulses in a digital signal to reproduce an analog signal. This method of modulation works best in the operation of electrical appliances, for instance, motors, via varying the average power delivered to the load. The operation of PWM is attributed to the fact that it can produce an efficient switch operation with less harmonic distortion in the output signal. As power systems move towards more digital control methods, the need for efficient PWM methods becomes increasingly evident, particularly in variable frequency drives and other motor control applications. PWM system design and implementation require proper care of the modulation methods involved. For instance, the Sinusoidal PWM and Third Harmonic Injection PWM are common methods that achieve maximum output current quality and minimum harmonic distortion within three-phase inverter systems. As discussed in PWM literature, for instance, Dahono et al. (2010), output current harmonics analysis is key to enjoying the performance of various modulation methods. These methods enable precise control of voltage and current waveforms to drive motor and other electrical apparatus smoother. With the design of a well-working PWM inverter, not just reliability but also the power transfer system's efficiency is enhanced as part of the general trend towards sustainable energy technologies. Developments in digital control technologies have also enhanced PWM system performance. Integration of Field Programmable Gate Arrays (FPGAs) in PWM inverter design facilitates sophisticated modulation techniques with real-time processing. The facility allows precise control of the switching frequency as well as adaptive response capability to changes in the load, thereby improving system responsiveness. In motor control, the use of self-sensing estimators, as in Belie et al. (2013), shows the feasibility of the replacement of traditional position sensors with enhanced system reliability at reduced expense. With further progress in such technologies, integration of PWM techniques with control systems using digital electronics has the capability of promoting innovative alternatives in sampling and pulse modulation engineering to pave the way for more effective and stable electrical drive systems (Figure 4.8).

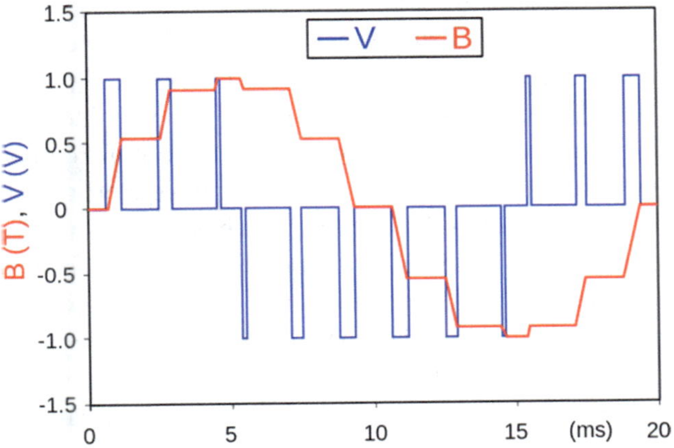

Figure 4.8. *Pulse width modulation (PWM).*

Source: https://www.analogictips.com/wp-content/uploads/2017/04/Fig-2-640px-PWM3-level-copy.jpg.

4.2.3. Pulse Position Modulation (PPM)

PPM is one of the major methodologies in digital communication, which captures energy efficiently and transmits data securely over a variety of channels. In this form of modulation, data is encoded by modifying the pulse positions in time, thereby allowing multiple bits of information to be transmitted in a single pulse interval. By positioning pulses in well-controlled positions, PPM reduces bandwidth needs tremendously and also enhances resistance against noise interference simultaneously. This performance is particularly realized in cases of high data integrity, where standard methods are guaranteed to fail. PPM therefore presents itself as a solution for specific applications such as optical communications, where it is paramount to maintain high signal fidelity in hostile environments (Lev et al., 2021). Pragmatic advantages of PPM are also evident through its alignment with advanced modulation schemes, resulting in enhanced performance measures. One such example is the hybrid PPM–GMSK scheme, a combination of the useful characteristics of PPM with Gaussian minimum shift keying. As demonstrated in current studies, the hybrid scheme has been able to effectively address issues of atmospheric turbulence, notably in optical wireless communication systems. With analytical bit error rates and outage probabilities being derived by studies, it is evident that the PPM–GMSK arrangement outperforms traditional modulation approaches like 2-PPM and minimum shift keying (MSK) both in terms of coverage distance and reliability. This verifies the multi-purposeness and adaptability of PPM in contemporary communication systems, demonstrating its potential to cater to the demands of current data transmission needs (Sahoo et al., 2018). Within the broader background of sampling and pulse modulation engineering, PPM is not only a signal of progress in transmission efficiency but also a reminder of the significance of integrated coding schemes minimizing distortion. The integration of designed source-channel coding with PPM has the potential to drastically reduce quadratic distortion and make it a possible solution for high bandwidth situations. Recent research indicates that such composite techniques have the potential to surpass existing modulation techniques, achieving maximum orders of decay in fidelity of the signal. With a focus being laid on a synergy between source quantization techniques and coding techniques, future research into PPM can offer new solutions towards reducing energy consumption and optimizing overall communication efficiency, making it a cornerstone of modern paradigms for signal transmission (Lev et al., 2021) (Figure 4.9).

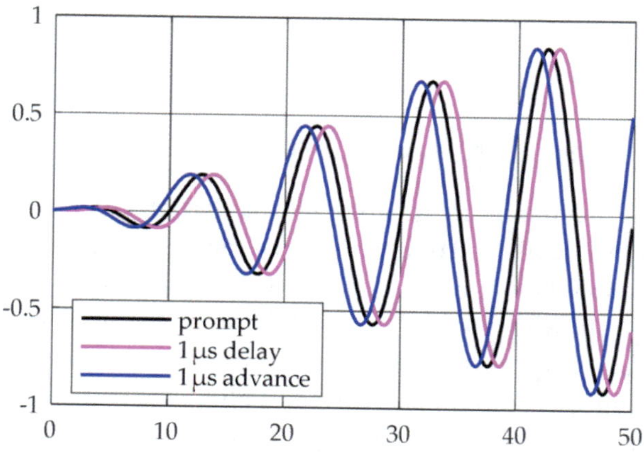

Figure 4.9. *Pulse position modulation (PPM).*

Source: https://www.researchgate.net/publication/343645598/figure/fig2/AS:11431281391604801@1745332913891/Pulse-position-modulation-PPM-signal.tif.

4.3. APPLICATIONS OF SAMPLING IN DIGITAL SIGNAL PROCESSING (DSP)

Applications of sampling in DSP are numerous and pivotal to the future communication world. Practically, sampling allows continuous-time signals to be converted into discrete sequences so that information is processable and transmittable in digital form. A key role in the process is played by the Shannon-Nyquist theorem, which formulates the conditions for perfect reproduction of signals without aliasing. In other uses, especially in audio and video processing, high-fidelity sampling methods are employed to ensure high fidelity. In music streaming services, for instance, proper application of sampling rates significantly enhances the listeners experience by eliminating distortions due to improper sampling techniques. This emphasis on correctness illustrates the significance of sampling practice in signal integrity (Eldar et al., 2011). Besides, DSP advancements have introduced new sampling techniques such as sub-Nyquist sampling targeted at minimizing ADC rates below the traditional Nyquist rates. The area holds a massive potential in resource-constrained environments, such as satellite communication systems where power consumption and bandwidth usage are priorities on the list. By bridging theory and practice, engineers are devising methods that take advantage of signal modeling in subspaces. This overlap allows for the evolution of DSP techniques from theoretical concepts to actual, hardware-driven implementations that enhance the effectiveness of signal processing. This transition from theory to application is particularly important in areas like telecommunications, where challenges such as noise and interference of signals must be resolved without wasteful use of resources (Eldar et al., 2011). In the context of narrowband satellite communications, sampling transcends theory and into the realm of complex real-world challenges. The software-defined receiver dissertation illustrates how efficient digital downconversion methods can counteract excessive levels of noise and large Doppler effects faced in deep-space communications. The study seeks computationally efficient but signal fidelity under adverse conditions. In addition, the actual applications of these algorithms—emulated in environments like MATLAB Simulink—demonstrates their effectiveness in processing real signals received from satellites. By exploring such advanced applications of sampling, it is realized that continued research and development in DSP can greatly contribute to communication technologies in providing new solutions to signal processing.

4.3.1. Audio Signal Processing

The field of audio signal processing is at the core of modern technology, allowing sound signals to be dealt with effectively and interpreted. The field encompasses various techniques that can be employed to sample, filter, compress and modulate audio signals, which enhance their quality and make them applicable in various fields. Audio sampling allows us to depict continuous sound waves in discrete terms that the digital system can process and examine. Sampling principles, for instance, the Nyquist theorem, emphasize the need for careful frequency selection to ensure the proper sampling of audio signals and not the loss of the same. These fundamental principles of audio signal processing directly influence communication, entertainment, and environmental sensing technologies, the width and worth of the field in everyday usage. The intersection of audio signal processing with pulse modulation has resulted in revolutionary methods of transmitting and receiving audio data. Pulse modulation makes it possible to encode sound signals in a modulated form that can be successfully transmitted across various channels of communication. Pulse Coded Modulation (PCM) and Delta Modulation (DM) are some methods that utilize the beneficial aspects of audio processing and modulation to offer clearer and more precise reproduction of sound.

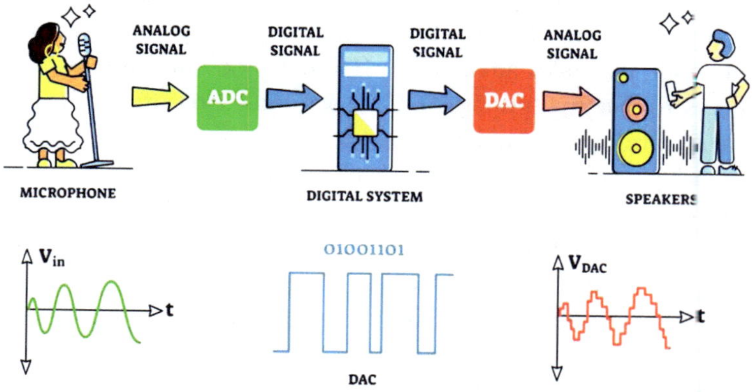

Figure 4.10. *Audio signal processing role in modern technology.*

Source: https://i.ytimg.com/vi/JwDJ29ZjzWE/maxresdefault.jpg.

Besides, new developments have indicated how low-cost personal computer setups can utilize pulsatile modulation methods for a broader class of applications, including digital signal analysis and remote sensing. For instance, speaker and microphone systems can detect objects and perform simple measurements through carrier signals that are modulated through pseudorandom noise, providing useful learning information about the use of such concepts (Campbell et al., 2011). The applications of audio signal processing go beyond the usual, creating new prospects for dealing with security issues in digital communications. The development of new methods of hacking, like acoustic covert channels, poses challenges that require a strong basis in the understanding of audio signal transmission channels. Recent studies identify that attackers are able to exfiltrate sensitive data from air-gapped computers via attacking audio emissions from internal hardware components, thereby revealing vulnerabilities within seemingly secure systems (Guri et al., 2020). Such a situation creates an imperative to develop innovative methods in audio signal processing that can identify, discourage, and defend against such security threats. Audio-gapped systems—systems that muffle audio hardware for security purposes—are therefore emerging as a counter-defense strategy against such threats. Through the growth of our understanding regarding audio signal processing in relation to security technologies, researchers are able to contribute towards more secure communication infrastructures (Figure 4.10).

4.3.2. Image Processing

Image processing is a fundamental tool in most engineering fields, particularly for sampling and pulse modulation. Imaging technologies have evolved, and it is now possible to move away from traditional multi-pixel systems to new single-pixel cameras. This not only reduces costs, but also makes maximum use of the life of imaging devices to render them more applicable in such fields as medical diagnosis and industrial inspection. Utilization of single-pixel cameras, as depicted in more recent work, has shown that it is actually possible to create high-resolution images using optimized modulation geometries. With algorithms designed for efficient post-processing, these systems have transformed the sampling and inspection of images, and this has influenced greatly the likelihood of impact on future imaging technology and devices and applications and their roles in shaping society. Apart from hardware advancements, image processing performance also relies on effective data-handling techniques, namely compression schemes that have bandwidth features control. For pulse modulation, it still remains critical to examine various methods of data compression techniques that optimize the performance of imaging systems without compromising inherent image quality. Such schemes can dramatically reduce data size processed without compromising image output integrity. New developments in bandwidth compression techniques tailored for video signals illustrate the continued drive to enhance image data storage and transmission. Coupling these methods with contemporary sampling techniques, engineers are now in a position to improve system responsiveness and efficiency by orders of magnitude—essential building blocks for real-time use in active environments. Additionally, development in terahertz (THz) imaging illustrates the evolving nature of image processing technology. The unique nature of THz radiation opens up new real-time imaging opportunities across different applications pending fulfillment of the current instrumentation constraints. Single-pixel camera capability in imaging THz images demonstrates an important milestone toward efficient image sampling. Employing novel techniques like noise elimination by carrier-wave modulation, research has considered successful performance in high-speed image capture while maintaining minimal complexity in post-processing. This is part of a wider trend in image processing engineering where the emphasis is increasingly being placed on developing convenient, lightweight solutions that do not compromise on performance. As these technologies continue to evolve, they promise to take a range of imaging applications to new levels, transforming industries that depend upon accurate visualizations.

4.3.3. Data Compression Techniques

Compression techniques play a critical role in the efficiency of data transmission and storage, particularly if used within the domain of sampling and pulse modulation engineering. As the size of digital data continues to grow in size, grow exponentially, so too does the relevance of compression techniques become obvious. Compression not only diminishes the quantity of data to be transmitted, making bandwidth efficiency greater, but also lowers storage capacity requirements. Methods like lossless and lossy compression serve different purposes since lossless compression ensures that original data is not changed, whereas lossy compression discards more information. Their impact can be felt in several applications, such as multimedia content and real-time communication, to enable the transparent implementation in systems that depend on the quick processing of large amounts of data. These methods must be applied and interpreted to achieve an optimum solution for contemporary engineering. There arises the significance of bandwidth usage in interaction with data compression methods pertaining to pulse modulation and sampling. Pulse modulation schemes commonly employed are PCM, DM, and others, although they are handicapped due to bandwidth constraints of the communication channel. Effective data compression techniques, therefore, can reduce some of these limitations by encoding information more succinctly at the source. The application of compression algorithms, particularly for video signals demonstrates how combining these technologies can significantly enhance sampling system performance. Crawford et al. (Ingels et al., 2025) explain various compression schemes that may be utilized in pulse modulation systems to achieve optimum transmission of large data volumes without sacrificing signal quality. This complementarity, in addition to improving transmission speeds, allows more advanced applications of signal processing to be implemented. Furthermore, the development of software-defined radar systems denotes the importance of incorporating innovative data compression schemes into modern communication systems. The flexibility offered by SDR allows for the generation of waveforms with arbitrary profiles and the implementation of sophisticated processing schemes, which can imply remarkable advantages in radar performance. Experimental findings indicate that innovative structures, like those employing P3 phase-codes, can take advantage of sophisticated data compression techniques to maximize signal purity and minimize operational bandwidth requirements. Utilizing bandpass sampling techniques in combination with successful compression, engineers are able to break the conventional state-of-the-art performance barriers, as supported by recent research (Kernec et al., 2017). This integration of the compression in pulse-modulated engineering improves the current technology and also offers possibilities in new application as in communication, imaging, and radar systems.

4.4. APPLICATIONS OF PULSE MODULATION

Applications of pulse modulation are far-reaching and significant in various fields of engineering and technology. One of the key applications is digital communications in which pulse modulation techniques such as Pulse Code Modulation (PCM) are used extensively for converting analog signals into digital signals. The digitalization ensures the signal is less susceptible to noise and interference during transmission. Moreover, pulse modulation methods like Pulse Amplitude Modulation (PAM) and Pulse Position Modulation (PPM) are also used in telecommunications to optimize bandwidth efficiency and improve transmission efficiency. These techniques enable effective data transfer in applications ranging from ground to satellite communication, thereby reflecting the vital role of pulse modulation in making modern communication systems operational and reliable. Apart from telecommunications, pulse modulation also has a central role to play in multimedia applications, particularly audio and video signal processing. Techniques such as PCM underlie digital sound playback and recording systems with high fidelity in the reproduction of sound. The techniques allow efficient coding of audio signals, thus allowing easier storage and transmission without degradation of sound. Similarly, video encoding algorithms are subjected to pulse modulation, wherein visual information is conveyed in MPEG and JPEG forms. Advances in integrated circuit technology accelerated these applications further, enabling the development of high-end multimedia systems utilizing pulse modulation for enhanced performance. In this way, not only does pulse modulation enhance the quality of multimedia, but it also drives innovation in digital media services. New technologies are also utilizing pulse modulation techniques for new applications beyond traditional uses. For instance, emerging advances in quantum optics utilize pulse modulation to generate squeezed light, with profound implications for quantum computing and secure communications technology. As mentioned, generation of squeezed light via spontaneous four-wave mixing in integrated nanophotonic devices demonstrates the remarkable technological advancement in the manipulation of quantum states, with the implication being for practical applications in continuous variable quantum technology (Collins et al., 2020). Further, in remote sensing technology such as Synthetic Aperture Radar (SAR), pulse modulation techniques are employed to detect small vibrations of targets, and thus the application of pulse modulation is reflected in high-accuracy measurement and monitoring tasks (Atwood et al., 2012). The continued quest for pulse modulation techniques is likely to give still more applications, incorporating these methods even further into the fabric of high-end engineering and technology.

4.4.1. Telecommunications

The development of telecommunications is deeply interlinked with that of advances in signal processing and modulation methods, specifically in sampling and pulse modulation engineering. It is a daunting task in modern systems to transmit information over long distances at high efficiency without any distortion or quality loss. By researching long-haul optical links, scientists have mentioned linear and nonlinear impairments that impede signal integrity. Recent studies promote the exploitation of time-frequency packing with low-order constellations as a feasible means of enhancing spectral efficiency. The approach of investigating the trade-off between system constraints and modulation formats has a great potential to solve theoretical and practical issues related to spectral efficiency, and it represents a powerful substitute for mere utilization of high-order modulation formats (Colavolpe et al., 2014). Pulse modulation techniques, particularly trellis-coded modulation, play a critical role in optimizing transmission capability in telecommunications. Adaptive trellis modem structure is indicative of a down-to-earth solution for optimizing power and spectral efficiency. The innovation integrates various modulation forms like Phase Shift Keying (PSK) and Quadrature Amplitude Modulation (QAM) with robust digital baseband processing. The adaptability allows systems to adjust dynamically, which enhances performance in different communication conditions. Capability for modulating styles like BPSK, QPSK, and 16-QAM with up to 20 megabits per second illustrates the possibility of tailoring telecommunications systems to counter new demands with less interference and optimal bandwidth usage (Budinger et al., 2025). As the speed of telecommunication growth picks up, the use of higher sampling rates and methods of pulse modulation will be key to overcoming the inherent limitations of digital communications. By refining these processes, engineers can move forward toward higher data rates and greater reliability, which will become essential for the larger Internet of Things (IoT) and future networks. The fusion of traditional analog techniques with cutting-edge digital technology fuels ongoing investigation for improved signal transmission and preservation of quality in conversion processes. The more information that is transmitted, the greater the potential to utilize adaptive coding and modulation technologies will impact the effectiveness and success of international communications, setting new standards in what can be accomplished through modern telecommunications infrastructure.

4.4.2. Radar Systems

Radar systems are a critical aspect of modern technology since they employ electromagnetic waves in various applications such as surveillance, navigation, and geological survey. The systems convert the radiated signals into embedded information and thus can detect and characterize distant objects. Radar operations involve transmitting bursts of radio waves and analyzing the reflected echoes, which give information regarding distance, speed, and even material composition. At the core of the efficiency of radar systems lies sampling of the returned signals. Proper sampling techniques enable the capture of essential data without loss, which is vital for proper signal processing and interpretation in real-time or near-real-time operations. Radar data processing utilizes advanced techniques of converting the analog echoes into usable data. For example, charge-coupled device (CCD) systems enhance the quality of the data processing. The CCD circuits perform well in executing basic operations such as input sampling, presumming, range correlation, and azimuth correlation in the analog domain. This capability is crucial for use in synthetic aperture radar (SAR) applications, where image formation relies significantly on the quality of these operations. In single-look and multiple-look imaging, CCD-improved radar systems establish an intensity of accuracy and detail that is essential in missions ranging from military surveillance to environmental monitoring (Arens et al., 1977). The type of ground material as measured by radar systems is instrumental to diverse applications ranging from analyzing subsurface properties and geophysical surveys. The capacity of radars to measure resistivity and magnetic permeability reveals important information about the physical conditions of the ground. Resistivity change can reflect levels of moisture, temperature, and other environmental factors affecting reaction of the material to electromagnetic waves. These results become increasingly significant as they inform engineers and scientists about the underground conditions, facilitating activities such as resource exploration to environmental conservation. The integration of such functions reveals the sophistication of radar technology and the importance of sampling and pulse modulation methods in effectively capturing and analyzing the hidden data (N/A, 2025) (Figure 4.11).

Figure 4.11. *Radar systems.*

Source: https://www.alliedmarketresearch.com/blog/Blog_banner_image/7qchN2e2Rz.jpeg.

4.4.3. Control Systems

Control systems constitute the central part of modern engineering practices, particularly for sampling and pulse modulation engineering applications. The control systems are engineered to manage and regulate dynamic processes so that the outputs can respond appropriately to varying inputs with the desired performance conditions remaining intact. By this definition, control systems are of the open-loop and closed-loop forms, with the closed-loop systems having feedback mechanisms to adjust actions based on discrepancies between desired outcomes and real outcomes. The use of feedback systems is critical in applications that require real-time adjustment and precision, hence their particular interest in use in digital signal processing applications. It is the understanding of rules that govern such systems, from stability, controllability, and observability, which becomes paramount for engineers seeking to develop successful models that successfully interpret the complexities of real operations. In signal processing, the use of sampling and pulse modulation techniques heavily relies on sophisticated control techniques. The efficiency of such techniques relies heavily on the ability of the control system to manage timing and synchronization. As an example, the sampling theorem specifies that a signal be sampled at least double its highest frequency to prevent distortion. As a result, control systems are utilized for ensuring sampling rates are properly tuned, with fidelity preserved and minimal critical information loss. By incorporating stronger control mechanisms, engineers can maximize system performance based on pulse modulation for the transmission of digital data. This integration enhances the signal quality as well as issues related to aliasing and quantization errors, further highlighting the position of control systems in communications reliability. Additionally, new developments in control technologies such as high-speed analog-to-digital converters (ADCs) characterize the convergence of control systems with modern signal processing techniques. These ADCs are key in the process of converting continuous signals into digital form, making the sampling and modulation efficient. Current studies indicate that the application of redundancy encoding techniques in such systems can be utilized for performance enhancement via the removal of potential errors and enhanced data integrity (Carroll et al., 2025). In addition, the mechanization of spacecraft communication systems on theoretical terms illustrates the application of control techniques in managing the data handling subsystems effectively (Hartman et al., 2025). It is proof of the flexibility of control systems, illustrating their relevance from terrestrial applications to space expedition endeavors. The continued advancement of control techniques will yield still further fruits in terms of sampling and pulse modulation technology, defining the fields dynamic personality.

4.5. ADVANTAGES AND DISADVANTAGES OF SAMPLING

The process of sampling is a basic technique in various fields, including pulse modulation engineering, in which sampling serves a significant function in efficient transmission as well as signal analysis. One of the biggest advantages of sampling is that it can quantify and detect continuous signals through discrete representation, thus enabling processing and storage of massive datasets while ensuring that essential information is not compromised. For instance, as highlighted in CERN studies, advances in data acquisition systems demonstrate how sampling techniques enable the handling of more luminosity data at the same high level of fidelity; the Quadrature Amplitude Modulation (QAM) system ran with data rates of approximately 3–4 Gbit/s under optimum conditions. This efficiency in handling data not only speeds up the process but also simplifies the complexities of continuous signal transmission, and hence the importance of sampling as a vital technique for today's engineering applications. However, sampling is accompanied with severe drawbacks that must be overcome to avoid pitfalls in signal representation. One major drawback is the aliasing effect that occurs when the sampling rate is lower than sufficient to capture the variations in signals accurately. If the signals are sampled below the Nyquist rate, valuable information may be lost or inaccurately represented, leading to distorted outputs that can severely affect the success of data retrieval and analysis. Also, as exemplified from the development of high-resolution imaging devices, even advanced techniques such as heterodyne detection are prey to these issues if sampling protocols are not under proper control. Finally, data sampled integrity depends on the selection of an appropriate sampling strategy, reinforcing the necessity of stringent planning and operation in pulse modulation engineering. In new technologies, sampling strengths and weaknesses are brought to the fore in ensuring accuracy and precision in data gathering. New technologies such as the high precision simultaneous full-field acquisition range imagers described in recent research employ complex sampling methods to achieve sub-millimeter accuracy over extended fields of view. These advancements demonstrate that even though sampling can make huge contributions and breakthroughs in fields ranging from physics to imaging, it also demands prudence with respect to the manner in which the sampling frequency may influence the overall system performance. The interfacing between different sampling techniques and their application to projects like that being pursued at CERN illustrates an integrated view of samplings function, towards ongoing research and development to further these instruments towards the pursuit of future technological innovation within pulse modulation engineering (Figure 4.12).

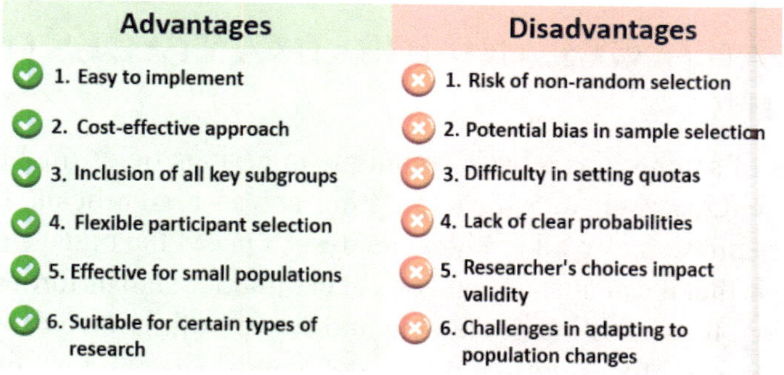

Figure 4.12. *Advantages and disadvantages of sampling.*

Source: https://cdn.educba.com/academy/wp-content/uploads/2024/01/Quota-Sampling.png.

4.5.1. Benefits of Digital Representation

The transition from analog to digital format in the area of pulse modulation and sampling introduced seminal breakthroughs in the realm of communication technologies. The simplest advantage, yet one of the most pivotal ones, lies in the format of digital representation as far as processing and transmitting data are concerned. Digital signals are easy to process using algorithms, which facilitates enhancements such as error correction and advanced methods of signal processing. This results in improved fidelity and transmission strength through numerous channels of communication. For instance, the ability to process information in digital format enables the implementation of sophisticated modulation methodologies, thereby enabling higher data rate achievement and systems implementation under stringent performance constraints, as symbolized by recent developments in programmable baseband modulators that can support multiple modulations, for instance, QPSK for premium wireless standards (Chakrabarti et al., 2010). Another crucial advantage of digital representation is the enhanced functionality it offers in data storage and retrieval of information. Digital representation makes it possible to compress data with little loss in quality, thus achieving maximum storage capacity.

This not only reduces the cost of operation in data handling but also achieves a maximum rate of data transmission. For instance, digital signals can be stored in a variety of formats that allow for rapid retrieval and access to process them, particularly where programs have to respond with real-time analysis of data. The inclusion of irregular nonuniform sampling schemes has also been shown to attain similar or even superior channel capacities, once again testifying to the efficacy of digital mechanisms to maximize data throughput without interference. This representational flexibility in the digital domain is a contrast to analog systems which suffer from performance scalability and data fidelity constraints (Chen et al., 2014). The benefits of digital representation extend beyond technical performance, such as interoperability and standardization across communication networks. Digital formats provide a common structure in that devices from various manufacturers can communicate with one another without restrictions.

This standardization is essential in a more interconnected world, in which devices are frequently dependent on exchanging vast amounts of data between incompatible systems. In addition, the digital world is constantly changing, enabling quick integration of new technologies and protocols that maximize communication capacity. From the insight on recent advances in software-defined radio architectures, digital representation not only offers scalability but also low-cost reconfiguration and parameterization, helping to circumvent legacy systems' limitations. The potential of transformation from digital representation thus justifies its centrality in driving the limits of sampling and pulse modulation engineering (Figure 4.13).

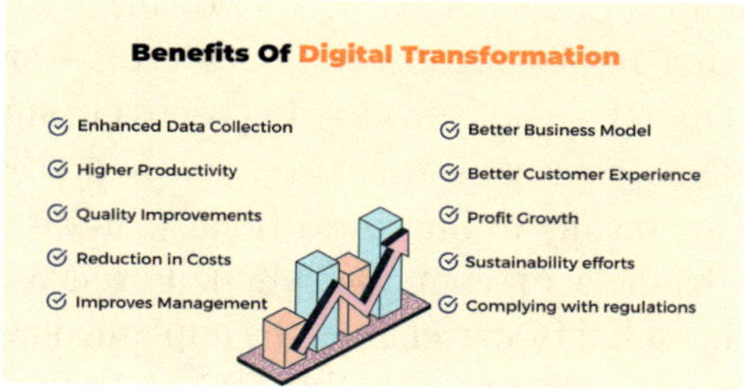

Figure 4.13. *Benefits of digital representation.*

Source: https://quixy.com/wp-content/uploads/2022/12/Benefits-of-Digital-Transformation.png.

4.5.2. Limitations of Sampling Techniques

Sampling techniques are at the core of digital signal processing but are riddled with limitations that, in return, affect considerably the fidelity of the recovered signal. A key among these is the issue of aliasing, which occurs when a signal is sampled below its Nyquist rate, leading to distortion and loss of essential data. This phenomenon can cause misrepresentation of the original signal, particularly in systems where fidelity is extremely crucial, e.g., audio or medical imaging. Moreover, even when the sampling rate satisfies the Nyquist criterion, engineering facts like quantization errors can introduce noise and inaccuracy. These problems are demonstrated in the scenario of high-order applications, such as pulse modulators, where the proper simulation of signals plays a vital role (Paul et al., 1961). Apart from aliasing and quantization error, the use of sampling techniques can also impose tight constraints on the system's overall efficiency. For example, the choice of sampling frequency not only dictates the data rate but also affects the signal processing algorithms employed thereafter. Higher sampling rates may result in too much data, for which more advanced compression algorithms are required that can, subsequently, further complicate processing and analysis. Similarly, hardware limitations, e.g., the limitation of analog-to-digital converters, can frustrate the operation of sampling schemes. As seen in pulse modulation engineering research, limitations of current modulation methods, like conventional current-regulated delta modulators, can exacerbate these issues and cause high ripple or error at different velocities, pointing towards the sensitive character of sampling in engineering procedures (Howe et al., 2005). Apart from purely technical limitations, sampling technique constraints also involve practical considerations, like resource accessibility and system design restrictions. In cases where rapid processing is paramount, like in the case of real-time systems or embedded systems, system resource usage versus sampling rate trade-offs are extreme. For instance, increasing the sample rate can enhance temporal resolution but only at the cost of increased latency or resource utilization. This delicate balance calls for engineers to optimize sampling methodologies while honoring all operating conditions. Lastly, the recognition and remediating of such deficiencies are key to optimizing the effectiveness of sampling methods in pulse modulation engineering and determining the reliability of output signals (Paul et al., 1961).

4.5.3. Trade-offs in Sampling Rate

In Engineering Sampling and Pulse Modulation, sampling rate selection is an essential trade-off that necessarily affects system function and data integrity. Higher sampling rates preferentially produce more accurate replicas of the original signal, enabling recording of fast dynamics and fine detail easily missed. Although this fidelity enhancement is paid for in terms of higher levels of data and processing load,. Systems operating at higher sampling rates are necessary for robust storage and higher-level processing components, which can significantly boost operation costs. Therefore, a careful assessment of requirements to analyze and system capability becomes an overriding consideration in the selection of an optimum sampling rate that can fulfill fidelity while remaining efficient. Lower sampling rates reduce volume of data, rendering storage and processing needs simpler while raising efficiency. They can, however, introduce aliasing effects, which tend to misrepresent the high-frequency signals.

This is particularly crucial for application in radar and audio engineering, where fidelity of signals has direct implications on the result of operations. One example is in radar technology for the exploration of the lunar surface in which the adequacy of the data recorded at various rates determines the quality of surface mapping systems. Goldberg et al. (2025) propose that precise adjustment of the sampling rates is vital to avoid critical information loss in the lower-resolution data streams, which will ultimately affect mission objectives and data precision. In addition, integration of new analysis methods can offset the consequences of reduced sampling rates, leading to enhanced overall system performance. Modern systems have the tendency to incorporate sophisticated algorithms that can interpolate or restore missing data points, allowing for functional use of reduced sampling rates without causing a substantial loss in reliability. In wireless technology-based sensor networks, use of intelligent algorithms allows for remote classification and monitoring of conditions, as observed by Baker et al. (2009). Such strategies stress that while sampling rate compromises are extremely applicable, advancement in signal processing techniques can offer opportunities for optimizations of system design. Meeting the innate sampling rate compromises in a well-balanced fashion is essential to efficient, dependable, and economical engineering solutions.

4.6. ADVANTAGES AND DISADVANTAGES OF PULSE MODULATION

The advantages of pulse modulation are tremendous, particularly in telecommunications and signal processing. One of its main advantages is that it is efficient and resistant to noise. Pulse modulation techniques such as Pulse Code Modulation (PCM) and Differential Pulse Code Modulation (DPCM) enable the transmission of signals over a long distance with dependability and are a significant component of modern-day communication systems. Moreover, pulse modulation is easy to combine with digital circuits and facilitates enhanced data compression and transmission rates. One such success is that with Quadrature Amplitude Modulation (QAM), good data rates of around 3–4 Gbit/s have been obtained in high-energy physics applications, reflecting the capability of pulse modulations to handle immense data volumes while maintaining signal integrity (Amaral et al., 2010). Therefore, the inherent benefits of pulse modulation make it the application of choice in most technological applications with the objective of enhancing overall communication system reliability. Irrespective of this, there are drawbacks to pulse modulation, which tend to reduce its efficacy under certain conditions. Among the initial challenges is the complexity of hardware involved, which can be associated with higher cost and larger power demands. This holds particular truth when the case is viewed in terms of intricate modulation schemes based on sophisticated signal processing equipment and algorithms. Such complexity can be a significant impediment in cases necessitating high accuracy as well as rapid processing. Further, susceptibility of certain pulse modulation techniques to quantization noise and timing errors can lead to degraded performance. As illustrated in high-accuracy applications, such as the simultaneous full-field acquisition range imager through amplitude modulation, any timing error would result in measurement error and even dangerous consequences in real critical situations (Carnegie et al., 2006). Thus, while pulse modulation offers its advantages, these disadvantages must be considered very cautiously on implementation. In addition to the technical challenges involved in pulse modulation, practical applications typically involve compromises between system performance and operational convenience. For example, the employment of effective error correction techniques is important where there is the possibility of high rates of error in pulse modulation systems. These, however, would add levels of complexity and delay that undermine the speed advantages generally provided by pulse modulation. Moreover, its installation on existing infrastructure may involve major upgrades to accommodate the high-end technology necessary to optimize performance. This transition can be cost-prohibitive and incurred against resistance to change in proven systems. Therefore, stakeholders must weigh the advantages of pulse modulation against these natural limitations, ensuring that the chosen solution fits the operational goal and technological capabilities of their own specific context.

4.6.1. Efficiency in Bandwidth Utilization

In Sampling and Pulse Modulation Engineering, bandwidth efficiency is the ultimate criterion for the optimization of communication systems. Optimization is particularly critical in consideration of the needs for increasing data transmission over a range of media. The use of high-level modulation methods, e.g., Quadrature Phase Shift Keying (QPSK), is a demonstration of developments in data rate optimization at the cost of the use of bandwidth. By optimizing the use of existing bandwidth, these schemes enhance higher throughput without the need for additional bandwidth assignment. The impact of software-defined radio (SDR) architecture parameterization highlights one of the most important achievements in developing more flexible and efficient systems, also justifying how today's technologies are able to greatly enhance bandwidth efficiency at low complexity and operational expenses (Chakrabarti et al., 2010). Continuing in this vein, the examination of narrowband radio communication systems reveals another level of bandwidth effectiveness. These systems are subject to strict limitations and specifications, such as the European ETSI EN 300 113, which defines limited bandwidth specifications. Subject to extreme restriction, signal processing technology advancements and modulation methods enable substantial amounts of data to be transmitted. Use of novel algorithms in signal synchronizing and most importantly modulation filtering methods allows for significant reduction of inter-symbol interference (ISI), improving overall communication performance. Consequently, these advancements allow for more effective use of narrowband channels with quality preserved and data rate optimized (Bobula et al., 2011). Additionally, examination of the inclusion of digital processing in modulation schemes elucidates current trends aimed at spectrum efficiency. The application implications of mixing linear and nonlinear modulation methods lead to increased data rates and reduced power dissipation in narrowband channels. Designers are increasingly creating adaptive digital modems that not only satisfy regulatory compliance but also enable the sustainability of high-performance systems in competitive markets. With industries adopting efficient communication solutions, the use of advanced algorithms and filters in modem design is an absolute determiner of excellent operation. Overall, signal modulation and sampling technologies with upcoming technologies can potentially reverse limitations towards bandwidth utilization, creating the threshold for enhanced advancements in communication systems (Chakrabarti et al., 2010; Bobula et al., 2011) (Figure 4.14).

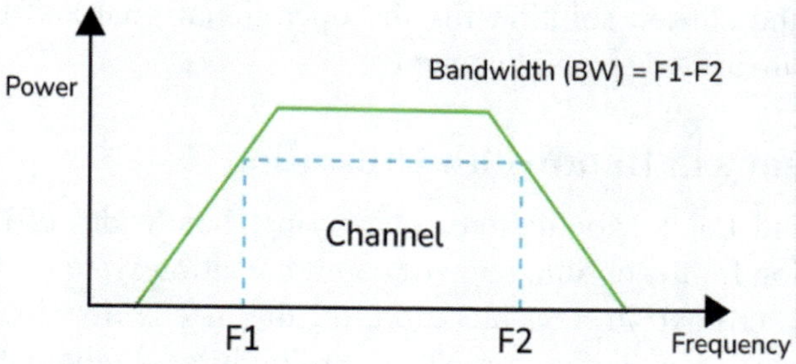

Figure 4.14. *Efficiency in bandwidth utilization.*

Source: https://media.geeksforgeeks.org/wp-content/uploads/20240702165307/Bandwidth-Efficiency.webp.

4.6.2. Complexity of Demodulation

Demodulation introduces a set of complexities that are of paramount importance in the field of sampling and pulse modulation engineering. As signals make their way back from modulated state to baseband representation, the techniques and algorithms used can play vital roles in the fidelity and efficiency of the communications system. Interesting to note is the fact that there are diverse architectures for demodulation with varying complexities and performance characteristics. The Coherent vs. non-coherent demodulation methods serve as an example to this effect; while coherent demodulation is more precise, it is plagued by phase mismatch and synchronization problems. Non-coherent approaches such as the transmitted reference receiver in Bevilacqua et al. (2005) are accompanied by low data-rate application-convenient designs but at the cost of overall performance. Advances in digital signal processing have also focused more on the effectiveness of demodulation methods. The Shannon-Nyquist theorem, or theory of sampling, is the prevailing principle that fuels these developments, emphasizing a proper sampling rate to prevent aliasing. But emerging sub-Nyquist sampling methods provide an exciting alternative, using mathematical models to enable a good representation of the signal without necessarily reducing the complexity of analog-to-digital conversion. As explained in Eldar et al. (2011), such approaches not only promise to simplify hardware implementation but also pave the way for more robust demodulation techniques that can be optimized for a variety of practical applications from communications to sensor networks. In the end, the simplicity of demodulation in sampling and pulse modulation engineering is a matter of balancing performance, usability, and price in system design. As new technology emerges, it becomes a test for designers to consider the significance of demodulator architecture choices, sampling topologies, and resulting integrated circuit complexities. Ongoing explorations of algorithmic performance vs. circuit functionality trade-offs, as highlighted in many studies, e.g., the ones referenced in Bevilacqua et al. (2005) and Eldar et al. (2011), are significant. This debate not only promotes the understanding of existing methodologies but also inspires innovation in novel demodulation techniques, again underlining the necessity for ongoing research and development within a rapidly evolving field.

4.6.3. Noise Immunity and Signal Integrity

The concept of noise immunity plays a central role in ensuring signal purity in electronic communication. In pulse modulation and sampling engineering, noise has the potential to complicate the quality of the signal received and lead to misinterpretation and system failure. Some methods employed by engineers to enhance noise immunity include error correction codes and differential signaling. These methods offer good answers to combat both external and internal noise sources that may arise during the transmission process. The importance of designing integrated circuits, as shown in the design of a 3-micron CMOS test chip for the Combined Release and Radiation Effects Satellite (CRRES), captures the strict efforts to ensure maximum noise tolerance in practical applications (Blaes et al., 2025). Without adequate noise management methods, reliability and performance of digital communication systems can be severely impaired, resulting in a constant necessity to conduct research here. Signal integrity relates closely to the concept of noise immunity, as it is a factor by which the quality of signals as they propagate through transmission media can be measured. With digital communications, one has to ensure a clean distinction between the logical '0' and '1' in the presence of interference. Signals tend to be distorted by various factors such as crosstalk, impedance mismatch, and environmental influences that can take a toll on the sampling process. By employing advanced modulation techniques and circuit design, engineers attempt to enhance signal quality, thus ensuring higher data transmission fidelity. This assertion is also supported by various innovations, including new patents that have documented advancements in circuit design towards signal integrity improvement. These advances reveal the continuous evolution of technology in compensation for the continuous problems of noise and interference (N/A, 2025). The intersection of signal integrity and noise immunity defines the critical balance required in modern communication systems. With technology advancements advancing sampling and pulse modulation engineering to new levels, pursuing greater data rates necessitates careful attention to these factors. System designers are now faced with the challenge of having to design circuits not only that will work as reliably in ideal conditions but also under conditions of noise. Due to this, there has been the application of simulation tools and real-time testing practices that quantify the performance of systems under varying environments. As a result, the field is undergoing a paradigm change in which integrated design approaches—like those used in CMOS technologies for applications like CRRES—are on their way to becoming the norm. Such approaches combine noise-reduction methods into the circuit topology itself, which is an active approach to addressing the issues of maintaining signal integrity in the presence of ever-increasing expectations placed on communications systems (Blaes et al., 2025).

SUMMARY

The research into Sampling and Pulse Modulation Engineering reveals these underlying principles as pivotal in today's communications and data processing. The rigid techniques employed in sampling permit the accurate mapping of continuous signals to discrete representations. This is exemplified through various projects, such as the model tests on the GEOS 3 radar altimeter. The data so obtained reinforce the challenge involved in designing efficient sampling procedures to ensure fidelity and integrity in signal conveyance. These procedures are ultimately not technicalities but are actually crucial determinants to decide the quality of modern technological communications, opening up ways for enhancement in many applications from radar altimetry to telecommunication (Clary et al., 2025). Apart from that, pulse modulation techniques, particularly sampling, increase the functionality of communication systems since data is allowed to be transmitted across various bandwidths effectively. Techniques such as Discrete Fractional Fourier Transform (DFRFT) have been proposed as powerful techniques for extracting information from modulated signals. As shown in present research, the method not only addresses the issue raised by noise but also enhances the estimates' accuracy regarding vibrations in synthetic aperture radar systems. Such enhancement illustrates the incorporation of sampling with modulation methods such that parameters once considered problematic to assess can be analyzed now, effectively maintaining the relevance of novel methods in the field (Atwood et al., 2012). Lastly, the interplay of sampling and pulse modulation techniques underscores the crucial nature of these engineering principles in advanced signal processing. In the future, the direction of the industry will be based on grasping and implementing these concepts in solving challenges and requirements in communications. The results derived from experimental tests, such as those related to radar altimeters, show significant improvements in calibrating sampling processes and modulation techniques. The continued evolution of these engineering concepts not only holds the potential to make existing systems more robust, but also has the potential to introduce innovations that will shape the telecommunications future. As such, a thorough grasp of Sampling and Pulse Modulation Engineering will become more important to future engineers and scientists that will try to remain ahead of the curve in terms of technological advances.

The fundamental principles of signal processing of the field of Sampling and Pulse Modulation Engineering should be known. One of the significant aspects is the relationship between sampling rates and fidelity of the reproduced signal. A signal should be sampled at least twice the highest frequency component of the signal, as per the Nyquist theorem, to avoid aliasing. This is the underlying design principle of most communication systems and is of paramount importance to enable faithful reconstruction and representation of the signal. Moreover, sampling rate selection has the potential to reduce the bandwidth requirement greatly, as well as system efficiency overall. It is thus the foundation of engineers' knowledge, as it makes possible the appropriate choice of sampling techniques that balance fidelity and resource usage in practical scenarios. Also of major importance in this field is the subject of pulse modulation techniques, techniques of encoding data onto carrier signals. These techniques, such as Pulse Amplitude Modulation (PAM) and Pulse Width Modulation (PWM), enable effective data transfer across different media. They not only provide noise resistance but also enhance the ability to compress data transmitted Various modulation methods have been demonstrated through numerous studies to yield bandwidth utilization efficiencies, especially in video signal transmission systems, as examined in bandwidth compression of video signals (Ingels et al., 2025).

Such pulse modulation method flexibility renders them indispensable in modern telecommunications where data rates are high and dependability is needed to operate effectively. Finally, advances in compression schemes are another critical field of progress in Sampling and Pulse Modulation Engineering. Compressive sensing techniques aim to minimize the bandwidth for transmission with the protection of the original signal's quality. For instance, consideration of source parameters of digitized analog data emphasizes the necessity of selecting appropriate algorithms capable of optimizing the quality-size trade-off (Ingels et al., 2025). In addition, with the rising prominence of digital communication, engineers have to keep adjusting to new technologies advancing data processing without sacrificing efficiency. This constant development requires a wide understanding of current practices and upcoming trends so that engineers are well-prepared to meet today's challenges of communication requirements.

REFERENCES

1. Amaral, L., Dris, S., Gill, K., et al. (2010). Feasibility of using bandwidth efficient modulation to upgrade the CMS tracker readout optical links. http://arxiv.org/abs/1010.1691 (accessed on 12th September 2025).
2. Arens, W. E. (1977). Charge-coupled device data processor for an airborne imaging radar system. https://core.ac.uk/download/pdf/10327526.pdf (accessed on 12th September 2025).
3. Atwood, T., Beach, R. J., Doerry, A. W., et al. (2012). SAR-based vibration estimation using the discrete fractional Fourier transform. https://core.ac.uk/download/286085257.pdf (accessed on 12th September 2025).
4. Baker, P. C., Judd, M. D., McArthur, S. D. J., et al. (2009). Development of an integrated low-power RF partial discharge detector. https://core.ac.uk/download/9023840.pdf (accessed on 12th September 2025).
5. Bevilacqua, A., Blazquez, B., Blazquez, F., Carbonelli, C., Carbonelli, M., Cassioli, D., Chen, J., et al. (2005). Implementation aspects of a transmitted-reference UWB receiver. https://core.ac.uk/download/pdf/11370300.pdf (accessed on 12th September 2025).
6. Blaes, B. R., Buehler, M. G., Jennings, G. A., et al. (2025). Product assurance technology for custom LSI/VLSI electronics. https://core.ac.uk/download/pdf/42840358.pdf (accessed on 12th September 2025).
7. Blu, T., Stantchev, I., Pickwell-MacPherson, E., et al. (2020). Real-time terahertz imaging with a single-pixel detector. https://core.ac.uk/download/323057912.pdf (accessed on 12th September 2025).
8. Bobula, M. (2011). Contribution to efficient use of narrowband radio channel. https://core.ac.uk/download/30310997.pdf (accessed on 12th September 2025).
9. Brongersma, M., Chen, F., Fong, S. W., et al. (2016). Picosecond electric-field-induced threshold switching in phase-change materials. http://arxiv.org/abs/1602.01885 (accessed on 12th September 2025).
10. Budinger, J., Hoerig, C., Tague, J., et al. (2025). Flexible digital modulation and coding synthesis for satellite communications. https://core.ac.uk/download/pdf/42814834.pdf (accessed on 12th September 2025).
11. Bunya, G. K., & Wallace, R. L. (2025). Programmable rate modem utilizing digital signal processing techniques. https://core.ac.uk/download/pdf/42827049.pdf (accessed on 12th September 2025).
12. Campbell, J. F., Flood, M. A., Hodson, W. D., et al. (2011). A low-cost remote sensing system using PC and stereo equipment. https://core.ac.uk/download/pdf/10564848.pdf (accessed on 12th September 2025).

13. Carnegie, D. A., Conroy, R. M., Cree, M. J., et al. (2006). The Waikato range imager. https://core.ac.uk/download/29197540.pdf (accessed on 12th September 2025).

14. Carroll, C. C., & Mott, H. (2025). A study of high-speed AD and DA converters using redundancy techniques: Interim report, May 10, 1963–May 9, 1964. https://core.ac.uk/download/pdf/85251333.pdf (accessed on 12th September 2025).

15. Chakrabarti, I., & Hatai, I. (2010). Multi-standard programmable baseband modulator for next generation wireless communication. http://arxiv.org/abs/1009.6132 (accessed on 12th September 2025).

16. Charbonnier, B., Gaudino, R., Greiss, I., et al. (2011). Recent results from the EU POF-PLUS project: Multi-gigabit transmission over 1 mm core diameter plastic optical fibers. https://core.ac.uk/download/pdf/11424445.pdf (accessed on 12th September 2025).

17. Chen, Y., Eldar, Y. C., & Goldsmith, A. J. (2014). Channel capacity under sub-Nyquist nonuniform sampling. http://arxiv.org/abs/1204.6049 (accessed on 12th September 2025).

18. Clary, J. B., & Hayne, G. S. (2025). Initial test results using the GEOS-3 engineering model altimeter. https://core.ac.uk/download/pdf/42878095.pdf (accessed on 12th September 2025).

19. Collins, M. J., Gerrits, T., Helt, L. G., Lavoie, J., et al. (2020). Broadband quadrature-squeezed vacuum and nonclassical photon number correlations from a nanophotonic device. http://arxiv.org/abs/1904.07833 (accessed on 12th September 2025).

20. Dahono, P. A. (2010). Design and analysis of three-phase three-level PWM inverter. https://core.ac.uk/download/12342426.pdf (accessed on 12th September 2025).

21. De Belie, F., Gabriel, F., Lataire, P., Neyt, K., et al. (2013). High-frequency issues using rotating voltage injections intended for position self-sensing. https://core.ac.uk/download/55798771.pdf (accessed on 12th September 2025).

22. Eldar, Y. C., & Mishali, M. (2011). Sub-Nyquist sampling: Bridging theory and practice. http://arxiv.org/abs/1106.4514 (accessed on 12th September 2025).

23. Eldar, Y. C., Gedalyahu, K., & Tur, R. (2010). Multichannel sampling of pulse streams at the rate of innovation. http://arxiv.org/abs/1004.5070 (accessed on 12th September 2025).

24. Gao, H., Lv, T., Wang, C., Yang, Y., et al. (2016). A belief propagation-based framework for soft multiple-symbol differential detection. http://

arxiv.org/abs/1608.01296 (accessed on 12th September 2025).

25. Goldberg, G. M., Hallmark, W. C., & Myers, R. H. (2025). Lunar contour mapping system /lucom/ final report, 5 Aug. 1964 - 18 Mar. 1965. https:// core.ac.uk/download/pdf/85255346.pdf (accessed on 12th September 2025).

26. Guri, M. (2020). POWER-SUPPLaY: Leaking data from air-gapped systems by turning the power-supplies into speakers. https://core.ac.uk/ download/579863256.pdf (accessed on 12th September 2025).

27. Hartman, D. J., Kahnke, J., & Kirk, R. J. (2025). Conceptual mechanization studies for a horizon definition spacecraft communications and data handling subsystem. https://core.ac.uk/download/pdf/85247138.pdf (accessed on 12th September 2025).

28. Hou, B., Hou, L., Liang, S., Marsh, J., et al. (2018). 1.55 μm AlGaInAs/InP sampled grating laser diodes for mode-locking at THz frequencies. https://core.ac.uk/download/154430638.pdf (accessed on 12th September 2025).

29. Howe, D., Wipasuramonton, P., & Zhu, Z. Q. (2005). Improved current-regulated delta modulator for reducing switching frequency and low-frequency current error in permanent magnet brushless AC drives. https:// core.ac.uk/download/58401.pdf (accessed on 12th September 2025).

30. Ingels, F. M., & Shahsavari, M. M. (2025). A study of data coding technology developments in the 1980–1985-time frame, volume 2. https:// core.ac.uk/download/pdf/42871670.pdf (accessed on 12th September 2025).

31. Le Kernec, J., & Romain, O. (2017). Performances of multitone for ultra-wideband software-defined radar. https://core.ac.uk/download/80698176. pdf (accessed on 12th September 2025).

32. Lev, O., & Khina, A. (2021). Energy-limited joint source–channel coding via analog pulse position modulation. 2021 IEEE Information Theory Workshop (ITW). https://www.semanticscholar.org/paper/5ec9d70d6cd-3b57689c8e02a7456fce333c000f (accessed on 12th September 2025).

33. Messerschmitt, D. G. (2011). Interstellar communication: The case for spread spectrum. http://arxiv.org/abs/1111.0547 (accessed on 12th September 2025).

34. N/A. (2025). NASA patent abstracts bibliography: A continuing bibliography. Section 1: Abstracts (supplement 39). https://core.ac.uk/ download/pdf/42817116.pdf (accessed on 12th September 2025).

35. N/A. (2025). Subsurface sounders. https://core.ac.uk/download/ pdf/42886158.pdf (accessed on 12th September 2025).

36. Paul, R. J. A. (1961). Hybrid methods for function generation. https://core.ac.uk/download/29409681.pdf (accessed on 12th September 2025).
37. Sahoo, P. K., Prajapati, Y. K., & Tripathi, R. (2018). PPM- and GMSK-based hybrid modulation technique for optical wireless communication cellular backhaul channel. IET Communications, 12, 2158–2163. https://www.semanticscholar.org/paper/37f011d11ff819cd5422ea3deb13d16323b42e62 (accessed on 12th September 2025).
38. Záplata, F. (2016). Advanced Algorithms for Satellite Communication Signal Processing. https://core.ac.uk/download/30311082.pdf (accessed on 12th September 2025).

CHAPTER 5

DIGITAL COMMUNICATION FUNDAMENTALS

Contents

INTRODUCTION

The application of digital communication has shifted the manner in which organizations and individuals communicate, ushering in a new era of speed, convenience, and immediacy. This shift is increasingly fueled by technological advancements, which have displaced barriers and allowed individuals to get in touch with anyone anywhere at any time. As a result, messages can be delivered through numerous different channels including e-mail, social media, and instant messaging. Such a shift not only facilitates individual communication but also enables businesses to operate more efficiently, with an integrated virtual workplace. But the growing use of online communication opens up basic concerns regarding information security, privacy, and misinterpretation chances. Awareness of these forces is critical, insofar as they necessarily shape the way digital communication is perceived and utilized in personal as well as business contexts. Besides, digital communication involves several practices and skills applicable to effective interaction in an online setting. The ability to convey ideas simply and effectively using digital media is more applicable in an information over-saturated age.

Figure 5.1. *Digital communication.*

Source: https://realbusiness.co.uk/wp-content/uploads/2016/06/shutter-stock_1463151758.jpg.

For instance, the ability to apply visual communication, such as infographics and digital storytelling, is significant in addressing people and enhancing comprehension. As highlighted by the continued evolution of technology, schools have integrated training in these core skills into their curricula. With this respect, integrating the conventional and computer-based methods is essential; it complements the native digital setting and validates the relevance of core skills in areas like manual drafting and technical communication, where precision and aesthetic appeal remain relevant (McLaren et al., 2007). This academic blend forms a solid base for students to excel in all aspects of digital communication. In addition, the impact of digital communication extends beyond individual interactions but impacts broader social trends and mores. Social media sites have spread a new paradigm, where anyone can broadcast content worldwide.

many times disseminating professional and personal communication. This democratization of information sharing encourages diverse voices yet possesses the ability to amplify misinformation and polarization. It is hence important critical media literacy that has become extremely necessary in evaluating the authenticity of online content. Technology and cross-fertilization of communications also invoke consideration of ethical responsibilities since people have to cope with issues like consent, respect, and the persistence of digital traces. By examining these complexities, it is clear how communication dynamics developed while at the same time honoring the ongoing challenges that digital spaces pose in order to develop meaningful and effective exchanges. Having cognizance of these various factors will ultimately guide both users and creators to more responsible digital communication environments (Figure 5.1).

Definition of Digital Communication

Digital communication, therefore, refers to the conveyance of information electronically through digital technologies primarily. The process has transformed the way individuals, organizations, and societies interact with one another. When comparing the previous modes of communication that predominantly utilized analog methods such as print and face-to-face communication, digital communication offers a host of platforms such as social media, email, and websites that guarantee real-time and global connectivity. The speed and performance of such devices have not only given personal communication greater access but also changed the workplace, as teams are able to work together from very distant locations. The value of digital communication is in its capacity to bridge distances and create relationships, reflecting its crucial position within modern social arrangements and institutions, as is clearly seen in the field of education where digital sites are being used with growing frequency for learning and participation. Beyond enhancing the availability and promptness of communication, digital communication has also significantly impacted the content and form elements of message transmission. The convergence of multimedia functionality—enabling text, audio, video, and graphics to come together—has transformed information presentation as well as consumption. This multi-disciplinary approach raises engagement by reaching various learning styles and requirements, particularly relevant in learning contexts. The emphasis on interactive and participatory communication models encourages collaborative learning experiences, leading to greater sharing of ideas. Moreover, it unveils the opportunity for distance learning and open educational resources that can facilitate accessibility and information democratization. As is evident with the arrival of MOOCs and other online learning resources, the trend illustrates the impact of digital communication on modern pedagogy, as noted in Mallmann et al. (2018). Moreover, the character of online communication extends to instant connections and content delivery alone but encompasses collaborative as well as participatory practices as well. Technology and communications convergence has brought about innovations in collaborative tools making collaboration more efficient and knowledge dissemination more effective. For example, electronic platforms facilitate real-time collaboration across domains and geographical distances, allowing several teams with different backgrounds to collaborate on shared goals. This is particularly critical in product design, where the integration of digital communication can enhance processes to be more efficient and seamless. As noted in Segonds et al. (2014), successful collaboration during the initial stages of design can significantly impact product development cost and results. Lastly, digital communication should be understood not only as a medium through which information is distributed but also as a key component in the development of innovation and collaboration between various fields.

Importance in Modern Society

In addressing the importance of digital communication in today's society, there is a need to examine its groundbreaking impact on face-to-face communication and communal structure. Information and communication technologies have redefined the way individuals interact, from the traditional face-to-face communication to a newer, more fluid, virtual style of interaction. The transformation not only connected the world as a whole but also created new frameworks in interpersonal relations. But with that ease comes the challenge of maintaining the fundamental rights of individuals within the online realm. As aptly stated in the recent research, the equilibrium between individual freedoms and governmental control is central in securing democratic values from potential misuse that could be witnessed in this novel digital reality (Saparbekova et al., 2025). Its influence is immense, shaping institutional trust and feelings of privacy as individuals negotiate the subtleties of today's communication spaces. Furthermore, digital communication is an essential tool for learning and skills development, adapting to the evolving needs of today's workplace. Educational institutions have in modern society adopted digital technology to develop collaborative cultures, stimulate critical thinking, and foster many varied competencies that underpin success in the global community. The European Union emphasizes the integration of digital literacy in education systems, viewing such proficiency as key to active citizenship and democratic life (Karaoulas, 2025). With expanding digital spaces, learning approaches move away from purely technical proficiency towards social and emotional intelligence as well. Such a comprehensive pedagogy lies at the heart of empowering individuals to live in the digital era with social responsibility. By fostering resilience and adaptability, schools have a central function in the development of capable people prepared to meet the challenges of rapid technological advancements. Additionally, digital communication has an expansive role in key areas like activism, public debate, and social change. Digital communication gives voice and power to individuals and groups in mobilizing support for whatever cause they wish, which enables grassroots movements to influence actual change. This democratization of information and involvement has revolutionized the approach to problems in society, giving voice to otherwise marginalized communities. Yet more use of digital channels also brings with it challenges of disinformation and concerns of digital divide that can render such movements futile. To provide that electronic communication remains a vehicle for positive social transformation involves ongoing effort to promote media literacy among the citizenry and push for counseling values centered around ethical information exchange. An educated citizenry is then essential so as to realize the maximum potential of electronic communication to push society forward while maintaining public discourse integrity in the modern era (Figure 5.2).

Figure 5.2. *Importance of digital communication in modern society.*

Source: https://fastercapital.co/i/Digital-Skills--Mastering-Digital-Skills--A-Digital-Nomad-sToolkit--The-Importance-of-Digital-Skills-in-the-Modern-World.webp.

Overview of Digital Communication Technologies

The development of digital communication technologies has radically altered how people and organizations interact and exchange information. At the forefront of this change has been the spread of the internet, the infrastructure behind most forms of communication such as email, instant messaging, and social media sites. The online environment enables users to communicate with one another at vast distances virtually in real time, both personally and professionally. Moreover, the revolution of mobile communication technologies also expanded accessibility and availability to be broader and more convenient so that connectivity is not limited by place anymore. These technologies, therefore, not only enhance efficiency but also establish a culture of collaboration that shapes existing societal norms and expectations. Another critical point of digital communication technologies is they enable improved information dissemination and sharing of knowledge. Sites such as blogs, wikis and forums allow people to express ideas, post studies, and comment on various subjects in the present moment. Democratization of information empowers people with access to vast knowledge that previously belonged to intellectual or professional categories. Moreover, organizations utilize these technologies for marketing and customer engagement, and data analytics are employed to refine their outreach efforts. As businesses develop to adapt to these changes, they must also deal with their ensuing challenges, such as data security and disinformation management that have taken greater importance in the modern digital age. Ongoing development of digital communications technologies raises consequential issues of further innovation in the same area. For example, the project to create an entire competency model for information technology education, as discussed by the Association for Computing Machinery and the IEEE Computer Society, is a case in point regarding how curricula that are able to keep pace with these evolving technologies are necessitated. Initiatives like IT2017 aim to equip graduates with skills and knowledge needed in order to be ready for future opportunities in a global economy. As digital communication progresses, it will be necessary for schools to keep pace with the developments and incorporate associated competencies in their curricula. This will help future professionals adequately respond to the challenges of digital communication in the dynamic environment (N/A, 2016; Alrumaih et al., 2016).

5.1. HISTORICAL CONTEXT OF DIGITAL COMMUNICATION

The history of digital communication reveals an environment shaped significantly by technological innovation and cultural transformation. Beginning in the later half of the 20th century, the advent of the personal computer brought with it unprecedented information accessibility. This break away from traditional means of communication enabled greater individuals to become involved in the creation and distribution of content. The proliferation of emails and online forums opened the door to interactive communication sites that would expand into social media networks later on. This change brought with it not only new means of sharing information but also the democratization of knowledge, wherein voices from every background started getting heard. As contested, such innovations developed a culture that leaned towards innovation, influencing anything from social relations to teaching practices, as reflected in the manner in which resources for enhancing communication abilities in various contexts evolved (McKelvey et al., 2011). It is also conceivable to analyze the trend towards new communication technologies in cultural terms of heritage and societal demands. While museums and cultural centers began to embrace digital platforms, they attempted to enhance visitor engagement by incorporating experiential methods that intermingled physical artifacts and digital storytelling. The introduction of interactive technologies allowed these institutions to redefine their roles in the preservation and transmission of culture. For instance, the development of augmented and virtual reality software made new forms of interaction possible with ancient artifacts, creating narratives beyond physical visits. This paradigm pointed out the necessity to join ancient modes of communication with present digital means, as those experiences can compound cultural accessibility and trigger ongoing argument over the meaning of the heritage (Falco et al., 2017). Moreover, the evolution of electronic communication reflects broader social forces and intersections of technology, and it is within that context that individual stories and collaborative possibility are important. The history of academics and authors within the academic tradition reveals how individual stories can catalyze broader transformation of communication practice, fueling innovation that inspires critical thinking and creativity in students. The repetitive cycle of publication, like the online communication world, requires an acute balance between innovation and staying within well-defined norms. As the stories progress, they provide fundamental insights into the operational challenges of producers working to uphold the integrity and ingenuity of their work. This symmetrical enmeshment of individual experience and technological systems thus points up the necessity of constant adaptation and exploration within the ever-evolving environment of digital communication (McKelvey et al., 2011).

5.1.1. Evolution from Analog to Digital

The transition to digital communication was a watershed moment in the history of communication technology. Analog systems initially dominated, with continuous signals that represented a wide range of physical phenomena. However, analog system shortfalls were soon to become evident, from vulnerability to noise to deterioration over distance, and wasteful compression of data. With the innovation of digital technology, a new paradigm came in where data is represented as discrete signals that allow for more clarity, stability, and efficiency. This evolution not only enhanced storage and transmission capacity but also allowed for the formation of sophisticated algorithms for error detection and correction, which further amplified digital communication's advantage over its analog counterparts. The evolution of digital communication systems initiated a revolutionary growth in mobile telecommunication, which revolutionized the interaction of people with information and technology. Parallel to the development of digitization, wireless technology emerged as a vital initiator of this process, offering mobility as well as effectiveness. The emergence of technologies such as Wireless Intelligence Networks (WIN) led to smarter, interactive communication systems that could support the needs of increasing numbers of mobile subscribers. As referred to in some of the literature, for instance, a description of modern telecommunication networks the transition from fixed wire systems to dynamic wireless applications is a landmark in responding to consumer needs and a way of overcoming the challenges associated with the mobile industry (Adesuyi et al., 2014). Development in digital communications has paved the way for breakthrough technologies, best exemplified by the introduction of 5G wireless networks. With even more sophisticated features such as millimeter-wave frequencies and high antenna arrays, 5G seeks to accommodate increasing demands for data-greedy applications without sacrificing power-efficient use. As argued in ongoing research, the use of beamforming technology significantly increases network capacity and energy efficiency compared to earlier generations such as the 4G (Deruyck et al., 2017). This transition from digital to analog is not merely technological change but also changing user expectations and societal demands since communication networks must adapt to support a more interconnected and information-driven society (Figure 5.3).

Figure 5.3. Evolution from analog to digital of communication technologies.

Source: https://vle.btvi.edu.bs/moodle/pluginfile.php/252765/course/overviewfiles/analogvsdigital.jpg.

5.1.2. Key Milestones in Digital Communication

The development of digital communication can be traced out in a series of significant milestones that have literally transformed the way individuals and organizations communicate. One of the most significant was the creation of the Internet at the close of the 20th century, which enabled instant, global communication on an unimagined scale. According to this, systems such as TCP/IP enabled standardization in data transmission, paving the way for diverse forms of digital communication, from email and instant messaging. This revolution not only transformed individuals' as well as businesses' communications but also necessitated a rethinking of governance and regulatory frameworks because the virtual realm began bleeding into the established patterns of human rights and privacy. As such, critical analysis of the implications of such changes has become unavoidable, which underscores the necessity of aligning legal frameworks for nurturing safe and just digital spaces as identified in Pyroha et al. (2025). Another defining landmark in digital communication is the advent of mobile technology, spearheaded by the smartphone. These technologies democratized access to communications, essentially placing connectivity power into the hands of individuals worldwide. The dissemination of cell-based applications changed run-of-the-mill functions such as calling, messaging, and social networking so that instantaneous communication is now possible no matter what the geographical distance. In addition, the rise of social media platforms has redefined social relationships, to not only facilitate interpersonal communication, but also serve as a means for political mobilization and public discourse. The change has raised significant ethical issues about data privacy, misinformation, and consumer agency, prompting more debate about regulatory frameworks to protect users online, as Pyroha et al. (2025) states. Looking ahead, the estimated launch of Sixth Generation (6G) cellular technology is a turning point in the history of digital communication. Designed to be deployed by 2030, 6G is poised to provide ultra-high levels of data transmission rates, enabling a network framework that can communicate virtually instantaneously across the world. With advances such as three-dimensional connectivity and optical technologies being integrated within the wireless space, the architecture of future digital ecosystems will have to undergo a radical change. It will provide for the increasing requirement of bandwidth and reliability that exists in modern-day communications, supplementing smart cities and the Internet of Things (IoT). It is imperative that industry leaders and scientists prepare for such developments by taking into account the implications and potential applications of 6G and ensuring there is equitable access and regulation, as emphasized in.

5.1.3. Impact of the Internet on Communication Methods

The discovery of the internet has transformed means of communication, ushering in a new paradigm in communication patterns with which people and organizations now interact with each other. Prior to the internet, communication was largely limited to face-to-face speaking, written mail, and fixed-line telephony, which had the effect of limiting the tempo and reach of communications. With the advent of digital media, the tempo of communication has increased manyfold, and real-time communication over long distances is made possible. This revolution has not only facilitated individual communication but has also produced a seismic impact on business communication. Companies now have the opportunity to leverage email, instant messaging, and video conferencing to enhance teamwork, automate processes, and eliminate geography. So, the internet has created an international marketplace of ideas and opportunities, transforming structurally the notions of interpersonal communication and organizational communication. In addition, the internet's contribution to communication has brought about a democratization of information dissemination. Social networking websites are today's public squares, where individuals can express their views and comments to huge numbers of people in the majority of situations without the constraints of the conventional media. This has empowered the users to the level where they can be active participants in the communication process rather than passive receivers. The ability to create and exchange content permits people to argue, coordinate social movements, and spread knowledge to communities. Concurrently, this democratization has had negative effects, including the spread of misinformation and the creation of echo chambers. With the internet transforming communicative practices, it is more critical now than ever before for users to apply a critical eye to the quality of information and the credibility of sources in this abundant, online world. In examining the broader scope of internet communication, one can understand how it is revolutionizing city spaces, if only in the context of smart cities. The application of internet technologies to city planning and management allows for enhanced communication between citizens, local governments, and service providers. Online forums enable residents to report crimes, receive services and participate in decision-making, allowing for a sense of civic engagement and transparency. However, the wisdom that is obtained from these interactions with data also allows city planners to optimize resources and improve public services. The construction of smart cities also has its accompanying vices, including privacy concerns and the digital divide between those connected and those not connected. Since cities continue to evolve through incorporation of technology, the dynamics between internet-based communication and public engagement require particular attention and deliberation about city living and city administration in the future (Xu et al., 2018; Albino et al., 2015) (Figure 5.4).

Figure 5.4. *Impact of the internet on communication methods.*

Source: https://img.freepik.com/premium-photo/transformative-impact-internet-communication_1088377-237.jpg.

5.2. TYPES OF DIGITAL COMMUNICATION

Under the umbrella of digital communication, various platforms and means have emerged, fundamentally changing how individuals exchange information. Social media is one of the most pervasive means of digital communication comprising an array of apps such as Facebook, Twitter, and Instagram. Social media enables people to create, distribute, and interact with information in real time, creating a worldwide web of connection. The social character of social media provides a setting in which there is feedback instantly and interaction among members, spawning communities of like-minded interests. This platform for interaction not only promotes individual communication but also becomes a site where companies and organizations speak to their constituency. The growing reliance on social media points towards its impact on communication tactics, enabling individuals to unlock their potential for self-expression and professional image-building, which is of crucial significance in the networked world of today. Email is another significant genre of digital communication that remains the backbone of working-level communication. Despite newer modes of communications entering the scene, email provides a structured and formal system of information exchange. It allows users to convey complex messages, transmit documents, and maintain records of exchange, thereby being an essential tool in business settings. Furthermore, email communication is characterized by asynchronous involvement, where the recipient can read at their convenience, thereby balancing instant and reflective communication. The ever-relevant nature of email is further cemented in that it meets today's demands, including augmented security measures and compatibility with many productivity programs.

Figure 5.5. *Types of digital communication.*

Source: https://clearinfo.in/wp-content/uploads/2023/04/Common-types-of-digi-tal-communication.png.

As companies increasingly use email for communication both within and outside the company, it is significant to understand its use and etiquette in order to interact efficiently in the virtual space. Instant messaging applications are a fast and informal means of virtual communication that have gained immense popularity over the recent years. Applications like WhatsApp, Slack, and Microsoft Teams provide users the luxury of sending messages immediately, allowing them to make choices and discuss matters quickly. These applications support both one-on-one and group communication and are most valuable in group projects and team environments. The addition of multimedia functionality, such as video calls and document sharing, improves the richness of communication and allows for the successful conveyance of complex information. As organizations embrace distant and international teams, instant messaging has risen to become a core driver of maintaining contact and productivity, redefining the manner in which collaboration is conducted in the era of digitization. The influence which these platforms have on people and business interactions belies the shift in communication dynamics in contemporary society, leaning towards the instant and the efficacious (Figure 5.5).

5.2.1. Text-Based Communication

Text-based communication forms the backbone of on-line interaction, enabling individuals to share ideas, emotions, and facts without the overt cues of face-to-face communication. As more use is made of textual modes—emails, messaging apps, and social media—it becomes important to understand the implications of this mode of communication. The absence of non-verbal cues may facilitate misunderstanding or misinterpretation, especially in the case of lacking context. However, it also leaves space for asynchronous communication, allowing the users time to reflect more thoroughly on their response. This is particularly critical in leadership, where creating a digital culture will require smooth and transparent communication approaches that can bridge the gap between different stakeholders, as recent studies on e-leadership and digitalization suggest (Cortellazzo et al., 2019).

Figure 5.6. *Text-based communication.*

Source: https://static.vecteezy.com/system/resources/previews/000/518/121/original/vector-man-and-woman-with-smartphones-concept-illustration-texting-messaging-chatting-social-media-customer-assistance-dating-communication.jpg.

Moreover, online social interaction also finds its definition in text-based communication, particularly through SNSs. Individuals utilize them to create social relationships around shared interests, and these end up creating active flows of social support and ideas. However, this process creates compulsive use behaviors, akin to behavioral addiction, as evidenced by recent empirical research (Daria et al., 2017). The psychological implications of such action highlight the twoness of text-based communication: it can build community and connection yet also carry dangers of over-dependence and addiction. It is crucial to realize these facets for individuals and organizations grappling with the complexities of virtual presence and social relationship. Finally, the evolution of text communication in the age of technology requires an examination of its ethical effects, primarily on privacy, transparency, and responsibility. As individuals make more intimate information available through texts, digital manners blur the lines between what is appropriate. Leaders and organizations have to campaign for ethical communications that protect user information with a culture of respect and trust. Digital platforms employed for text-based interaction need to be guided by ethical values to prevent exploitation and misuse. Digital communication, leadership, and ethics should unite to provide a sustainable environment for people to engage meaningfully and safely (Cortellazzo et al., 2019). The need for constant research in this regard remains very high since it helps ensure that the advancements in digital communication are for the benefit of society in general (Figure 5.6).

5.2.2. Audio and Video Communication

The arrival of audio and video communication has transformed the digital communication paradigm, reshaping the way information is produced and consumed. Along with the dissemination of the internet, the channels for the transfer of audio and video, such as podcasts, streaming services, and video conferencing applications, have become mainstreamed into everyday life. It not only amplifies user engagement but also makes content creation available to everyone, so every voice has an opportunity to be heard. Signal processing techniques have become critical to optimizing the quality of such communications because they can isolate desired audio and video signals from unwanted interference and noise. Stochastic process-oriented methods, as discussed in relevant literature, illuminate the mathematical frameworks underlying such developments to enable engineers to create systems that are capable of dealing effectively with unwanted signals in audio and video formats (Etten et al., 2006). With the emergence of digital communication, incorporation of micro protocols into audio transmission systems has evolved as a key area of research. One of the best examples of this trend is network steganography, where sensitive information is concealed in seemingly harmless digital information. Inclusion of micro protocols within digitized audio signals delivers new avenues for enhancing reliability, performance, and security in transmission. These integrated channels accommodate comprehensive feature sets, including dynamic routing and effective data transmission without attracting undesirable notice. Experimental findings endorse the feasibility of such design practices, demonstrating the potential to improve overall audio communication systems through ensuring secure and efficient data transfer (Keller et al., 2015). These advances go to demonstrate how researchers are ever on the lookout for innovative solutions that address both technical constraints and user needs in audio and video communication. Ultimately, the future of audio and video communication will continue to be controlled by advancing technology and the mathematical foundations of signal processing. As engineers more and more employ sophisticated methods of filtering and optimizing media signals, users can expect clearer and more efficacious communications experience. The potential of digital communication in various sectors—be it workplace, schools, or social interactions—will likely expand further, driven by the increased usage of artificial intelligence and machine learning technology. The future map reveals an exciting convergence of theoretical expertise and practical knowledge, stressing the importance of having a solid command of stochastic processes in the formulation of dependable communications systems (Etten et al., 2006). Audio and video communication will remain at the core of linking far-flung regions with the incessant innovations (Figure 5.7).

Figure 5.7. *Audio and video communication of digital communication.*

Source: https://proavdc.com/wp-content/uploads/2023/12/Audiovisual-Communication.jpg.

5.2.3. Social Media Platforms

The advent of social networking websites has dramatically changed the world of online communication, touching not just individual interactions but also reshaping the tasks of various professional disciplines. Sites such as Twitter, Instagram, and Snapchat are integral tools for communication that offer real-time information dissemination and community interaction. The rapid development of these media has forced communication professionals, like journalists, to rethink their methodologies to suit shifting audience expectations. With further development of these platforms, it remains necessary to understand the subtleties of their functioning and role for individuals willing to manage this fast-paced world effectively. These characteristics are traditionally emphasized by teachers in learning models, identifying the manner in which social media has infiltrated the culture of modern communication behaviors, particularly with respect to social trends and cultural discourse (McLean et al., 2020). Besides that, the functions of social medias extend beyond communication; they significantly shape diversity and social issues of justice, catalyzing debates on these essential matters across communities. The platforms' functionality opens up the space for the voice of the marginalized to become visible and raise the demand for equity and representation lacking in traditional media.

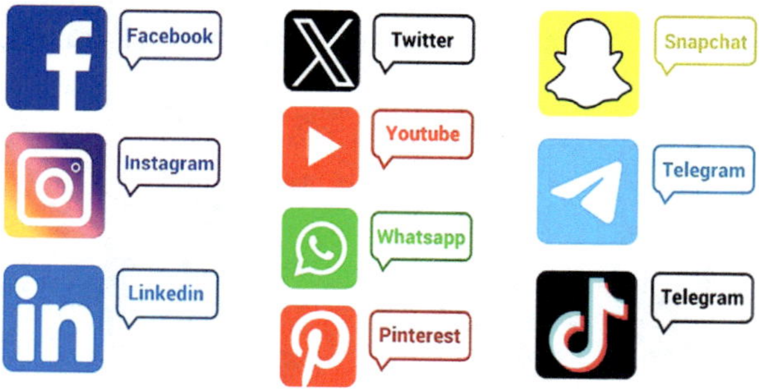

Figure 5.8. *Social media platforms.*

Source: https://www.creationinfoways.com/blog/uploading/1799955929.jpg

This has also created online communities not only campaigning for social change but also bestowing a sense of belonging on their audiences. Critical analysis of such sites reveals the challenges that digital inequality poses, making the case for equitable participation in web-based communities. Thus, academic research into social media websites must address their complex nature as tools for empowerment and marginalization, placing emphasis on the need to develop strategies for equitable access and engagement in these online environments (McLean et al., 2020). Finally, the fleeting nature of social media phenomena makes it hard to understand its overall effect on society. As platforms keep changing their algorithms and features, the users and creators are compelled to change their approach as well, in order to remain relevant and engage people. These changes underscore the necessity for successful content that speaks to people, thus enhancing the importance of digital literacy and strategic communication skills in this ever-changing landscape. In addition, with the increase in influencers and going viral of information, information dissemination dynamics have shifted significantly, leading to new authorities and forms of credibility. Social media academic writing therefore has to address these new trends and how they are accommodated in the practice of communication so that students and professionals are taught in-depth about the platforms governing our digital lives (McLean et al., 2020) (Figure 5.8).

5.3. DIGITAL COMMUNICATION TOOLS AND TECHNOLOGIES

The growth in digital communication tools and technology has revolutionized the face of inter-personal relationships and organizational life. Instant messaging software, video conferencing software, and social media websites are some of the tools that have made real-time communication possible across distances, thereby wiping out geographical distances. This real-time feature of digital communication has turned out to be both beneficial and detrimental. Individuals and organizations can communicate more easily now, but such reliance on these technologies usually causes problems of information overload and decreased face-to-face interaction. Furthermore, digital literacy is placed in greater demand because individuals now need to employ various platforms with varying degrees of technical expertise needed. Adopting these tools in everyday processes of communication points towards understanding their functioning and effects on effective and successful interactions. Moreover, the emergence of telemedicine technologies is just one example of the revolutionizing role that digital communication plays in the field of specialized professional domains such as medicine. As more of these new tools become integrated into healthcare systems, nurses and healthcare professionals are being pushed into roles that require the construction of new digital competencies. Telehealth not only expands the access to care, but also remakes the nurse-patient relationship generally, especially with regard to remote consultations. Nurses are expected to evidence leadership and soft skills in the use of these technologies for successful patient needs response, as has been highlighted in a scoping review of assessing the evolving nursing role in the COVID-19 pandemic (Isidori et al., 2022). The movement towards digital sites clarifies the necessity for systematic models of training that address technological know-how as well as mainstream clinical capability. Beyond supplied areas such as healthcare, the influence of digital communication tools extends to broader societal frameworks, particularly disaster management and infrastructure resiliency. The most vital infrastructure that enables continuation of societal functions is increasingly dependent on novel digital applications that facilitate real-time monitoring and decision-making. Technologies like the Internet of Things (IoT) and artificial intelligence can provide valuable information regarding asset condition and emergency response to natural disasters, enabling overall resilience (Argyroudis et al., 2021). However, the potential of such technologies is not being exploited since there is a lack of cohesive strategy and integrated policies to guide their deployment. By promoting a multidisciplinary response to the adoption of these technologies, the stakeholders can create robust infrastructural systems that not only withstand adversity but also safeguard socioeconomic stability in an uncertain world (Figure 5.9).

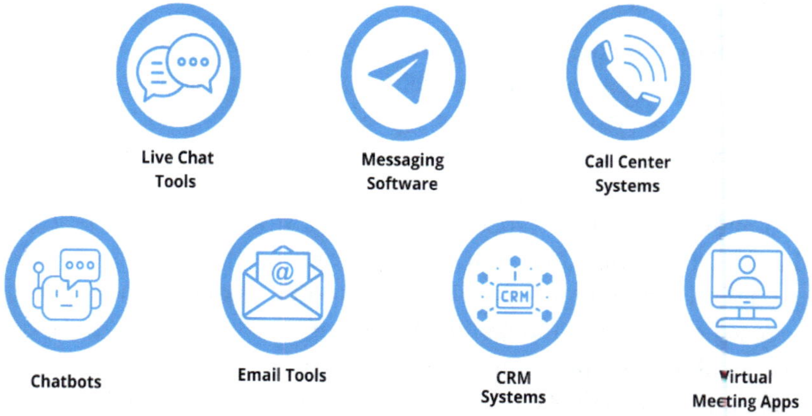

Figure 5.9. *Digital communication tools and technologies.*

Source: https://www.regpacks.com/wp-content/uploads/2023/09/image15.png.

5.3.1. Email and Instant Messaging (IM)

Email and instant messaging (IM) are two cornerstones of electronic communication that have revolutionized personal as well as professional communication. Email, launched with the onset of the internet era, is an official medium for message exchange, documents, and files across the globe. Its asynchronous nature allows the users to send and receive messages at their convenience, thereby allowing varying work hours and time zones. In comparison, instant messaging offers the immediacy and conversational mode closer to face-to-face communication, allowing real-time chat through text, picture, and even video call. Although email may remain the preferred mode of professional communication and formal discussion, IM has emerged as a convenient tool for quick questions and informal chat between colleagues, friends, and family members, creating a blend of formality and informal interchange that characterizes modern communication environments. The versatility brought about by both email and instant messaging has transformed office dynamics to its core, encouraging more cooperation and efficiency. For instance, instant messaging sites enable groups to communicate quickly, leading to quick decision-making and real-time feedback, and this can be very helpful in speed workplaces. On the other hand, emails are used in instances of elaborate discussions, assigning tasks, and documentation, serving as an archive of valuable information that one can refer to later.

Figure 5.10. *Email and instant messaging (IM).*

Source: https://www.techdonut.co.uk/sites/default/files/email-calanders-and-instant-messaging-400255669.jpg.

These electronic communication tools also support telecommuting by bridging geographical distances, enabling employees to remain connected and work in concert even with miles of distance between them. Not only does this make people more productive but also creates a culture of inclusiveness because individuals can work from diverse places, making it possible for opportunities in divergent input and creative thinking. The integration of email and instant messaging into work flows poses serious issues regarding management of electronic communications. While convenience on these websites promotes them to be used regularly, it also leads to an excess of messages that the users have to navigate through on a daily basis. Organizations are understanding the importance of controlling the usage of such tools so that there is no overflow of communication and management. Having clearly defined rules for the context most suitable to each medium—informal or formal communication or rapid-exchange messages—can diminish complexity of interaction, thereby preventing important information from being overlooked. Furthermore, as established in mobile commerce and communication practice studies (Hsieh et al., 2014; Thatcher et al., 2011), understanding the bases of digital communication becomes crucial for effective business management because it has direct implications on future opportunities and outcomes (Figure 5.10).

5.3.2. Video Conferencing Software

Video conferencing has revolutionized the process of communication in most sectors, institutions, and social activities with the advent of information and digital technology. Unlike other modes of communication, which depend on words spoken or written, video conferencing enables one to see and hear others at the moment. This enables greater participation and understanding, giving a sense of presence crucial for collaboration. As companies adopt remote work models, enterprise-quality video conferencing needs have grown, redefining team meetings, idea-sharing, and relationship building dynamics. Seamless integration of such solutions makes it easy to enable flexible workflow, which is important for maintaining productivity across varying work environments. Thus, video conferencing has become an essential tool in the world of connectivity today, enhancing the level of communication to meet the demands of a global audience (Perkins et al., 2014). The impact of video conferencing software is beyond organizational uses, with long-term impacts on learning pedagogies as well. With the growth of distance learning, teachers more than ever before are utilizing such platforms to deliver interactive lectures and forums. This transformation not only offers more convenient access but also facilitates cooperative learning through the connecting of students from different geographic locations. Video conferencing in this context acts as a key facilitation tool, bridging spaces that traditional learning media cannot bridge. It becomes a platform where students can interact with other students and instructors, significantly enhancing the overall learning process. As suggested by contemporary studies, availability of Web 2.0 tools such as video conferencing is a significant component in determining collaborative spaces for students, promoting more collaborative learning space as reported across different cohorts (Ferreira et al., 2011). Additionally, software program ability in video conferencing permits it to be combined with other communication technologies, which maximizes its effect. Integrated messaging, screen sharing, and file transfer are some of the functionalities that constitute an end-to-end communication environment that accommodates varied learning and work styles. Such flexibility is very worth it for collaborative work projects, where team members can collaborate on documents concurrently with video-based communication. Such features play a key role in maintaining project momentum and can potentially yield better outcomes than conventional methods of communication. In a pledge for more user-friendly features and interfaces, video conferencing applications continue to become better, not only promising to make professional communications more effective but also reinventing the education industry. As basic digital communication fundamentals advance, embracing video conferencing is vital to develop efficient interactions and achieve collaborative goals in most scenarios (Perkins et al., 2014).

5.3.3. Social Networking Applications

Social media applications have revolutionized interpersonal interaction and communication of people with one another in the virtual space. Facebook, Twitter, and Instagram are not only channels for human interaction but also significant sites of information exchange and community formation. The advent of such applications has facilitated users to generate and share information within seconds with a worldwide audience, thereby making the playing field level for accessing information. But this shift has profound social consequences, not least the establishment of a culture of instantaneity and constant connectivity which can create both positive and negative social processes. The frenetic pace of development of social media has encouraged researchers and practitioners to examine the effects of these platforms on human action, on relationships, and indeed on the broader organizational systems within which they are situated. The Covid-19 pandemic put an unprecedented reliance on social networking apps into overdrive as humans and organizations adjusted to confinement and working from home settings. During this times of crisis, the utility of these platforms attracted even more attention as they leveraged them as central interfaces for professional engagements and human interactions. As George et al. (2020) points out, the sudden shift towards remote work emphasized the necessity for quality communications instruments and, on the other hand, revealed both the opportunity and the limitations of digital communication. The challenge of maintaining interaction and collaboration in a distributed environment highlighted fresh assumptions about work procedures and coordination schemes. While organizational restructuring of communication strategies resulted in questioning social media functions in collaboration and innovation, concerns were raised regarding how virtual spaces are best able to anchor innovative practice amidst ongoing technological evolution. Furthermore, the development of smart cities through increasing dependence on digital communication for sustainability of urban living even more accentuates the shifting context of social networking uses. With cities embracing technology in their infrastructure, the role of social media in facilitating civic engagement and community resilience has been a timely debate. Albino et al. (2015) are of the view that the concept of a smart city encompasses more than one dimension, such as participatory government and social interaction. Social networking websites empower citizens to speak out and engage in urban planning, entrusting power to people to shape local policy. This shift raises the question of equity of access and participation in digital space, particularly among minority populations. Thus, as social networking finds its place as part of smart urban initiatives, one should examine the implication for social integration and the future ability of cities to shape a networked and activated citizenry (Figure 5.11).

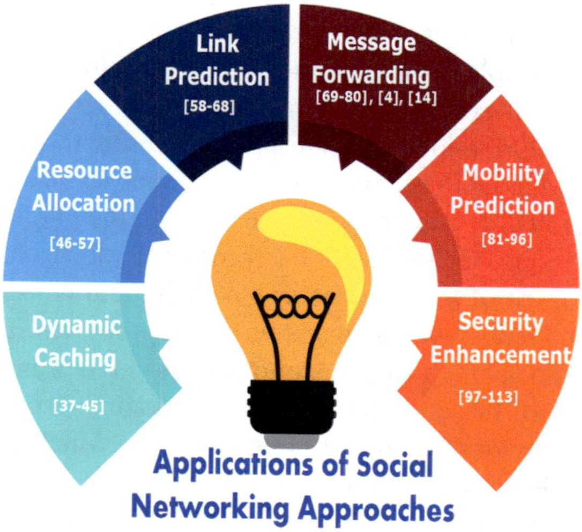

Figure 5.11. *Social networking applications.*

Source: https://www.researchgate.net/profile/Dush-Nalin-Jayakody/publication/327821353/figure/fig4/AS:674720508829696@1537877313141/Applications-of-social-networking-approaches-These-applications-consist-of-dynamic.png.

5.4. THE ROLE OF DIGITAL COMMUNICATION IN BUSINESS

The evolution of digital communication has redefined business relationships, affecting everything from internal communication to customer relationships. Organizations no longer rely on old forms of communication but rather utilize all sorts of digital media—emails, social networks, and instant messaging—to facilitate seamless communication among teams and with outsiders. This shift enhances collaboration and facilitates timely feedback, which is indispensable in a fast-paced environment. Besides, the ability to communicate across remote geographical locations diminishes collaboration hurdles, enabling organizations to adopt global operations. This merging based on technology supports skill management tenets, where abilities crafted using technology can enable organizational effectiveness in generating collective aims, emphasizing the importance of strategic utilization of skills in a competitive environment (Huda et al., 2025). Therefore, electronic communication is not only a method of communication but also a strategic resource in business development. In addition to internal communication facilitation, online platforms are also core to how businesses engage with customers. Through social media, businesses can communicate with their audiences, provide answers, and foster community, leading to an interactive customer relationship experience. This two-way interaction fosters customer loyalty and trust, seriously impacting brand perception and customer retention. Companies also leverage analysis from online interactions to give greater insight into customer behavior, segmenting their marketing efforts to accommodate individual needs and wants. Such an effort not only produces revenue but also engenders innovation through feedback mechanisms that can inform product development. As businesses navigate this world of cyberspace, the use of effective communication strategies becomes increasingly vital in order to thrive in a constantly evolving technology-driven market (Pineda et al., 2024). Strategic use of electronic communication also finds its application in organizational decision-making processes since the exchange of information is more crucial than ever. The speed and efficiency of electronic communication allow for rapid decision-making, which allows the manager to respond rapidly to market trends and operational problems. As businesses opt for transparency and openness through communication, they stand to benefit from increased employee involvement and motivation. Moreover, having electronic communication in a skills-managed environment facilitates the development of a knowledgeable workforce that can respond to new challenges. Training and ongoing education on digital communication tools underpin these skills, highlighting the importance of companies investing in ongoing learning as a growth strategy. Ultimately, understanding and leveraging the role of digital communication is essential for businesses that seek to thrive in a more globally connected world (Huda et al., 2025).

5.4.1. Enhancing Collaboration and Productivity

With the current era of digital communication, promoting collaboration and productivity is central to the objectives of organizations seeking to leverage the powers of today's communication technologies. Part of working toward this objective is maintaining generational diversity in the workplace. According to recent studies, digital literacy differences across generations have the potential to become communication gaps that inhibit collaboration and productivity at large. For instance, age workers may experience challenges emanating from their low trust in the application of digital technologies or limited training exposure. Institutionalizing ongoing digital competences training and mentorship initiatives, as suggested by Ezurike (2025), generates an environment embracing these generation gaps. These interventions not only equip all workers with empowerment but also enhance collective capabilities, leading to improved teamwork and productivity among diverse teams. Moreover, the application of digital collaboration platforms becomes obligatory to meet the needs of a virtualized work culture.

Figure 5.12. *Enhancing collaboration and productivity.*

Source: https://vestalhub.com/wp-content/uploads/2024/04/Enhancing-Collaboration-Maximizing-Team-Productivity-with-HubSpot-Service-Hubs-Team-Management-Features.jpeg.

The performance of virtual teams relies on a number of factors, including leadership, trust, and structured communication practices that become crucial in bridging the challenges of remote work arrangements. Findings included in Hoque et al. (2025) indicate that organizations that meet these aspects strategically can establish high-performing virtual teams that are capable of surpassing traditional productivity levels. By embracing the appropriate digital means and proper channels of communication, teams are capable of coordinating, sharing knowledge, and remaining active despite geographical distances. This strategic approach not only enhances collaboration but also stays teams on target and aligned to organizational objectives, ultimately contributing to performance and innovation. Lastly, encouraging a continuous improvement culture and feedback is essential for sustaining collaboration and productivity within an online setting. Enabling open communication where team members feel secure in expressing their ideas and issues results in a learning- and adaptable-conducive environment. As companies struggle to unravel the nuances of internet communication, routine feedback cycles can help identify areas for development and celebrate success. In establishing programs that promote group consideration, paired with the knowledge provided through studies like those mentioned in Ezurike (2025) and Hoque et al. (2025), places importance on empathy and understanding in leadership. By prioritizing learning culture and support, organizations can drive their collaborative efforts to their highest potential, resulting in sustainable productivity and an aligned workforce that is more capable of addressing future challenges in a rapidly digitalized world (Figure 5.12).

5.4.2. Marketing and Customer Engagement

The rise of electronic communication has radically transformed the marketplace and customer interaction process, and organizations have to employ novel methodologies to reach their target audiences successfully. As consumers increasingly employ digital media for accessing information, firms are compelled to personalize their marketing approaches to foster interactivity and faith. Electronic media such as social networks, websites, and phone apps provide immediate interface with consumers, increasing the effectiveness of marketing campaigns. For instance, Hajj and Umrah tour operators can utilize these sites to post open information, thereby establishing customer trust and credibility (Ahmadi et al., 2025). Through concentration on two-way communication and interactive dialogue, companies can not only sell their services but also establish loyal customers who feel valued and heard. Furthermore, the effectiveness of e-commerce campaigns often relies on the ability to segment and gain information regarding customers' behavior. RFM model is a very practical tool for this purpose, enabling companies to measure their customers by Recency, Frequency, and Money. Through this practice, companies are able to identify certain customer groups and target marketing according to each specific choice habits and needs of those sets. For instance, a coffee shop like The Soul can use RFM analysis to plan specific promotions for various consumer segments ranging from loyal users to new consumers (Aldyah et al., 2024). These targeted approaches not only increase engagement but also result in customer satisfaction because shoppers tend to respond positively to promotions based on their individual experiences and preferences. Lastly, effective marketing and customer engagement in today's age are reliant on a strong commitment to transparency, personalization, and interactivity. Firms must realize that forming and maintaining customers' trust and loyalty is a continuous process requiring regular effort and dynamic adjustments. Efforts such as leveraging AI chatbots for time-sensitive customer interactions and social media analytics to gauge customer sentiment can significantly enhance the customer experience as a whole. As has been observed from the Hajj and Umrah services findings, organizations that practice interactive and transparent communication are most likely to cultivate successful relationships with customers, establishing loyalty and long-term retention (Ahmadi et al., 2023). Therefore, complete understanding of digital communication fundamentals is necessarily the responsibility of organizations aspiring to master the nuances of modern marketing and customer relationships (Figure 5.13).

Figure 5.13. *Marketing and customer engagement.*

Source: https://www.vedia.ai/blog/wp-content/uploads/2019/10/customer-engagement-benefit-your-business.jpg.

5.4.3. Remote Work and Virtual Teams

The rapid development of electronic communication technologies has transformed the workplace in a fundamental way with the result that telecommuting is now widely practiced and virtual teams have come into existence. This transformation is largely driven by the demand for flexibility in the modern workforce and the preference for work-life balance on the part of employees. As companies realize the advantages of having remote teams, they are investing heavily in technologies that facilitate communication, collaboration, and project management at geographically dispersed locations. This transformation demands not only the use of digital tools but also a keen understanding of virtual leadership, a key competency in this new work environment. To address this changing landscape, most companies are endeavoring to have in place a culture that encourages remote work, where employees are able to remain engaged and productive despite the physical separation of members from each other. Successful communication is the basis of successful remote work and virtual teams. In the absence of face-to-face interaction, breakdowns and misunderstandings among team members can happen if not managed in an active manner. Virtual teams must build strong communication systems that use several web platforms to develop transparency and harmony in their collaborative work. Secondly, leaders are equally crucial in developing an open and transparent communication environment in which all members feel free to share their ideas and grievances. Among these, most significant findings of recent studies emphasize the requirement of efficient communication techniques and digital tools for enhancing virtual leadership abilities. Such techniques not only create feelings of belongingness among telecommuters but also enhance team performance as well as innovation. With remote employment becoming increasingly popular, various elements are responsible for its propagation and impact on work environments. Organizations are increasingly recognizing the likely cost savings, increased talent availability, and worker satisfaction associated with remote work arrangements. The benefits of these also have associated challenges such as the need for employees to get necessary skills in order to perform adequately within a digital platform. Employees must be capable of traversing through multiple communication outlets and create self-management skills to thrive in an independent work environment. Bringing these elements together points to the importance of creating a flexible and robust workforce amidst change. Hence, understanding the dynamics of remote working and virtual teams is imperative in leveraging electronic communication to improve organizational performance and the health of employees (Disman et al., 2023; N/A, 2008).

5.5. CHALLENGES OF DIGITAL COMMUNICATION

In the rapidly evolving environment of computer-mediated communication, the merging of innovative technologies raises serious challenges that might hamper fruitful collaboration and interaction. Despite the possibilities of technologies such as Building Information Modeling (BIM) and advanced data analytics, computer-mediated platforms become immersed in uncertain flows of communication. Malaysia's building sector is a prime example, where technology adoption resistance, fueled by a lack of digital expertise, stifles the improvement in effectiveness and efficiency. As Gheda et al. (2025) notes, such obstacles are typically in the shape of excessive costs and low standardization processes, which can undermine the success of digital initiatives and lead to delay or failure. Consequently, organizations must recognize these challenges as critical drivers that must be addressed in their digital transformation for them to develop a culture of effective communication.

Figure 5.14. *Challenges of digital communication.*

Source: https://www.vmcdn.ca/f/files/villagelife/sponsored-content/bazoom-digital-communication-insights.jpg;w=960.

Secondly, application of digital communication tools is also beset with an event known as alarm fatigue, particularly for high-risk environments such as healthcare. As discussed in studies of systems like Digital National Early Warning Score (NEWS), while such technology has the potential to enhance patient safety and streamline communication amongst healthcare teams, it also has the potential to generate a cacophony of alert messages that could overwhelm staff. The key is achieving equilibrium between the necessity of monitoring and the potential for desensitization to alert, which can lead to severe gaps in patient care. As Noya et al. (2025) indicates, the hesitation to adopt such technologies is typically based on issues of culture and resource availability, which compel organizations to re-strategize their adoption strategies. In addressing these challenges, there must be an overall strategy with a focus on user training, technology literacy, and robust infrastructure to ensure that web-based tools enhance rather than constrain effective communication. In addition to cultural and technical hurdles, digital communication is typically also plagued by information overload challenges. As more and more organizations rely on numerous digital platforms for collaboration, employees may be confronted with an information avalanche that results in lost productivity and heightened tension. The problem can stifle workflow and create barriers to effective communication, since essential messages become lost amidst too much content. In addition, the lack of consistency in these various channels can multiply miscommunications and misinterpretations and make integrated communication a formidable task. In an attempt to combat these difficulties, organizations must establish mechanisms that align communication processes and establish well-defined protocols for information flow. In overcoming the information overload barrier and establishing systematic channels of communication, businesses can leverage the entire potential of digital communication while reducing its inherent shortcomings (Figure 5.14).

5.5.1. Miscommunication and Misunderstandings

The subtleties of online communication have profoundly changed the character of interpersonal communication, yet in so doing also heighten the consequences of miscommunication and misunderstanding. With communication transferring from person-to-person encounter to text media, tone, context, and cultural allusions become susceptible to failure to translate. This is particularly common in settings where the actors hold differing cultural affinities, each of which will carry individual preconceptions and assumptions regarding language. The presence of misinformation on social media further complicates this dynamic, as is clear from the "TT Refugee" incident, where there were widespread false presumptions in terms of a crisis that were widely spread before being resolved through repeated communication. The double-sided nature of social media—as a means of disseminating falsehoods and as a means of counteracting these misperceptions—underscores the nuances involved in online communication and consequent misunderstandings that can arise in multicultural communication (Ji et al., 2025). In clinical settings, communication can have devastating consequences, particularly among those populations that face linguistic and cultural constraints. The migrant women's encounter of gaining access to maternity care demonstrates the manner in which language may lead to not only miscommunication but also to adverse health effects. Recent qualitative research has identified that the integration of digital translation technology, including the Swedish-Arabic dialogue support, has proved successful in enhancing interaction among Arabic-speaking women and midwives. This device provided a formal method of sharing information and improved the understanding of complicated medical issues. Moreover, it enabled women by limiting their reliance on interpreters, enabling them to obtain the correct and timely treatment. All these advancements are significant in addressing the miscommunication that can occur among varied health environments and emphasize the significance of culturally sensitive digital communication (Bitar et al., 2025). The effects of miscommunication extend beyond personal relationships, influencing public opinions at large and fostering polarization. In the digital realm, rapid dissemination of misinformation can lead to massive misconceptions among communities, inducing polarization and suspicion. Social networking websites, while they facilitate intercultural dialogue, can also strengthen misconceptions when untended. As evident in the provided case, public discourse can transform fundamentally by way of public engagement, in which users themselves negate misinformation and facilitate a more critical understanding of matters. The interactive dynamic among technology and human communication calls for an active stance to stop miscommunication, with emphasis on digital literacy and critical reading of sources. Through creating spaces that encourage open discussion, online communication may also become an instrument of unification instead of segregation, eventually promoting cross-cultural understanding and collaboration (Ji et al., 2025).

5.5.2. Privacy and Security Concerns

In the modern era of communication through digital media, issues of privacy and security are more than ever. With users putting greater trust in digital media for domestic and working affairs, the risk of unauthorized access to sensitive data escalates by thousands. Cybercrime, data breaches, and identity theft are some of the common perils that violate users' privacy. Individuals have a tendency to downplay the value of information belonging to them, which reaches its climax in a casual approach towards online security. This casual approach may result in catastrophic consequences, including financial loss and reputation loss. Therefore, developing an awareness about privacy rights and security measures is of immense significance to use these online forums safely. Education that improves digital literacy has the potential to empower the users to control their digital identity, thus addressing some of the security concerns of digital communication. Technologies have improved and brought forth opportunities along with challenges regarding privacy and security. Technologies such as artificial intelligence and blockchain technology have the potential to revolutionize security measures. In particular, AI and blockchain technology convergence holds the prospect of automating verification processes and increasing transaction security. In addition to increasing operational efficiency, this collaboration is also addressing privacy challenges by creating open and tamper-proof systems. Smart contracts and advanced encryption techniques allow companies to protect confidential data and facilitate trust in online transactions. Ongoing innovation of such technologies spells a bright future for secure communication but poses the need for careful consideration of ethical concerns as well as the need for robust regulations to guide their use (Narwal et al., 2024). With the digital communication taking place across many domains, it is essential to recognize and address ethical and legal concerns regarding privacy and security. The intersection of these issues increasingly shapes public policy and public discourse. Users want organizations to have good practices in place to protect their information but always question these organizations' commitment to privacy. In addition, differing international data protection standards make it difficult for multinational firms to comply, and may expose users to abuse. To reduce these concerns, frameworks must be established which will address all the aspects and strike a balance between the privacy rights of users and security needs. With a joint effort by policymakers, technology innovators, and teachers, efficient steps can be formulated that ensure maximum digital security with the basic aspects of online privacy intact in the digital ecosystem (Pandey et al., 2024) (Figure 5.15).

Figure 5.15. *Privacy and security concerns.*

Source: https://www.columbiasouthern.edu/media/udobzpao/privacy-issues-in-cybersecurity.jpg.

5.5.3. Digital Divide and Accessibility Issues

The term used to refer to the very broad difference between those with easy access to digital technology and those without such access is the digital divide. This difference is becoming more pronounced as day-to-day life increasingly relies on digital communication with ramifications for education, employment, and social involvement. There are a number of determinants that explain this divide, including socioeconomic position, place, and educational level. Residents of rural or underprivileged populations have limited access to the internet and resource deficits, which may hinder being in the virtual world in a useful manner. Such disparities thus threaten to further accentuate the prevailing inequalities, forming a vicious circle in which marginalized communities become increasingly isolated from the virtual world. Conquering these rudimentary problems of accessibility requires an interdisciplinary approach to ensure equal access to digital communication platforms and tools. Initiatives to combat the digital divide ought to target innovative strategies that facilitate accessibility. Public-private partnerships play a vital role in enhancing connectivity and resources. For instance, community access to broadband in local communities has been identified as a key factor in bridging barriers to technology adoption. By virtue of public-private partnerships, city governments can fund the development of infrastructure and applications responsive to the particular needs of their constituencies. This is in line with Knight's 2009 call for e-government and increased transparency via open technology. Greater connectivity not only facilitates individual empowerment but also enables enhanced community engagement, creating a more participatory online community where everyone gets heard. Moreover, accessibility consideration is not a matter of mere connectivity; usability and functionality of digital instruments are equally critical. Telecare technologies, as an example from the health sector, are a case of both promise and challenge in addressing issues related to accessibility. The technologies offer remote monitoring services that can enhance care delivery, but there are issues in adoption. Ethical concerns regarding surveillance, education of technologies to accommodate individual requirements, and effects on the relationship between caregivers and patients must be critically assessed. In telecare debates, more work must be done to comprehend how technology is perceived and used by practitioners who are working with these innovations. Ensuring their accessibility and equity requires a continuous conversation concerning their practical applications and the experiences of users in real-world contexts (Gant et al., 2011; Eccles et al., 2013).

5.6. FUTURE TRENDS IN DIGITAL COMMUNICATION

As online communication evolves, there are certain trends emerging that foretell the future of growth in this category. One of them is the increasing application of artificial intelligence (AI) and machine learning in communication platforms.

These technologies not only are enhancing user experience with personalized content recommendations but also are mechanizing communication processes across various channels. For instance, chatbots driven by artificial intelligence are becoming a common feature in customer service, allowing businesses to respond more efficiently. This captures a larger trend towards automating online communication, wherein the goal is to provide instant response and respond to customer needs in real-time, enhancing interaction and satisfaction ultimately. Among the most critical domains that are defining directions for future trends in online communication is growing focus on security and privacy. More content being put online, there are fears of information leaks and unauthorized use, which are driving encryption technology and secure communications protocols. Network steganography, a mode of hiding information in regular network traffic, highlights the danger posed by rogue parties seeking to exploit vulnerabilities in electronic communications (Caviglione et al., 2014). Companies and individuals are becoming more aware of such risks, and this is propelling the movement towards adopting cybersecurity measures to protect individual data. This is not just an indication of transforming society to place greater importance on privacy but also an indication that any new communication technology will have to incorporate robust security as a default part of its design. Also, access to communication digital technologies is anticipated to expand significantly driven by rising levels of internet penetration and technology price decline. Penetration of mobile phones and cheap devices has allowed developing world people to access the digital space, thus transforming their communication mode. Development of mobile data connectivity, such as 5G technology, will further develop the speed and quality of digital communication (N/A, 2016). As more and more individuals gain access to these technologies, opportunities for richer collaboration platforms and richer forms of communication will emerge. This development has the potential to introduce new modalities of creation and exchange of content, enabling richer interaction and more inclusiveness in worldwide patterns of communication. Last but not least, the digital communication future will be shaped by a meeting of technological innovation, consumer demand for security, and a priority for accessibility.

5.6.1. Artificial Intelligence and Automation

With AI and automation, the world of digital communications has been revolutionized with a complete shift in the way organizations engage with their stakeholders. The technologies allow for real-time responses, increase customer engagement, and automate processes, creating efficiencies previously unachievable. For instance, AI-driven chatbots are able to handle countless customer inquiries simultaneously, providing instant feedback that increases customer satisfaction while reducing operational expense. The incorporation of automation into digital communication enables companies to work throughout the day and night, making them available at all times. This not only improves customer service but also creates a responsiveness and agility culture within organizations. As firms are adopting these technologies increasingly, they are also being asked to redesign their communication strategy to leverage the full potential of AI so that their digital communication remains meaningful and impactful in an environment of competition. Moreover, application of these emerging technologies has numerous implications for organizational process and the function of employees. As efficiency is boosted by AI and automation, they also initiate a paradigm shift in the workforce, with traditional roles either redefined or rendered obsolete. Workers now have to join hands with AI systems to perform tasks that require a blend of human judgment and machine precision, rather than performing repetitive tasks manually.

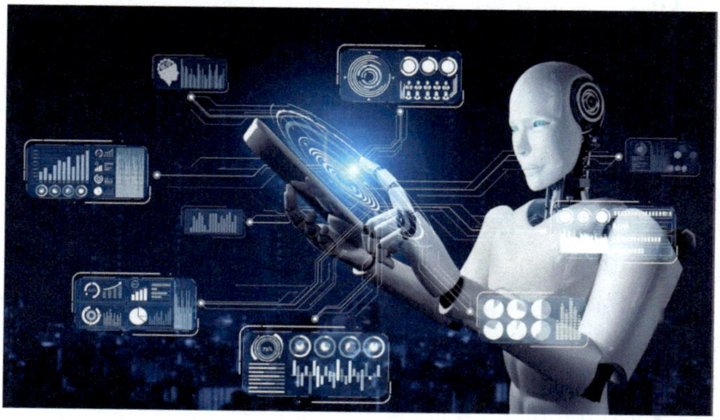

Figure 5.16. *Artificial intelligence and automation.*

Source: https://octavius.ai/wp-content/uploads/2023/10/1_k4IDvabuf0rgLkMRnl-3hcA-scaled.webp.

This partnership necessitates an employed staff that is proficient not only in their own job roles but also in technology-driven modes of communication. For this reason, organizations are investing in training programs to equip the employees with the capabilities needed to shift towards these alterations. The inter-industry models of business process reengineering (BPR) can be a front line model for integrating AI tools into organizational operations, optimizing flows to capitalize on automation advantages and address potential drawbacks for job roles and workplace culture (Popoola et al., 2024). However, the rapid growth of AI in digital communication provides no margin for trivializing serious ethical and legal issues that should not be avoided. Applications of AI technologies, such as generative models, have been the source of controversy regarding bias, privacy, and disinformation. With platforms like ChatGPT proving successful in generating text that can sound human-like, organizations need to grapple with the ethical implications of using AI to generate content that could inadvertently replicate biases or disinformation and, in the process, affect public trust. Furthermore, transparency of AI processes becomes prominent when stakeholders increasingly demand explanations of how decisions are made by automatic means. The discussion of regulatory options that can address these issues becomes prominent in an effort to make AI used in communications accountable. With increased development in the conversation on AI, further studies are necessary in addressing the complexity involved in its integration into electronic communication, with regards to ethics, responsibility, and the possibility of beneficial and adverse effects of its widespread use (Yogesh et al., 2023) (Figure 5.16).

5.6.2. Virtual and Augmented Reality Applications

The advent of virtual and augmented reality (VR/AR) technologies marks the beginning of a new era in the evolution of digital communication, dramatically changing how individuals interact with information and one another. Through the immersion of users within richly interactive contexts, these technologies allow for a more intuitive and interactive experience, particularly in learning and work environments. For example, immersive simulations provide students with the capacity to practice skills in the real world within a simulated, controlled environment, enabling experiential learning. This is particularly relevant in areas such as medicine and engineering, where practical application of what has been learned is paramount. With increasing demand for new modes of communication, the work of VR/AR is still on an incline, indicating the shift in the digital landscape and the possibility of these technologies redefining the norms of communication in various fields of study. Furthermore, as VR/AR technology develops further, so too does integration within the Metaverse, hinting at a new horizon of digital communication. Visual semantic structures are taking the lead in this revolution, leveraging progressive multimedia information to enhance the virtual interaction experience. According to new studies, these structures employ VR/AR-enabled headsets to generate semantic data that not only enhances the efficiency of communication but also focuses on sustainability in digital environments. Through effective use of resources, these systems are designed to minimize energy requirements while simultaneously improving user interaction. This dual focus on performance and sustainability reflects a greater understanding of the ecological implications of technological advancement and emphasizes the necessity for sustainability in the domain of digital communication. Although the optimistic advances in VR/AR technology promise a lot, there is still challenge in utilizing their full potential in various sectors. The integration of such technologies into education systems for instance, requires an extensive understanding of pedagogical models as well as the digital literacy of students and teachers. Recent studies indicate that while digitalization has an appeal in higher education, available hurdles in the shape of poor training, resource shortages, and resistance to change can hinder proper implementation. The ongoing research on building structured models of digital integration will overcome these challenges such that VR/AR applications complement traditional learning while promoting digital literacy. On the whole, overcoming these obstacles will usher in a better utilization of VR/AR, maximizing channels for communication, and developing collaborative learning spaces that will contribute to teaching in the modern age (Jagatheesaperumal et al., 2025; Kovalchuk et al., 2025).

5.6.3. The Rise of 5G and Its Implications

The introduction of 5G technology is a monumental leap in digital communication with promises of unprecedented speed, efficiency, and connectivity. In contrast to its predecessors, 5G utilizes advanced radio technology and greater bandwidth, which has data transfer speeds 100 times faster than the 4G. This also increases the number of devices that can connect simultaneously by three times, catering to the growing Internet of Things (IoT) ecosystem Accordingly, consumer experiences will be revolutionized in incredible ways, supporting seamless streaming, virtual reality applications, and enhanced mobile gaming. Even more significant, though, is that 5G implications reach beyond consumer convenience; industries are poised to leverage the abilities of 5Gs to drive operations, streamline processes, and develop innovative services, thereby positioning organizations to better manage the increasingly digitalizing business environment and address evolving consumer demands, foregrounding its pivotal role in shaping future communication futures. The effect of 5G technology is widespread, speeding up developments that drive innovation and transform life on a day-to-day basis. In medicine, for instance, 5G's low latency characteristic enables real-time remote patient monitoring and telemedicine services, likely to revolutionize patient care, access, and outcomes. With data-intensive procedures enabled, physicians can use augmented reality and artificial intelligence to guide surgery, improving precision and safety. With this, other sectors such as transportation will also be forced to experience monumental change, with smart vehicles relying increasingly on 5G for dynamic communication with the environment. This communication revolution will result in smarter traffic control and enhanced safety measures, leading to an overall remaking of city living. The power to change operational paradigms further proves 5Gs ubiquity across many disciplines, pointing towards its pivotal position in the ongoing process of digital communications. The social implications of 5G technology are profound, cutting across social interactions, economic structures, and cultural trends of the globe. Through widespread connectivity, individuals will notice themselves transforming the manner in which they utilize technology and interact with each other; prospects for cooperative projects, learning, and information exchange spread exponentially. Moreover, organizations will continue to adopt 5G technology in order to further optimize customer interaction tactics, drive data analytics, and enhance organizational performance as a whole. However, this progress has its potential drawbacks; concerns over data privacy and security come up since huge amounts of data are being passed across networks. Balancing innovation with ethics will be critical in navigating the effects of 5G technology. As such, 5G onset not only heralds a new era of communication but also challenges policymakers, institutions, and individuals to consider the holistic impact on society at large, for its revolutionary possibilities come ever closer to reality (Figure 5.17).

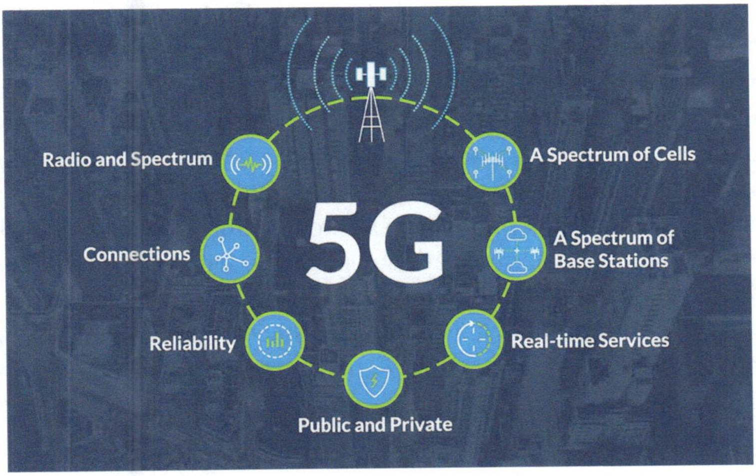

Figure 5.17. *Rise of 5G and its implications advanced radio technology.*

Source: https://www.arm.com/-/media/global/solutions/5g/arm-5g-infrastructure-graphic.jpg?revision=4157f53b-7b23-4d6f-aa5e-541f122b4afb&h=587&w=1280&hash=0542DF888642B74708F3A144F670EAEF.

SUMMARY

Lastly, the study of digital communication fundamentals discovers the intricacy of tradition and contemporary times. The difference between manual technical drafting and computer-aided design (CAD) indicates a deep insight in this regard. As a survey with various stakeholders, such as professional CAD users and school students, identifies, experience acquired from tradition is required skill such as problem-solving, creativity, and precision. Even in our age of electronic communication, students evidence a strong preference for traditional methods of drafting, suggesting that an integrated approach—maturity training both in traditional and electronic skills—can better prepare individuals to address the multi-faceted issue of modern communication technologies (McLaren et al., 2007). Also, the value of knowledge in digital communication can be derived in the area of cybersecurity as seen through progress in network steganography. This method, wherein information is conveyed in stealth within innocuous digital messages, poses formidable challenges to security. The advent of malicious entities employing steganography mirrors the necessity for studying in detail the potential as well as risks of digital media. Hence, creating sound foundational knowledge of digital communication is essential to establishing sound plans for countering such threats efficiently (Caviglione et al., 2014). Ultimately, thorough analysis of digital communication fundamentals directs towards the immediate need to embrace a well-balanced strategy stressing both traditional competencies as well as new innovation. Through establishing a learning environment that recognizes the importance of various modalities of communication, we prepare successive generations with the ability to handle complexities in a growingly dynamic digital age. With this two-faceted understanding, not only do individuals become flexible in transferable abilities, but their entire pool of communication resources is enriched too. Emphasizing interconnectedness between different communication modes is required in an effort to train well-informed and responsive practitioners capable of bridging the difference between traditional and digital modes in their practice.

In critiquing the topography of digital communication, there is an essential need to identify the ongoing pertinence of long-standing skills against technological advances. The integration of experiential methodologies in pedagogical processes, particularly in fields such as graphic communication, highlights a stark dichotomy between legacy skills and modern approaches. Studies have found that students performing manual traditional hand technical drawing do not reject digital technology but rather benefit from the core skills derived from these procedures. This perception calls into question the supposition developed by Prenskys thesis on digital natives, indicating that students appreciate the individual discipline, precision, and problem-solving skills gained by using standard drafting methods. Such findings underline the need for a differentiated curriculum that incorporates both digital technologies and key skills, and prepares students to navigate the intricate age of digital communication with a combined set of skills (McLaren et al., 2007). Moreover, the rise of Open Educational Resources (OER) and Massive Open Online Courses (MOOCs) is a paradigm shift in higher education, particularly in digital communication. These innovations make it possible to access knowledge and shared learning environments that transcend spatial distances. The emergent debate around OER and MOOCs necessitates an epistemological perspective to avoid theoretical confusion and maximize pedagogical application. It is argued by researchers that the integration of technology should be accompanied by a reconceptualization of education theory and economic interests as they determine the mobilization of open educational practice significantly. Finally, this conversation is not only about the availability of resources but also about broader pedagogical topographies that direct the teaching, learning, and use of digital communication in various contexts (Mallmann et al., 2018). Overall, therefore, the confluence of old competencies and new digital spaces represents a major locus for future research and pedagogic practice. Through the integration of knowledge both from traditional manual drafting approaches and the evolving OER and MOOC environment, teachers and policymakers can formulate strategies that preserve the worth of fundamental skills and provide space for innovation. This is a balance that leads to a culture of learning that supports critical thinking, creativity, and adaptability in learners. With the online space in flux, the challenge is how to balance the benefits of new technologies with skill-based learning that-readies the student for complex communication tasks. Thus, making the models of education in digital communication robust, inclusive, and reflective of the pluralistic nature of current learning environments becomes imperative for preparing students to address future challenges in the digital era (McLaren et al., 2007; Mallmann et al., 2018).

REFERENCES

1. Adesuyi, R. S. (2014). The future of mobile industry. https://core.ac.uk/download/19447812.pdf (accessed on 12th September 2025).
2. Ahmadi, E. S. A., & Sudahri. (2025). Analysis of communication strategies of Hajj and Umrah travel agencies in building pilgrims' trust through digital media: A case study of PT. Assunniyah Al-Jauhari. https://core.ac.uk/download/656179992.pdf (accessed on 12th September 2025).
3. Albino, V., Berardi, U., & Dangelico, R. M. (2015). Smart cities: Definitions, dimensions, performance, and initiatives. Journal of Urban Technology, 22, 3–21. https://doi.org/10.1080/10630732.2014.942092 (accessed on 12th September 2025).
4. Aldyah, T., Heikal, J., Muzakkar, M., Sugino, S., et al. (2024). RFM segmentation analysis for determine online marketing strategy: The Soul Coffee Mate case study. https://core.ac.uk/download/616688763.pdf (accessed on 12th September 2025).
5. Alexandrov, O. (2025). Role of educational technologies in the formation of innovative competence in future agronomy specialists. Cherkasy University Bulletin: Pedagogical Sciences. https://www.semanticscholar.org/paper/52e68551956a446648f57883206de851334db72f (accessed on 12th September 2025).
6. Alrumaih, H., Impagliazzo, J., & Sabin, M. C. (2016). An information technology competency model and curriculum. https://core.ac.uk/download/86427251.pdf (accessed on 12th September 2025).
7. Argyroudis, S., Mitoulis, S. A., Chatzi, E., Baker, J. W., Brilakis, I., Gkoumas, K., Vousdoukas, M., et al. (2021). Digital technologies can enhance climate resilience of critical infrastructure. Climate Risk Management, 35, 100387–100387. https://doi.org/10.1016/j.crm.2021.100387 (accessed on 12th September 2025).
8. Bitar, D., Oscarsson, M., & Hadziabdic, E. (2025). Midwives' perceptions of communication at antenatal care using a bilingual digital dialog support tool – A qualitative study. BMC Pregnancy and Childbirth, 25. https://www.semanticscholar.org/paper/d562496120aa45daa7d69dd3417d0eda7e324999 (accessed on 12th September 2025).
9. Caviglione, L., Mazurczyk, W., Meier, M., Wendzel, S., et al. (2014). Hidden and uncontrolled–On the emergence of network steganographic threats. http://arxiv.org/abs/1407.2029 (accessed on 12th September 2025).

10. Cortellazzo, L., Bruni, E., & Zampieri, R. (2019). The role of leadership in a digitalized world: A review. Frontiers in Psychology, 10. https://doi.org/10.3389/fpsyg.2019.01938 (accessed on 12th September 2025).

11. Dal Falco, F., & Vassos, S. (2017). Museum experience design: A modern storytelling methodology. https://core.ac.uk/download/153363274.pdf (accessed on 12th September 2025).

12. Deruyck, M., Joseph, W., Martens, L., Matalatala Tamasala, C., et al. (2017). Performance evaluation of 5G millimeter-wave cellular access networks using a capacity-based network deployment tool. https://core.ac.uk/download/84046715.pdf (accessed on 12th September 2025).

13. Disman, D., Pringgabayu, D., Sobandi, A., Waruwu, B., et al. (2023). The art of virtual leadership: Insights from a literature review. https://core.ac.uk/download/587917226.pdf (accessed on 12th September 2025).

14. Dwivedi, Y. K., Kshetri, N., Hughes, L., Slade, E., Jeyaraj, A., Kar, A. K., Baabdullah, A. M., et al. (2023). Opinion paper: "So what if ChatGPT wrote it?" Multidisciplinary perspectives on opportunities, challenges and implications of generative conversational AI for research, practice and policy. International Journal of Information Management, 71, 102642–102642. https://doi.org/10.1016/j.ijinfomgt.2023.102642 (accessed on 12th September 2025).

15. Eccles, A. (2013). The complexities of technology-based care: Telecare as perceived by care practitioners. https://core.ac.uk/download/17185487.pdf (accessed on 12th September 2025).

16. Ezurike, N. (2025). Bridging the generational gap in the digital age: Challenges and strategies for enhancing workplace cohesion and performance. British Journal of Multidisciplinary and Advanced Studies. https://www.semanticscholar.org/paper/e5d62c854187b031f4f26efebede86c5272ab785 (accessed on 12th September 2025).

17. Fayad, A., Cinkler, T., & Rak, J. (2024). Toward 6G optical fronthaul: A survey on enabling technologies and research perspectives. IEEE Communications Surveys & Tutorials, 27, 629–666. https://www.semanticscholar.org/paper/adc743738d5fb96f17bd2c1eb1593359e28e9385 (accessed on 12th September 2025).

18. Ferreira, G., & Wilson, T. (2011). E-learning and support tools for Information and Computer Sciences. https://core.ac.uk/download/106675.pdf (accessed on 12th September 2025).

19. Gant, J., & Turner-Lee, N. (2011). Government transparency: Six strategies for more open and participatory government. https://core.ac.uk/

download/71357252.pdf (accessed on 12th September 2025).

20. George, G., Lakhani, K. R., & Puranam, P. (2020). What has changed? The impact of COVID pandemic on the technology and innovation management research agenda. Journal of Management Studies, 57, 1754–1758. https://doi.org/10.1111/joms.12634 (accessed on 12th September 2025).

21. Gheda, M. L. M., Chung, H. W., Abdul Aziz, F. F., & Mohamad Nusran, N. F. (2025). Bridging the digital divide: Overcoming challenges in technology adoption in Malaysia's construction industry. Malaysian Journal of Social Sciences and Humanities (MJSSH). https://www.semanticscholar.org/paper/04c2e03072a0f18f626d46f53b6d4456b7051c24 (accessed on 12th September 2025).

22. Hoque, M., Zhang, X., Rahman, M. F. F., Zahra, B., & Hasnat, M. S. (2025). A comprehensive review of enhancing collaboration and performance in virtual teams. Review of Business and Economics Studies. https://www.semanticscholar.org/paper/d7db48d618ee8ba9732e3c72b401197a1ab60a5c (accessed on 12th September 2025).

23. Hsieh, C. T. (2014). Mobile commerce: Assessing new business opportunities. https://core.ac.uk/download/55330114.pdf (accessed on 12th September 2025).

24. Huda, A. N., Hamdani, U., Rahayu, W., & Winarno, A. (2025). Pengembangan manajemen skill pada wirausaha lokal: Studi kasus UMKM Bu Umar Cake. Jurnal Ekonomi, Bisnis dan Manajemen. https://www.semanticscholar.org/paper/de9bceeb69daa8bf2eeee2bfb01f6f1d65eddcb9 (accessed on 12th September 2025).

25. Isidori, V., Diamanti, F., Gios, L., Malfatti, G., Perini, F., Nicolini, A., Longhini, J., et al. (2022). Digital technologies and the role of health care professionals: Scoping review exploring nurses' skills in the digital era and in the light of the COVID-19 pandemic. JMIR Nursing, 5, e37631–e37633. https://doi.org/10.2196/37631 (accessed on 12th September 2025).

26. Jagatheesaperumal, S., Cui, Y., Ji, S., Yang, Z., Kang, J., & Zhang, Z. (2025). Visual semantics in metaverse and Web 3.0: Integrated learning and communication. IEEE Communications Magazine, 63, 107–1[illegible]. https://www.semanticscholar.org/paper/80e23df39cc0a380eb3383420dae3481c6c6aa66 (accessed on 12th September 2025).

27. Janoske McLean, M. L. (2020). JRSM 4910: social media & society (Syllabus). https://core.ac.uk/download/618310282.pdf (accessed on 12th September 2025).

28. Ji, Y., Zhao, H., Zhai, J., & Zhao, C. (2025). Cultural misunderstandings caused by misinformation and the mitigation strategies through cross-cultural communication on Rednote. Lecture Notes in Education Psychology and Public Media. https://www.semanticscholar.org/paper/59cf4952df29da9317641c8eb7211c42e82ce2e4 (accessed on 12th September 2025).
29. Karaoulas, A. (2025). The development of knowledge and skills in European education. International Journal of Research in Education Humanities and Commerce. https://www.semanticscholar.org/paper/5bb8271598e7767d1863d780c834c6c8a98d94cf (accessed on 12th September 2025).
30. Keller, J., Mazurczyk, W., Naumann, M., Wendzel, S., et al. (2015). Micro protocol engineering for unstructured carriers: On the embedding of steganographic control protocols into audio transmissions. http://arxiv.org/abs/1505.07757 (accessed on 12th September 2025).
31. Kovalchuk, V., Kovalchuk, A., Vovk, B., Derevyanchuk, O., & Rozhkov, O. (2025). Model of digitalization professional training of future vocational education teachers. Environment. Technology. Resources. Proceedings of the International Scientific and Practical Conference. https://www.semanticscholar.org/paper/673de429cdaf1b76ff109058fc27e2a1e3626713 (accessed on 12th September 2025).
32. Kuss, D. J., & Griffiths, M. D. (2017). Social networking sites and addiction: Ten lessons learned. International Journal of Environmental Research and Public Health, 14, 311–311. https://doi.org/10.3390/ijerph14030311 (accessed on 12th September 2025).
33. Mallmann, E. M., Mazzardo, M. D., Nobre, A., Nobre, H., et al. (2018). MOOC and OER: Identity management. https://core.ac.uk/download/162557582.pdf (accessed on 12th September 2025).
34. McKelvey, K. (2011). Turning an idea into a valuable teaching resource and research output. https://core.ac.uk/download/9991258.pdf (accessed on 12th September 2025).
35. McLaren, S. V. (2007). Exploring perceptions and attitudes towards teaching and learning manual technical drawing in a digital age. https://core.ac.uk/download/9016897.pdf (accessed on 12th September 2025).
36. Mourtzis, D., Angelopoulos, J., & Panopoulos, N. (2022). A literature review of the challenges and opportunities of the transition from Industry 4.0 to Society 5.0. Energies, 15, 6276–6276. https://doi.org/10.3390/en15176276 (accessed on 12th September 2025).
37. N/A. (2008). New Jersey's growing remote workforce and the skill

requirements of employers. https://core.ac.uk/download/71341901.pdf (accessed on 12th September 2025).

38. N/A. (2016). SciTech News, 70(4). https://core.ac.uk/download/83049835.pdf (accessed on 12th September 2025).

39. Narwal, R., Gupta, A., Gupta, A., & Kukreja, A. (2024). AI and blockchain: Transforming digital transactions. International Journal of Advanced Research in Science, Communication and Technology. https://www.semanticscholar.org/paper/8b589b31896a4662b57bd185acba9f2c80b8a0c9 (accessed on 12th September 2025).

40. Noya, F., Trisyani, Y., & Emaliyawati, E. (2025). The benefits, factors, and barriers faced in the implementation of digital national early warning score in hospitals: A scoping review. Journal of Nursing Practice. https://www.semanticscholar.org/paper/f1176705c7e139e842207240edd7d35e6f330da7 (accessed on 12th September 2025).

41. Pandey, H., Lunia, C., & Rathore, N. (2024). Navigating the digital maze: Privacy and security challenges in the era of cloud computing. International Journal for Multidisciplinary Research. https://www.semanticscholar.org/paper/b18320453eefe1d2af6a683608d2cab50c5a6f59 (accessed on 12th September 2025).

42. Perkins, R. A. (2014). Digital dissemination platform of transportation engineering education materials founded in adoption research. https://core.ac.uk/download/286607665.pdf (accessed on 12th September 2025)

43. Pineda, M., Jabba, D., & Nieto-Bernal, W. (2024). Blockchain architectures for the digital economy: Trends and opportunities. Sustainability. https://www.semanticscholar.org/paper/a9eb8c961bc7364180b3c5ed458e354b93349442 (accessed on 12th September 2025).

44. Popoola, O. A., Adama, H. E., Okeke, C. D., & Akinoso, A. E. (2024). Cross-industry frameworks for business process reengineering: Conceptual models and practical executions. World Journal of Advanced Research and Reviews, 22, 1198–1208. https://doi.org/10.30574/wjarr.2024.22.1.1201 (accessed on 12th September 2025).

45. Pyroha, S., & Kolb, O. S. (2025). Human information rights in the digital space. Uzhhorod National University Herald. Series: Law. https://www.semanticscholar.org/paper/02b7e5642e108d048738f52ce038118cac95abe1 (accessed on 12th September 2025).

46. Saparbekova, E., & Zhetpisov, S. (2025). The rule of law and digital transformation of public administration. Bulletin of the Karaganda

University "Law Series." https://www.semanticscholar.org/paper/8cd10a8fb233ec4bf72c881b2d191e67eb7e87a4 (accessed on 12th September 2025).

47. Segonds, F., & Veron, P. (2014). PLM and early stages collaboration in interactive design: A case study in the glass industry. https://core.ac.uk/download/29137497.pdf (accessed on 12th September 2025).

48. Thatcher, A. (2011). Managing social activity and participation in large classes with mobile phone technology. https://core.ac.uk/download/39668566.pdf (accessed on 12th September 2025).

49. van Etten, W. (2006). Introduction to Random Signals and Noise. https://core.ac.uk/download/pdf/11469291.pdf (accessed on 12th September 2025).

50. Xu, L. D., Xu, E., & Li, L. (2018). Industry 4.0: State of the art and future trends. International Journal of Production Research, 56, 2941–2962. https://doi.org/10.1080/00207543.2018.1444806 (accessed on 12th September 2025).

51. Zeng, Y., Zeng, L., Zhang, C., & Cheng, A. S. K. (2022). The metaverse in cancer care: Applications and challenges. Asia-Pacific Journal of Oncology Nursing, 9, 100111–100111. https://doi.org/10.1016/j.apjon.2022.100111 (accessed on 12th September 2025).

52. Zhumadilova, M., & Berkenova, R. (2025). Fundamentals of linvodidactic methodology for the formation of writing skills of students in other languages. Bulletin of the Eurasian Humanities Institute, Philology Series. https://www.semanticscholar.org/paper/b0e63aed734dbba8f1adfc64070d71d6f1d975ef (accessed on 12th September 2025).

CHAPTER 6

DIGITAL MODULATION TECHNIQUES

Contents

INTRODUCTION

The addition of digital modulation techniques has extensively transformed modern communication systems with the advent of faster and more efficient transmission. Digital modulation is a technique used to encode data onto the carrier wave so that the signals are transmitted through various media like optical fibers and wireless channels. This movement away from analog toward digital technology is characterized by enhanced performance, for example, resistance to noise and increased data rates that can be accommodated. Because contemporary networks call for increasingly large bandwidths, modulation methods such as amplitude modulation (AM), frequency modulation (FM), and phase modulation (PM) have been extensively created and optimized. The arrival of sophisticated signal processing technologies further complements these methods so that signal distortions can be mitigated and communication systems be run at optimal levels. In fiber optic communication, digital modulation techniques are of essential importance in overcoming inherent impairments with fiber transmission. The evolution of coherent detection systems has revolutionized digital signal processing, best illustrated by the possibility of utilizing advanced digital signal processing (DSP) algorithms. This is achievable through developments that allow improved spectral efficiency through multi-level modulation formats, managing linear distortions that limited performance earlier.

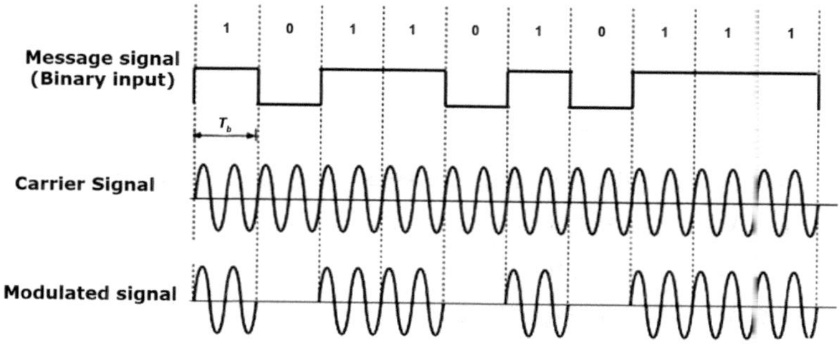

Figure 6.1. *Digital modulation techniques.*

Source: https://i0.wp.com/netizzan.com/wp-content/uploads/2023/05/Digital-modulation.webp?resize=719%2C405&ssl=1.

Furthermore, with improving telecommunication systems in the direction of higher data transmission speeds, e.g., 100 Gbps and more, new challenges arise, namely with respect to non-linear effects like the Kerr effect. Complete understanding of fiber nonlinearity and an operational plan of effective NLC techniques are of the utmost importance for optimal communication performance and system dependability, enhancing the role of modulation techniques in modern telecommunications (Amari et al., 2017). The rapid development of digital technologies further emphasizes the importance of effective digital modulation techniques in current use. Alongside the popularity of digital media, the environment of copyright and data protection has been fundamentally transformed. Compared to analog technology, which inherently limited the quality of copies, digital processes enable the reproduction of identical digital data without degradation, further aggravating piracy and problems with data integrity. To solve these issues, digital watermarking has been a key mechanism in intellectual property protection. By embedding information in digital material, watermarking preserves and protects original works against abuse. The integration of digital modulation and watermarking not only safeguards content but also reflects the mutual relationship between technological advancement and human issues, displaying the multifaceted consequences of digital modulation techniques (Serdean et al., 2002) (Figure 6.1).

Definition of Digital Modulation

Digital modulation is the techniques used to convert digital information to an analog signal suitable for transmission via various media. In telecommunications, it is crucial because a proper and efficient data transfer is required. There are digital modulation techniques that can be classified based on how they modulate the carrier wave, carrying the data signal. Amplitude Shift Keying (ASK), Frequency Shift Keying (FSK), and Phase Shift Keying (PSK) are some of the usual ones. All these techniques possess certain advantages in terms of bandwidth efficiency, noise immunity, and complexity of implementation. For instance, PSK can pack data very compactly without sacrificing much performance to noise, and hence it is a popular choice for all communication systems. One intrinsic characteristic of digital modulation is that it can be tailored to any application. With home networking, the use of digital modulation technologies such as eight-level pulse amplitude modulation (8-PAM) and orthogonal frequency-division multiplexing (OFDM) demonstrates how these technologies are optimized for high-speed, reliable data transmission. Both modulation techniques have already been successfully deployed to provide high data rates over some distance, demonstrating the real-world capabilities of these techniques. The technological advancements from these recent efforts, such as the POF-ALL and POF-PLUS projects, reflect the continued advancement in modulation technologies that enhance the performance and efficiency of digital communication networks (Abrate et al., 2010). Furthermore, the significance of digital modulation techniques extends beyond direct application; they form the basis of devising advanced communication networks. With telecommunication demands evolving, so do the modulation methods that form the basis of these systems. By way of example, the recent introduction of Galois-Division Multiple Access (GDMA) illustrates how newer schemes provide optimized bandwidth maximization and multiplicity of users on one platform for improved overall system performance. With the use of finite field transforms that provide optimized capacity usage, GDMA is a new approach to dealing with resources in more congested communication channels (de Oliveira et al., 2015). These advances underscore the pivotal role of digital modulation in enabling modern communication infrastructures to be capable of supporting current and even prospective data transmission demands.

Importance in Modern Communication

In the rapidly evolving environment of modern communication, digital modulation techniques are the backbone of effective information transfer. They ensure that data is transferred and received via various media, including wireless networks, satellite systems, and optical fibers. Techniques such as Quadrature Amplitude Modulation (QAM) and Frequency Shift Keying (FSK) increase the efficiency of the communication by enabling multiple signals to occupy the same bandwidth. As the need for broadband access and mobility throughout the world continues to increase, the significance of such techniques to improve bandwidth efficiency and signal integrity cannot be emphasized enough. Moreover, application of advanced modulation methods contributes a great deal towards increasing the overall strength and reliability of communications systems against interference as well as attenuation of signals, which is especially needed in highly congested urban centers where electromagnetic noise against signal purity is a major issue. Digital modulation methods directly affect a number of aspects of modern communication, primarily the wireless technology domain. The Burts-by-burst (BbB) adaptive systems are a case of a more sophisticated approach towards addressing the dynamic nature of wireless channels. Such systems have the ability to adjust the level of modulation based on real-time channel conditions, optimizing throughput and reducing latency. This capability is vital under signal quality-varying environments, where it becomes hard for traditional fixed-mode systems to keep pace. With the advent of digital communication, the integration of BbB-adaptive and symbol-by-symbol adaptive systems is capable of effectively compensating against channel distortions, leading to improved user experience. Furthermore, the combination of these techniques with diversity methods like Rake receivers and space-time block coding enhances reliability, particularly in contexts where users are subjected to compounded challenges like shadowing or multiuser interference, further solidifying the role of digital modulation in contemporary communication. Aside from functional efficiency, digital modulation techniques possess socio-economic importance which underscores their position in contemporary communication. Growth in mobile connectivity and Internet of Things (IoT) reflects the extent to which a society relies on advanced communication systems. By improving the dependability and efficiency of data transmission, these techniques facilitate innovations in important fields such as telemedicine, smart cities, and driverless cars. Standardization of electromagnetic compatibility (EMC) follows the pace of the advent of digital modulation methods as witnessed by attempts to update standards guiding wireless system coexistence. This path emphasizes as much the technical necessity as the social requirement to offer an interference-free communications environment in a more globalized world. Therefore, the continuous advancement in digital modulation is not merely technological advancement; it is component elements of modern communication infrastructure fueling economic growth and social development (Figure 6.2).

Figure 6.2. *Importance in modern communication.*

Source: https://fastercapital.co/i/BabyBells-and-Digital-Switching--Transforming-Communication-Systems--The-Backbone-of-Modern-Communication.webp.

Overview of Key Techniques

The discipline of digital modulation methods encompasses a variety of significant methods that enhance the efficacy of communication and integrity of data. Among them, amplitude modulation (AM), frequency modulation (FM), and phase modulation (PM) are inherent methods. Amplitude modulation varies the amplitude of the carrier signal to convey information, which, even susceptible to noise, is simple and cost-effective for low-bandwidth applications. Frequency modulation, on the other hand, changes the frequency of the carrier signal, with greater noise immunity, thus more appropriate for higher fidelity requirements. Phase modulation, being more complicated in nature, does provide a means of representing data through changing the phase of the carrier signal, allowing higher data rates and better bandwidth efficiency. The selection between these techniques is based on the specific needs of the communication link, bandwidth, and the desired degree of resilience against interference, making them key features of digital modulation techniques. These fundamental techniques have developed into sophisticated digital modulation techniques, primarily in the field of optical communications and wireless systems. Techniques like quadrature amplitude modulation (QAM) and orthogonal frequency-division multiplexing (OFDM) illustrate the synthesis of multiple modulation methods in a bid to maximize data transport. QAM utilizes both AM and PM in conjunction, enabling multiple bits to be carried per symbol and therefore maximizing spectral efficiency. OFDM breaks down the communication frequency into numerous orthogonal sub-carriers, each of which is modulated individually, improving resistance to multipath fading. This attribute is particularly useful in interference-rich environments or bandwidth-limited conditions to provide enhanced signal integrity. The sophisticated nature of these techniques underscores the ongoing evolution of digital communication strategies, adapting to the demands of modern applications within an evolving technology landscape. Furthermore, the application and development of these digital modulation techniques are largely dependent on advances in digital signal processing (DSP). DSP accounts for the mitigation of impairments in signal transmission, primarily in fiber optic communications. For instance, optical fibers nonlinear distortions and linear impairments, which can lead to impairment of transmitted signals' fidelity, are efficiently addressed by coherent detection techniques, supplemented by DSP algorithms. Combining modulation techniques with DSP enables augmented capacity and reliability through practices such as the creation of 100 Gbps WDM systems with multi-level modulation formats. As data demands keep expanding at breakneck speed research into optimizing techniques, including the compensation method for nonlinear effects, continues to be pivotal to the maintenance of solid and efficient digital communication channels. The correlation between DSP and modulation is thus a cornerstone in achieving performance in high-speed communications systems, pointing to the sophistication and intricacy of dominant technology trends (Serdean et al., 2002; Amari et al., 2017).

6.1. BASICS OF DIGITAL MODULATION

Digital modulation is a fundamental technique used in modern communication systems to facilitate efficient data transmission through various mediums. Converting digital signals into analog signals, modulation techniques enhance data transmission over long distances and minimize the impact of interference and noise. Modulation has different techniques available, including amplitude modulation (AM), frequency modulation (FM), and phase-shift keying (PSK), each with their own features and uses. For instance, PSK enables full phase change control and hence is very noise-tolerant, whereas frequency modulation is the benchmark against which one measures resistance to varying signal strength. Familiarity with these fundamentals is necessary in order to develop an understanding of more sophisticated ideas and applications in digital communications. The other major characteristic of digital modulation is the fact that it finds application in adaptive systems, particularly those which adapt instantaneously to variable channel conditions. For instance, burst-by-burst adaptive HSDPA systems demonstrate how digital modulation can dynamically adjust to signal quality changes.

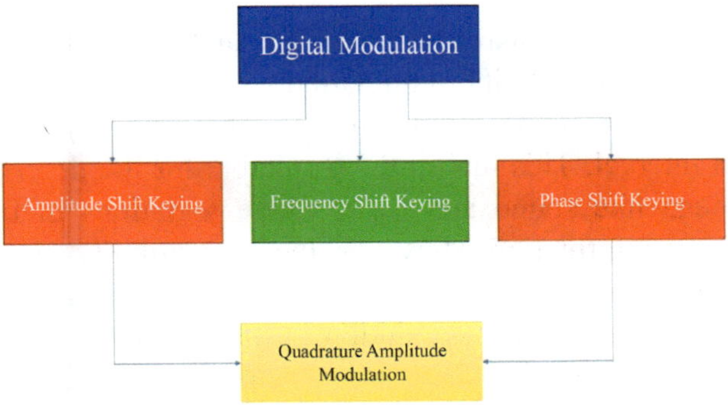

Figure 6.3. *Basics of digital modulation.*

Source: https://www.researchgate.net/profile/Diponkor-Bala/publication/348364945/figure/fig15/AS:978360746381315@1610270787739/Classification-of-Digital-Modulation.png.

Adaptive systems utilize techniques such as OFDM for data rate improvement and robustness in bad conditions. The reason why adaptive modulations are successful is that they can offset instantaneous short-term variations, thereby improving total performance. The changing characteristics of wireless technologies highlight the necessity to understand these changes, since they are an intrinsic part of future communication systems, which depend on sustaining elevated data rates and reduced error rates under conditions of quick change (Choi et al., 2007). The incorporation of sophisticated digital signal processing methods further enriches digital modulation. Techniques such as space-time coding and Rake receivers are now inexorable in countering multipath fading, a pervasive phenomenon in wireless communication. In reducing bit error rates and improving signal quality by spreading data across multiple transmission paths, these methods bear fruit. Additionally, with software applications increasingly being coded into new domains such as driverless cars, awareness of the complexity of radar systems becomes more significant. Radar uses principles of digital modulation to detect and track targets with a special emphasis on the widespread applicability of these techniques. An overview of the several industries such as automobile technology and telecommunication would make one understand the importance of digital modulation basics in shaping the future of technology and communication (Rummel et al., 2019) (Figure 6.3).

6.1.1. Analog vs. Digital Signals

In telecommunication, the distinction between analog and digital signals is at the heart of what defines digital modulation techniques. Analog signals have waveforms that are continuous and can take a theoretically infinite range of values over an interval. This makes them suitable for the description of real-world phenomena like sound and light, where variations in frequency and amplitude convey information. However, the susceptibility of analog signals to distortion and noise over transmission limits their effectiveness in modern communication systems. The revolution towards digital signals has thus grown in momentum, which convey information in terms of discrete values improving the integrity of the signal and allowing more effective processing of data. Digital signals, by contrast, quantify information by digitizing it into binary formats – distinct collections of 0s and 1s. Such a form of representation supports more efficient error detection and correction processes, thus ensuring transmission of data with increased fidelity. Digital modulation techniques such as Quadrature Amplitude Modulation (QAM), utilize these principles to achieve maximum bandwidth efficiency and deliver higher transmission rates for data. Such techniques are very useful in high-speed communication systems, where data volume and transmission rate become the topmost priorities. The study of digital modulation with QAM has brought positive results in certain applications, including in particle physics, where experimental data may overwhelm traditional systems and therefore demonstrate the real worth of digital signals in processing huge amounts of information. High-level digital modulation methods are emphasized in applications such as proposals for the upgrade of the Large Hadron Collider (LHC) at CERN. The planned Super LHC is to gain an order of magnitude in luminosity and thus produce much larger volumes of data (Albert et al., 1998). This escalation creates problems with data handling and processing rates and requires the deployment of efficient digital modulation methods. Laboratory sessions involving digital communications highlight the need for endowing telecommunications experts with proficiency for managing such conversions between analog and digital devices (Albert et al., 1998). The ongoing growth of digital signal processing has far-reaching effects across various fields, again underlining the need for ongoing research and improvement in digital modulation techniques, which ultimately decide the prospects of communication technology.

6.1.2. Modulation Process Explained

The modulation technique is a fundamental technique employed in digital communication systems to offer the capability to transmit data in various means of media, such as copper wires and fiber optics. Modulation is simply the process of varying a carrier signals' characteristics—amplitude, frequency, or phase—to represent information. Modulation offers means through which the signal is transmitted a long distance with little quality loss such that it can cross noise and interference barriers. Other modulation techniques, such as amplitude modulation (AM) and frequency shift keying (FSK), have their own trade-offs, advantages, and applications. Proper selection of the techniques is necessary to ensure optimum bandwidth efficiency as well as signal integrity in digital communications, particularly in high-speed networks like those employed in the latest fiber-optic technology. The development of coherent detection systems has revolutionized the process of modulation, particularly in optical communication networks. In coherent systems, DSP algorithms are used to correct for optical impairments and enhance transmission performance. In reference to the article, those systems support advanced wavelength-division multiplexing (WDM) standards that support increased spectral efficiency using multi-level modulation formats. They also handle linear distortions optimally, stressing the adaptability of the modulation process in current applications. The integration of coherent detection with new modulation techniques allows for significant progress in achieving high data rates, such as in 100 Gbps and beyond. With increasing pressures from larger data demands, it becomes more important to understand and balance the modulation process, especially with the additional complexity of fiber nonlinearity. It becomes an issue with advancing transmission technology, particularly with fiber nonlinearity, which is prominent at high input powers. As noted, the Kerr effect introduces an additional layer of complexity in the modulation process that needs advanced compensation techniques. A number of nonlinear compensation (NLC) methods have been proposed to overcome these limitations and improve transmission reliability in general. A good understanding of these methods is required for current and future WDM systems and how to strike a delicate balance between implementation complexity and performance trade-offs. Exploring the process of modulation in the context of these new technologies not only helps in the design of better systems but also highlights the key role of innovation in digital modulation techniques, leading to opening up new avenues for future advances in information communication technology (Figure 6.4).

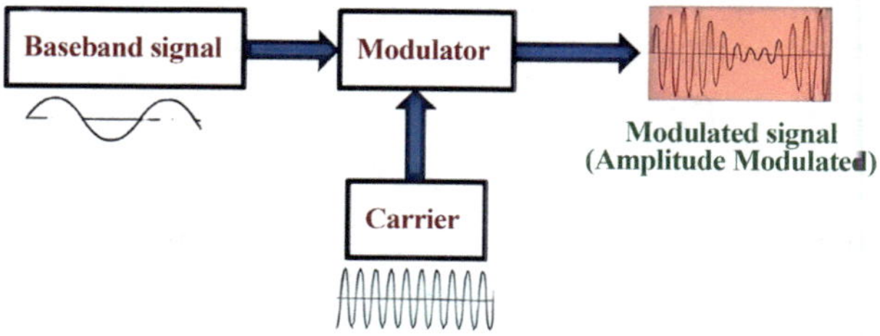

Figure 6.4. *Modulation Process Explained.*

Source: https://engineerstutor.com/wp-content/uploads/2018/09/2.jpg.

6.1.3. Key Parameters of Modulated Signals

In communication through digital means, analysis of the modulated signal is important in order to support effective data transmission. Bandwidth, power spectral density, and bit error rate (BER) are some significant parameters that contribute greatly towards the identification of performance of modulation techniques. Bandwidth determines the amount of data that can be successfully transmitted using a medium of transmission, with higher bandwidth supporting higher data rates. Conversely, a low bandwidth will generally cause data congestion and increase the likelihood of transmission errors. Power spectral density describes the way transmission power is allocated among different frequencies, which holds true very much for practical channels with noise and interference. Hence, these parameters make engineers more capable of designing best-suited modulation techniques for specific applications by balancing speed, reliability, and usage of resources, which ultimately results in better communication systems. Another significant modulated signal parameter is the signal-to-noise ratio (SNR) definition. SNR measures the amount of the signal of interest relative to the ambient noise, and its effect directly applies to the quality of data transmission. Larger SNR values equate to cleaner signals, which provide lower error rates and enhance the overall system performance. Techniques such as error correction coding and modulation diversity are employed to increase SNR, particularly in adverse environments with high interference. Further, modulation scheme development, e.g., multiple-input multiple-output (MIMO) systems, reflects the innovative strategies to maximize SNR and capacity. For instance, methods based on the application of genetic algorithm-aided optimization techniques have been established that are capable of providing enhanced performance in actual applications, particularly for multiuser MIMO-OFDM, as is evident from recent research publications. This ongoing advancement in modulation techniques indicates the need to understand thoroughly SNR and its implications on signal integrity and efficiency. Finally, the selection of modulation schemes also depends on the specific applications and transmission conditions involved. There are various environments—urban, rural, or indoor—each with different levels of multipath fading and interference that call for the use of particular modulation techniques. For instance, quadrature amplitude modulation (QAM) can be applied in situations with good conditions for high SNR and low distortion, while phase-shift keying (PSK) can be applied in worse-case scenarios to offer noise resistance. Engineers are able to modify their modulation methods in real time using this knowledge of the essential parameters and achieve optimal performance and dependability in different communication environments. Additionally, research into adaptive modulation techniques, which adapt modulation schemes dynamically based on measured channel conditions, presents an excellent area of future research and can result in next-generation systems that harmoniously combine advanced digital modulation techniques. The knowledge of key parameters is hence important for encouraging innovation and effectiveness in digital communication.

6.2. TYPES OF DIGITAL MODULATION TECHNIQUES

Digital modulation techniques form the cornerstone of modern communication networks, where information is efficiently mapped onto carrier signals to be sent over any transmission medium. Among these techniques, Amplitude Shift Keying (ASK), Frequency Shift Keying (FSK), and Phase Shift Keying (PSK) are the most popular ones. ASK varies the amplitude of the carrier signal with respect to the binary data being sent, and it is easy to implement but sensitive to noise interference. In addition, FSK modulates the carrier signal frequency to represent binary data and provide more noise resistance in data communications. PSK varies the phase of the carrier wave in representing information and providing an improved immunity against the degradation of the signal. These techniques satisfy various requirements of applications and thus lead to various applications in environments ranging from simple wireless communications to complex digital transmission systems. Continuing with digital modulation, Quadrature Amplitude Modulation (QAM) has gained widespread use due to its ability to send data on scarce bandwidth. QAM combines both amplitude and phase modulation, where greater than a single bit of information is sent with each change in the signal. This characteristic makes QAM highly beneficial for sending data at high speed, such as in digital television and broadband internet.

The cost, however, is added complexity in the process of demodulation and the requirement for sophisticated methods of coherent detection. Orthogonal Frequency Division Multiplexing (OFDM), on the other hand, is unique in that it can correct the problems of multipath propagation, which is very common in wireless communications. OFDM can pack data efficiently and is also immune to interference using many carrier frequencies in an orthogonal mode, making it the foundation of wireless communication standards of today, such as LTE and Wi-Fi technology. The impact of digital modulation goes beyond typical uses of communications. For instance, Phase Locked Loops (PLLs) are critical blocks in synchronization processes in various digital systems. PLLs are often utilized in mixed-signal systems, where they facilitate synthesizing stable frequencies needed for uniform operations in integrated circuit designs, particularly in System on Chip (SoC) implementations. Charge-Pump (CP) PLL design is a good approach to clock generation operations such that devices like Analog to Digital Converters (ADCs) and Digital to Analog Converters (DACs) can operate effectively. Such on-chip operations stress the importance of digital modulation techniques not only for accurate data transmission but also for efficient operation of integrated systems in more complex technological environments, as quoted in Burbidge et al. (2008). Thus, the study of these kinds of modulations is crucial for the advancement of digital communications and their applications within contemporary technology, like developments in mobile robotics, as quoted in Carnegie et al. (2011) (Figure 6.5).

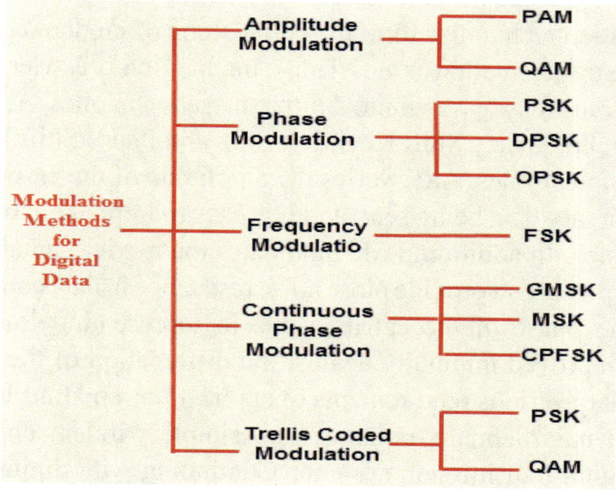

Figure 6.5. *Types of digital modulation techniques.*

Source: https://www.watelectronics.com/wp-content/uploads/types-of-digital-modulation1.png.

6.2.1. Amplitude Shift Keying (ASK)

ASK is a fundamental modulation technique in digital communication, known for simplicity and efficiency in encoding binary data. In ASK, the amplitude of the carrier signal is changed in proportion to the data being transmitted—typically two states: high amplitude for a binary 1 and zero or low amplitude for a binary 0. Due to the fact that this method features easy implementation, it is especially apt for application in various applications, particularly in low data-rate systems. ASK, however, is prone to noise and fading, particularly in situations where signal integrity is in doubt. These vulnerabilities can provide less optimal performance in regards to bit error rates, and hence more secure digital modulation techniques must be explored for uses where increased reliability in data transmission is required. One of the most limiting aspects of ASK is its performance under noise, particularly in real-time communication systems. As indicated by more recent studies, although ASK is advantageous in systems where bandwidth efficiency is prioritized, it suffers when it comes to maintaining signal quality, particularly in systems with time-varying channel states. For example, in conditions emulated in heterodyne systems, the bit error rate (BER) performance of binary amplitude shift keying (2ASK) has been found to be generally poorer than with other schemes like binary frequency shift keying (2FSK) or more sophisticated approaches like binary polarization shift keying (2PolSK), the latter of which was shown to exhibit better signal-to-noise ratio (SNR) performance under turbulence and phase noise (Lee et al., 2011). Hence, while ASK has some particular applications, its limitations emphasize the need for ongoing advancement in digital modulation methods to be useful for a wide range of applications. Further, the advent of low-power wireless communication protocols has revived interest in ASK, particularly in license-free industrial, scientific, and medical (ISM) frequency bands. Modern applications such as home automation utilize the protocols to enable multiple devices to be combined efficiently. As outlined in recent work, the robust framework provided by software-defined radio (SDR) technology enables rapid prototyping and system deployment utilizing ASK to enable adaptive flexibility on the basis of specific application requirements (Kumbhar et al., 2016). While ASK can be effective in such scenarios, for ensuring guaranteed connectivity, other error correction schemes or utilization of stronger modulation schemes must be used to ensure general system performance. Last but not least, as an appropriate aspect of digital modulation techniques, Amplitude Shift Keying remains relevant but with use in contemporary wireless systems having to be managed within its inherent limitations and in accordance with the evolving demands of communication technologies.

6.2.2. Frequency Shift Keying (FSK)

FSK is a technique of digital modulation in which data is recorded onto a carrier wave by varying its frequency. It is particularly applied where resistance to noise and interference is crucial. In FSK, the frequency of the carrier varies between discrete frequencies, which are proportional to digital data that is being sent. For instance, in a two-stage FSK system, two distinct frequencies are employed to encode the two binary digits 0 and 1. The primary advantage of FSK is its relative resistance to amplitude fluctuations and phase distortion, so it is particularly suitable for transmission over many other kinds of transmission medium, including radio and optical communications systems. Further, its capacity to adapt to different channel conditions enhances data transmission reliability, which is important in telecommunication and networking industries. Another stunning aspect of FSK is its bit error rate (BER) performance, especially how it fares compared to other modulation techniques.

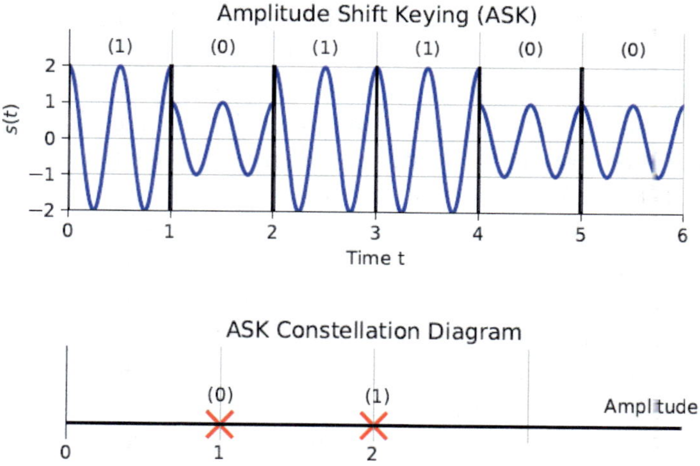

Figure 6.6. *Frequency shift keying (FSK).*

Source: https://wirelesspi.com/wp-content/uploads/2020/12/figure-ask.png.

It has been found through studies that FSK can at least match, if not outperform, other modulation techniques under certain circumstances, such as when noisy channels or atmospheric turbulence exist. This is well demonstrated in optical coherent communication systems wherein FSK has been compared with other schemes like binary polarization shift keying, and it has demonstrated that FSK is capable of yielding better signal-to-noise ratio (SNR) performance. The analytical results confirm the hypothesis that error performance of FSK schemes can be consistently reliable, enabling its use in real applications where data integrity is essential (Lee et al., 2011). Such performance measures are critical in designing effective communication protocols and optimizing the existing systems. Moreover, work on index-modulated systems has introduced new innovative techniques incorporating frequency shifting to enhance the performance of the traditional FSK. One example is the index-time frequency hopping (I-TFH) system, which has been proposed as a method to maximize power and spectral efficiency in optical communications. This technique consists of common modulation of signal frequency and timing with the purpose to reduce non-turbulent error probabilities. Unlike this, I-TFH outperforms traditional methods in terms of BER and power efficiency and reflects the dynamic transformation of digital modulation schemes (Escribano et al., 2019). The integration of these advanced methods with traditional FSK demonstrates continued innovation in the field and highlights the importance of FSK to modern communication technologies (Figure 6.6).

6.2.3. Phase Shift Keying (PSK)

PSK is one of the more powerful digital modulation techniques, simply due to the fact that it can effectively transmit data via a large variety of communication channels. By varying the phase of a carrier wave to convey digital information, PSK is able to achieve a high level of spectral efficiency. This approach is especially useful in situations where bandwidth is limited, as it is able to pack the highest number of bits per symbol. In particular, forms of PSK, including binary and quadrature phase shift keying, enable two and four bits per symbol to be sent, respectively, increasing the quantity of data that can be transmitted in a single broadcast cycle. This spectral efficiency makes PSK an attractive choice for applications ranging from satellite communications to newer wireless applications. PSK is not challenge-free, however, particularly when it comes to noise and interference. The biggest drawback with PSK systems is their phase noise sensitivity, which degrades the quality of the signal and the incidence of bit errors, particularly in adverse environmental conditions. Here, the effectiveness of PSK in real operation is quantified through performance metrics such as Bit Error Rate (BER). Analytical outcomes and simulation results prove that BER performance can be highly affected by factors such as atmospheric turbulence and non-ideal devices. For instance, research on coherent optical binary polarization shift keying has given insight into PSK sensitivity and performance compared to other plans and demonstrated the reliability of PSK in maintaining signal integrity under poor conditions (Lee et al., 2011). Incorporating advanced modulation techniques such as Space-Time Shift Keying (STSK) may further amplify the PSK system capabilities. STSK, as part of a greater Multiple-Input Multiple-Output (MIMO) strategy, aims to take advantage of spatial diversity to provide more reliable and greater capacity transmission. By having data spread across multiple antennas, STSK takes advantage of the spatial characteristics of the channel and can perform better than conventional modulation schemes. The combination of PSK and MIMO technologies not only makes the overall system more efficient but also more robust against channel degradation like fading and interference. As communication technology advances, PSK's incorporation into new approaches like STSK offers exciting possibilities to meet ever-increasing stringent requirements for high-rate, reliable information transmission in various applications (Chen et al., 2012) (Figure 6.7).

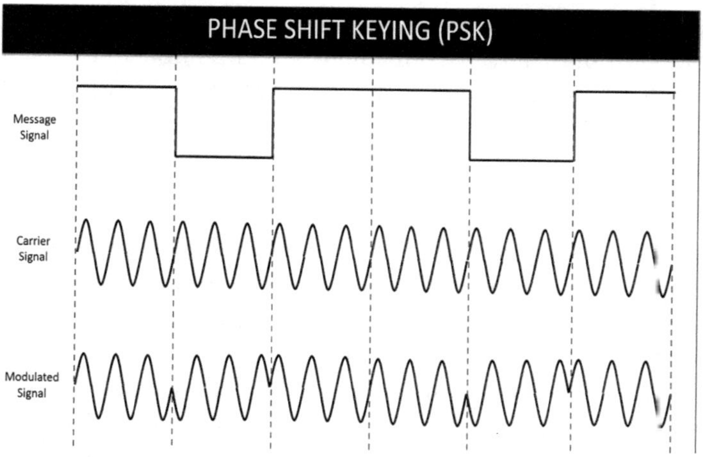

Figure 6.7. *Phase shift keying (PSK).*

Source: https://i.ytimg.com/vi/Ma-hAQsbYFI/maxresdefault.jpg.

6.3. ADVANCED DIGITAL MODULATION TECHNIQUES

In contemporary telecommunication, evolved digital modulation techniques have a key role to play in maximizing data transmission efficiency and quality. Not only are they essential to improved bandwidth usage, but also to improved channel robustness against channel distortions. Application of high-level modulation schemes, such as quadrature amplitude modulation (QAM) and higher-order modulation schemes, significantly boosts spectral efficiency. The transition from traditional modulation methods to these new alternatives is a result of the explosive demand growth for high-data-rate services. Furthermore their application provides enhanced performance in congested frequency bands.

an important factor for today's wireless communication systems. As the industry continues on the path toward next-generation networks, the understanding and implementation of these advanced modulation methods become increasingly essential, delivering communications to be both reliable and efficient under a broad range of operation environments. A key element of advanced digital modulation is the breakthrough of coherent detection systems, which employ digital signal processing (DSP) algorithms to revert the distortion brought about by optical fibers. As highlighted in current studies, coherent systems enhance transmission speeds by a substantial degree, enabling the current standards of wavelength-division multiplexing (WDM) technology. This shift towards coherent detection not only promotes spectral efficiency through multi-level modulation formats but also includes efficient solutions for handling linear fiber distortions. Consequently, the industry has experienced a remarkable boost in data rate, with systems operating at 100 Gbps and advancing towards speeds of 400 Gbps. The combination of future-generation modulation techniques and coherent detection platforms is transforming optical communications in equipping larger capacity networks to effectively accommodate increasing data traffic demands, such as in Amari et al. (2017). Furthermore, the alleviation of challenges brought forth by nonlinear effects in fiber optics remains a critical focus with increased transmission speeds. The Kerr effect, particularly at elevated input powers, produces significant nonlinear degradations that contribute to signal degradation. In order to counter these effects, many nonlinearity compensation (NLC) techniques have been proposed to improve the performance of high-speed systems. Integration of these techniques with advanced digital modulation techniques provides an even more robust communication system that is capable of compensating for the complexities of the real-world transmission environment. In particular, the development of methods that enhance modulation format classification through artificial neural networks is characteristic of best-practice applications within the discipline. This serves not only to enhance the level of accuracy in modulation classification but also to optimize the performance of systems in general, and in doing so, reflects the significance of continued research and refinement in this fast-developing area of telecommunications, as presented in Fathallah et al. (2018).

6.3.1. Quadrature Amplitude Modulation (QAM)

QAM is a sophisticated digital modulation technique that employs both amplitude and phase variations to transmit data at an effective rate through a channel of communication. It doubles the data transmission rate by modulating two carrier waves simultaneously, one of which is 90 degrees out of phase with the other. This two-way modulation scheme enables a more robust scheme of data encoding and is therefore particularly coveted for application in systems like digital television and broadband internet. QAM's ability to be configured in various forms ranging from basic ones like 16-QAM to more complex ones like 64-QAM and 256-QAM, each with greater data capacity but at the cost of heightened susceptibility to noise and signal degradation, renders it so coveted. The effectiveness of QAM mostly relies on the environment that it operates in. At best, higher-order modes of QAM like 64-QAM support more data rates and hence are more effective in cases of applications where data needs to be transferred at high speeds. With the increase in modularity, however, susceptibility to interference and noise increases.

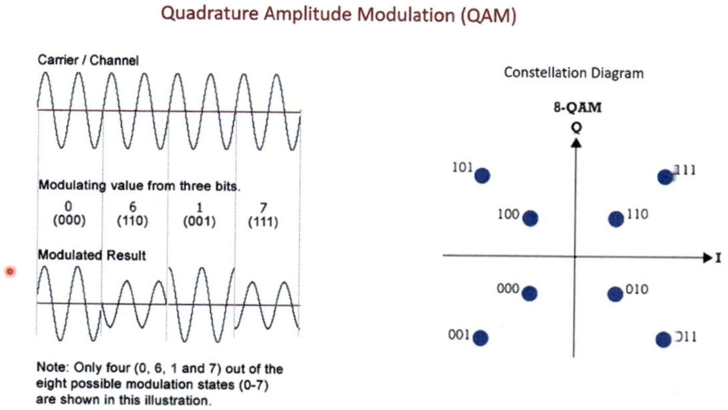

Figure 6.8. *Quadrature amplitude modulation (QAM).*

Source: https://i.ytimg.com/vi/zZX9A6h7upo/maxresdefault.jpg.

For example, lower-order modulation schemes like Binary Phase Shift Keying (BPSK) and Quadrature Phase Shift Keying (QPSK) are famously noted for their durability in hostile environments, and the same might explain why they are utilized in various critical communication environments where reliability is the overriding requirement. Experiments have demonstrated that lower order modulation systems can perform better than higher schemes at error rates under a variety of types of noise, such as in research that explores the viability of QAM for use in Orthogonal Frequency Division Multiplexing (OFDM) systems (Karimian et al., 2010). Enhancing the performance and efficiency of QAM systems, advanced methods like trellis-coded modulation (TCM) are employed. TCM integrates error correction coding with modulation in order to increase the overall data transmission reliability. The modular and flexible design of a Trellis

Modem Codec (TMDC) allows for flexible modulation formats like various QAM levels to accommodate the specific requirements of the channel. This is significant for systems where bandwidth is limited but data throughput is crucial. The seamless integration of coding techniques enhances both power performance and spectral efficiency, demonstrating increased development in QAM as a prominent digital modulation technique. Additionally, prototype and test-bed development is another step in the scientific research of modulation schemes, allowing extensive evaluation of their performance in diverse operational conditions (Budinger et al., 2025) (Figure 6.8).

6.3.2. Differential Phase Shift Keying (DPSK)

DPSK is a modulation method that has several benefits in the context of digital communications, mostly by making it possible to have strong data transmission through environments with fading and noise. Unlike conventional phase shift keying, DPSK sends information through the difference in phases between consecutive symbols. This aspect makes DPSK particularly beneficial for wireless and mobile communications, where it is not easy to maintain an unwavering reference signal. DPSK, by eliminating the need for coherent detection, lowers the complexity of the receiver without sacrificing too much in performance and is thus used extensively in a number of applications, such as satellite communications and optical data communication. DPSK systems can be considerably enhanced if coupled with advanced error-correction techniques. For instance, Turbo coding can significantly improve DPSK transmission's bit error rate (BER) performance. It is reported that the use of a Turbo-coded scheme with M-ary DPSK can provide high coding gains and hence greater data transfer efficiency. Differential Amplitude and Phase Shift Keying (DAPSK) combined with DPSK can also utilize the amplitude and phase data and deliver more throughput without compromising signal reliability. More specifically, research has demonstrated that a 64-ary DAPSK system has the potential to gain over 4.2 dBs more than a standard 64-DPSK system at equal output rate, and thus its intrinsic efficacy in relation to enhancing communication link quality under optimized conditions (Hanzo et al., 2011). The adoption of adaptive demodulation systems imposes an added complexity level on DPSK applications. Adaptive Demodulation (ADM) offers a general approach to signal recovery independently of channel state information at the transmitter. It is supported by a variety of differential schemes, including M-DPSK, which can modify decision boundaries adaptively in accordance with the received signal quality and thus achieve the best performance under varying channel conditions. It has been demonstrated that ADM systems are beneficial in Rayleigh fading channels, where natural environments tend to critically compromise signal integrity. With a focus on the regeneration of the most reliable bits through near-optimal implementations, such systems maximize spectral efficiency and minimize bit error rates, rendering high-rate communication both efficient and reliable (Abouei et al., 2010). As developments in DPSK continue, its synergy with adaptive techniques solidifies it as an integral part of modern digital modulation techniques (Figure 6.9).

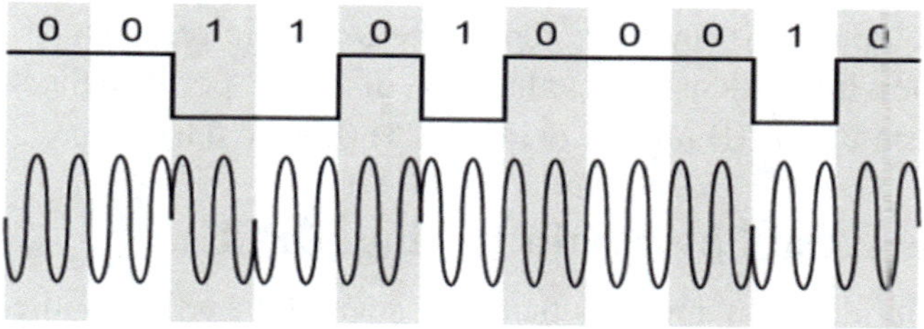

Figure 6.9. *Differential phase shift keying (DPSK).*

Source: https://electronicspost.com/wp-content/uploads/2018/06/dpsk-waveform.png.

6.3.3. Orthogonal Frequency Division Multiplexing (OFDM)

OFDM is a significant advancement in the area of digital modulation techniques, particularly for high-speed communication systems. Through the division of a single stream of data into multiple smaller streams sent together across different frequencies, OFDM manages to keep away from multipath fading and interference issues that other modulation techniques experience. This aspect is priceless, particularly in environments where signal degradation occurs due to obstructions or reflections. Moreover, OFDM's natural design allows for the support of overlapping subcarriers, which maximizes spectral efficiency at no substantial cost of inter-carrier interference (ICI), another critical factor in amplifying data throughput in situations of compressed bandwidth. The advantages of OFDM extend beyond robust performance to include its suitability to varied channel conditions, under which dynamic resource allocation becomes possible. In practical application, OFDM systems can be adjusted to change the modulation scheme on a subcarrier by subcarrier basis, boosting the net data rate based on the channel quality. This feature is particularly beneficial in applications such as wireless communication and broadband networks, where varying user demand necessitates flexible and effective system response. Recent studies elucidate the growing use of OFDM in wireless as well as optical transmission channels, citing its suitability in next-generation network architecture. Particularly, emerging technologies such as optical OFDM (O-OFDM) have been found to enable ultra-high-speed data transmission, thereby satisfying the growing need from the internet as well as data-intensive applications (Leenheer et al., 2013). Experimentation and research validate the effectiveness of OFDM in a wide range of applications, including those with high mathematical modeling content. OFDM capacity to stabilize and maximize communication systems for varying operating conditions is evident through the establishment and investigation of complex control systems such as Artificial Neural Networks for DC-AC converters applied in wind power generator systems. Findings indicate that the integration of sophisticated control methods can significantly enhance output stability and quality but further solidify OFDM as a leading digital modulation technique. With researchers and engineers still exploring and enhancing OFDM technology, they can unleash new solutions that meet the rapid pace of digital communication demands in the contemporary world (Buswig et al., 2011).

6.4. APPLICATIONS OF DIGITAL MODULATION

Digital modulation schemes have significantly changed the area of wireless communications and have made possible new applications across a broad range of areas. One such application is possible in high-speed data communication systems, where reliability and efficiency in digital modulation play major roles. Techniques such as Adaptive Orthogonal Frequency Division Multiplexing (OFDM) are now central technologies of the next generation wireless systems. Specifically, burst-by-burst adaptive HSDPA-type multicarrier systems utilize digital modulation to counter channel quality variations, achieving greater throughput than fixed-mode systems. The addition of diversity schemes like Rake receivers also enhances these systems by mitigating the adverse effects of fading, thus ensuring data integrity even in challenging environments. The importance of these advances depicts the significance of digital modulation in enabling efficient and reliable communications in more complex scenarios (Choi et al., 2007). Digital modulation is also used widely in multimedia broadcasting, which has experienced tremendous growth as a consequence of advances in technology. Quadrature Amplitude Modulation (QAM) is one of the methods that allow good quality video and audio to be sent through bandwidth-limited channels. As content consumption trends shift towards streaming platforms, the need for robust modulation methods that can handle varying channel conditions has become the order of the day. Digital modulation methods based on multiuser MIMO systems can enhance the user experience through the provision of high-definition content and low latency. Using adaptive modulation techniques, these systems will adaptively change modulation parameters based on real-time channel conditions of the communication channel, thereby optimizing quality and efficiency of multimedia transmission and adapting to user demand and environmental fluctuation. Such adaptability represents the quintessential significance digital modulation plays in maintaining high performance for broadcast applications (Hanzo et al., 2007). Furthermore, digital modulation technologies move into the Internet of Things (IoT) space where connectivity and efficiency are interlinked. IoT networks consist of devices spread across varied environments which must talk to each other, typically in regards to limited power consumption and bandwidth consumption. Techniques such as Frequency Shift Keying (FSK) and Phase Shift Keying (PSK) are being used within IoT technologies for secure data transmission with low energy expenditure. Furthermore, the use of genetic algorithms for multiuser detection maximization in MIMO-OFDM systems shows innovative synchronizing among modulation techniques and advanced computational methodologies to promote performance under congested network conditions. With the expansion of IoT, the need for successful digital modulation will also be greater, proving its central role in enabling an interconnected world (Choi et al., 2007; Hanzo et al., 2007).

6.4.1. Telecommunications

The development of telecommunications has played a critical role in the development and utilization of digital modulation techniques, which are significant for efficient data communication. Digital modulation forms a central pillar in information communication over various media, reacting to the diverse requirements of modern communication systems. For instance, digital modulation techniques such as Frequency Shift Keying (FSK), Binary Phase Shift Keying (BPSK), and Quadrature Phase Shift Keying (QPSK) are now essential solutions for telecommunications. Each of these approaches yields various strengths in terms of bandwidth efficiency, power consumption, and immunity to noise so that engineers can select the best one based on the specific application in question. The application of these techniques demonstrates the intricate interplay between theoretical concepts and practical means that characterizes the field of telecommunications, demonstrating the operation of their facilitating high-quality and reliable communication. As telecommunications technology develops, the necessity for robust and agile modulation techniques becomes more pronounced. New technologies, particularly in packet transmission at high speeds, have spurred the development of more advanced systems that are capable of efficiently handling dynamic channel variations. BbB-adaptive multicarrier systems reflect this trend as a demonstration of how adaptive modulation techniques can improve performance by adapting to real-time changes in signal quality. These systems utilize the concepts of orthogonal frequency division multiplexing (OFDM), with the ability to modulate symbols adaptively to provide higher throughput and lower bit error rates (BER) relative to fixed modulation systems. The application of such adaptive methods is crucial in addressing the challenges brought about by fading and interference, which can drastically distort the integrity of telecommunications networks. By employing techniques that adapt to the dynamic condition, the telecommunication engineers are able to establish a stronger and more effective communication network. Telecommunication and digital modulation techniques are further complicated by the inclusion of diversity techniques, which are employed to fight a range of channel impacts. Space-time coding and Rake receivers are a few examples that yield stronger signals by traveling through multiple paths over one channel. Such approaches are capable of reducing errors and improving the overall system performance efficiently, especially when affected by correlated fading or multiuser interference. However, the effectiveness of diversity techniques can be undermined under suboptimal conditions, so practitioners should have an idea about the dynamics of the communication environment. As telecommunication increasingly involves more advanced modulation and diversity methods, the challenge is to optimize these systems for different uses so they can serve the increasing demands for speed, reliability, and resilience in current communication. Research on these technologies is essential in order to continue the development of cutting-edge telecommunication technology (Figure 6.10).

Figure 6.10. *Telecommunications of telecommunications.*

Source: https://www.frost.com/wp-content/uploads/2019/08/Brazil-Telecommunications.jpg.

6.4.2. Broadcasting

Broadcasting has also improved dramatically with the advent of digital modulation techniques to significantly enhance the efficiency and quality of broadcast signals. Digital broadcasting is the transmission of video and audio content using digital signals rather than analog signals, allowing more data to be compressed and ensuring greater signal purity. By employing various modulation techniques, like Quadrature Amplitude Modulation (QAM), several channels can be broadcast combined on one frequency channel by broadcasters. This multichannel capability is important in achieving the fullest available bandwidth, hence offering the audience a variety of programming options. Besides, digital broadcasting can counteract signal attenuation with distance and interference, which are fundamental constraints on traditional analog systems. The nature of real-time digital signal processing is a fundamental focus in today's research and technology development of modern broadcasting systems. Diversity techniques need to be adopted in enhancing the reliability of digital broadcasting systems.

Figure 6.11. *Broadcasting the advancement of digital.*

Source: https://i2.wp.com/cerberussystems.co.uk/wp-content/uploads/2017/06/Global-Capital-Studio-3A.jpg?fit=2050%2C1536.

By employing transmitter and receiver diversity techniques, the impact of multipath fading can be minimized, which is likely to distort signal clarity in mobile reception. These techniques involve transmission of several signals simultaneously through different paths in a manner that at least some of the signals reach the receiver without significant loss, thereby improving the overall signal-to-noise ratio (SNR). Such optimizations are most important in systems that implement DVB standards, like DVB-H, where reception in high mobility is a prime requirement. The application of these diversity techniques reduces the infrastructure cost through diminishing the number of transmitter sites required while at the same time addressing environmental issues. Broadcasters are thus in a position to provide quality services that are appropriate for modern consumers (Bard et al., 2006). For digital modulation, the problem of system performance under varying conditions becomes pivotal to the role of robust channel modeling. Accurate models are needed in simulating receiver-dominated transmission environments with variables such as receiver speed and Doppler shifts. From literature, fading channels can be classified into mobile, portable, and fixed environments, each of which presents unique issues for signal integrity and reception quality. This nuance implies the need for adaptive modulation schemes that would be able to adapt to the dynamic nature of real broadcasting conditions. To facilitate these efforts, software written in tools such as MATLAB enables engineers to model and examine the influence of many parameters on signal distortion in particular, focusing on Bit Error Rate (BER) and Modulation Error Ratio (MER). These studies are basic to the enhancement of the performance of systems under DVB-H/SH standards in facilitating better quality mobile content delivery (Polák et al., 2013) (Figure 6.11).

6.4.3. Data Transmission

Data transmission is the fundamental component of modern digital communications systems, and it encompasses a number of techniques designed to efficiently convert and transmit information through a variety of different channels. One crucial aspect of data transmission relies on modulation methods, which determine how the information is coded for transmission. Digital modulation techniques play a critical role in guaranteeing the data is received and transmitted correctly and with efficiency, with consequences for signal quality, bandwidth usage, and immunity to interference and noise. The quality of these techniques dictates how well digital communication performs overall, particularly in bandwidth-limited areas where data rates need to be optimized. One noteworthy advancement in digital modulation techniques is the application of differential pulse code modulation (DPCM), where noteworthy potential for video data transmission has been exhibited. DPCM allows efficient compression of video data and hence enables transmission through bandwidth-limited channels. Research indicates that a DPCM-based algorithm can achieve broadcast-quality video transmission at remarkably low rates of 1.8 bits per pixel. This success underscores the value of effective encoding techniques in data transmission since they can provide high-quality content without having to call upon enormous bandwidth resources. These developments not only improve the user experience but also have a positive effect on the economic feasibility of digital broadcasting systems (Shalkhauser et al., 2025). The data transmission challenges come primarily from balancing quality, speed and complexity. For instance, while bandwidth-efficient digital video codecs offer tremendous gains in signal quality and data-rate reduction, these come with complicated hardware implementations. The complication poses challenges of wide adoption, particularly in analog-dominant environments. Nonetheless, with more advanced digital video compression algorithms demonstrating their superiority over traditional ones, the transition to digital formats has become greatly desirable. Both ground and satellite video service operators now turn to these integrated options for enhancing channel capacity. Copying the efforts of NASAs to deploy digital compression to transmit high-definition video from space also demonstrated this new direction, emphasizing how cutting-edge modulation methods are relevant in the emerging data transmission scenario (Bizon et al., 2025) (Figure 6.12).

Figure 6.12. *Data transmission is digital communication systems.*

Source: https://thumbs.dreamstime.com/b/data-transmission-channel-motion-digital-flow-transferring-big-181311171.jpg.

6.5. ADVANTAGES OF DIGITAL MODULATION

Digital modulation techniques possess various important advantages compared to their analog counterparts, particularly when it comes to signal integrity and noise tolerance. The first very important advantage is the higher ability to recover signals under noisy conditions. Digital signals can be processed using a variety of methods, such as error detection and correction protocols, which will identify and correct any errors introduced during transmission without compromising the whole signal. This is more critical when communication systems require fidelity and clarity to be preserved at all costs, like in broadcasting and mobile communications. Moreover, digital modulation supports power and bandwidth efficient utilization, as well as the capacity to accommodate multiple channels without interference. This robustness renders data transmission more secure and hence digital techniques preferred in most contemporary communication applications. Another major advantage of digital modulation is its capability to support high data rates. As described in the present evolution of communication technologies, e.g., the requirement of 5G services, the ability of carrying large amounts of data in an efficient manner is crucial. Digital modulation schemes can make use of complex schemes, such as Quadrature Amplitude Modulation (QAM), which increases data throughput significantly. These technologies allow for the provision of multimedia services with increased bandwidth and data capacities yet remain totally compatible with existing digital infrastructure. This advance of optical wireless communication-based technologies that capitalize on the vast unregulated band further enhances this advantage with the prospect of several times faster communications than available today as indicated by the growing volume of research in this area (Chowdhury et al., 2018). By enabling higher data rates, digital modulation is a bedrock of modern telecommunication. Further, digital modulation techniques are inherently capable of supporting advanced signal processing capability, which aids in further enhancing the quality of a communication system. Conversion to digital form allows for the application of sophisticated algorithms that enhance signal quality and reduce distortion. Techniques such as filtering and pulse shaping can do much to alleviate issues like intersymbol interference that weigh so much on analog transmission. As an example, using wind generator systems, where the control over output signals has to be highly accurate, digital modulation provides the platform on which effective control structures can be constructed through algorithms and simulations (Buswig et al., 2011). Also, digital modulation provides a platform for future innovation in wireless communication, opening the door to new services and applications that can be developed in the future, thereby making its importance more cemented within the purview of developing communication technologies.

6.5.1. Improved Noise Immunity

The digital modulation techniques applied in contemporary communication systems have significantly enhanced the noise immunity. The improved noise immunity is of very critical importance as it directly affects the reliability and clarity of transmitted signals, especially in environments with various types of interference. Techniques like Phase Shift Keying (PSK) and Frequency Shift

Keying (FSK) have turned out to be robust solutions in maintaining signal integrity amidst noise. For instance, coherent optical binary polarization shift keying (2PolSK) has also recorded spectacular improvements in bit error rate (BER) performance compared to standard schemes like binary amplitude shift keying (2ASK). The findings indicate that 2PolSK exhibits enhanced signal-to-noise ratio (SNR), meaning immunity to disruptions in the atmosphere, thus emphasizing the effectiveness of advanced modulation techniques in ensuring data integrity in challenging conditions (Lee et al., 2011). In addition to PSK and FSK, Spread Spectrum Modulation is also an important technique designed to combat noise and interference in digital communications. Spread Spectrum Modulation selects a signal closely approximating the noise-like pattern and operates using a much wider bandwidth than necessary. The built-in strengths of Spread Spectrum Modulation are its high resistance to radio-frequency interference (RFI), especially in locations with high levels of technological noise. By having its bandwidth increased, Spread Spectrum not only minimizes the effects of interference but is also highly sensitive in detection to natural noise sources. The utilization of pseudorandom sequences to create signals facilitates reduced ambiguity in signal detection, which further adds to enhanced overall robustness of communication (Messerschmitt et al., 2011). Spread Spectrum Modulation is therefore the ultimate example of the new methods embraced in digital communications to counteract issues of noise efficaciously. The significance of Improved Noise Immunity extends beyond signal clarity, as it is essential in many applications, including wireless communications, satellite systems, and underwater acoustic networks. The combination of sophisticated modulation techniques enables more stable connections to be made, making the transmission of data in a seamless manner even in adverse situations possible. The evolution of modulation techniques also continues to keep pace with technological advances and the increasing demand for increased data rates. Increased error correction capability with these modulation schemes provides additional insurance against noise, and efficient communication in more crowded spectral environments. Finally, the commitment to higher noise immunity using digital modulation techniques is essential to advancing the efficiency and effectiveness of modern communication systems, rendering them a mainstay of our information-based civilization.

6.5.2. Efficient Bandwidth Utilization

Optimized usage of bandwidth is a critical component of contemporary digital communication networks, particularly in the face of persistent high-speed data transmission demands. Digital modulation techniques are central to the efficient utilization of available bandwidth through the simultaneous sharing of the same limited spectral space by multiple signals. Of these methods, orthogonal frequency-division multiplexing (OFDM) is special in that it can divide a broad channel into numerous thin subcarriers of bandwidth. Not only does this make the spectrum more efficient but it also performs better against inter-symbol and inter-carrier interference. Such adaptability makes OFDM particularly well suited to fulfill demands placed by emerging communications technologies, especially with loads piling up onto networks.

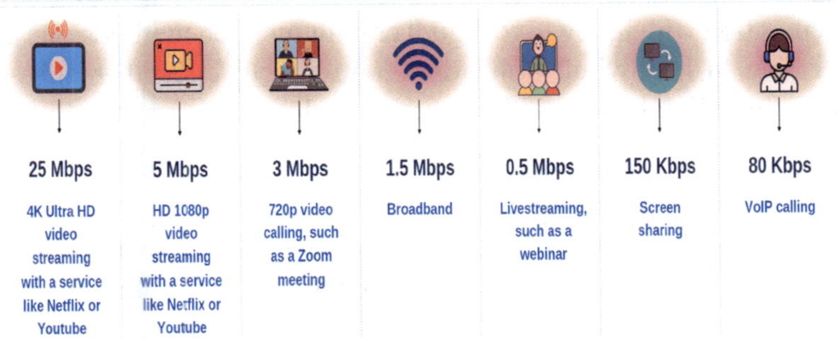

Figure 6.13. *Efficient bandwidth utilization is digital communication systems.*

Source: https://external-content.duckduckgo.com/iu/?u=https%3A%2F%2Fcdn.broadbandsearch.net%2F2022%2F11%2F1668139511489-2._Recommended_Bandwidth_for_Online_Activities_at_Home_and_in_Business.png&f=1&nofb=1&ipt=0dbd77d5e97fa321dc377df1334a49065b02c06a97166f5798f4e848cf9e169b.

OFDM implementation facilitates an improved bandwidth allocation method that is compatible with different applications but accommodates high data rates. Besides, the popularity of orthogonal frequency-division multiplexing in contemporary networks testifies to its importance within efficient bandwidth use in the field. As presented in studies on OFDM technologies, this method allows flexible allocation of spectral resources, accommodating multiband channel conditions and data rates with superior efficiency. Such flexibility enables the use of elastic optical networks, whose functionality can be dynamically modified according to the demands of the digital era. With high capacity to support a wide range of services and comparatively high-growing Internet traffic, OFDM facilitates effective bandwidth usage. The continued advancement of OFDM technology also holds potential for increased robustness and scalability of communication networks, thus making them capable of transferring higher data volume in the highly interconnected world. In renewable energy, optimum use of bandwidth exceeds the traditional in telecommunications. Application of digital modulation techniques in wind generator systems, as an example, illustrates the variety of such techniques in signal transmission optimization in alternative energy applications. Application of controllers, including those derived from mathematical models and artificial neural networks, maximizes the performance of DC-AC converters to stable output voltage and effective distribution of energy. This technology development not only maximizes the efficiency of power systems but also addresses the importance of modulation techniques to efficient control of frequency resources. By merging renewable energy generation with innovative digital communication methods, the potential for innovation in bandwidth utilization continues to expand, reaffirming the significance of optimal bandwidth allocation in different domains (Figure 6.13).

6.5.3. Enhanced Security Features

In today's world of digital communication, advanced security features are vital in safeguarding transmitted information as the threat landscape keeps evolving. With advancing digital modulation technology and hence developments, so also should the security environments preserving these communications develop and improve. It is important to install robust encryption methods; they ensure that even if a signal is intercepted, the data will be out of reach to inappropriately targeted entities. For example, chaotic masking modulation, which is constructed upon the foundation of chaotic systems, shows the possibility of integrating strong adaptive observer-based synchronization for strengthening security in real-time systems. This process not only maintains the security of the data but also provides secure transmission process integrity against any uncertainty, as discussed in Raoufi et al. (2006). Such methodologies bring to the fore the importance of advanced modulation techniques in achieving high levels of security in all kinds of communication channels. Moreover, the integration of machine learning (ML) algorithms with digital modulation techniques is a significant leap towards enhancing security factors. ML can deal with large sets of data to determine potential vulnerabilities and respond dynamically against attack vectors through modulating parameters or encryption keys in real-time. In a global scenario where networks must support an array of devices and services, this dynamic ability provides increased immunity to harmful endeavors. Hence, deployment in heterogeneous networks, cognitive radios, and Internet of Things (IoT) applications strengthens the security ecosystem, leading to a more trusted communication system. The variations in the contributions made by various kinds of ML algorithms, from supervised to reinforcement learning point towards the multidimensional perspective towards security in wireless networks, as per Chen et al. (2019). This synergy between latest technology and security solutions indicates the importance of innovation in adaptive modulation techniques. For the purpose of better comprehending the role of advanced security capabilities in digital modulation techniques, examining their significance in critical applications such as military and civilian use is a good example of their potential to transform. Maintaining confidential communication channels is paramount in military operations, where the need for encrypted communication has direct relevance on mission success. In the same way, in non-military use, where personal data about individuals is being communicated, robust security features must be employed in order to maintain user trust and regulatory compliance. Design and implementation of sophisticated modulation schemes, with adaptation of adaptive learning as well, represent a required foundation for obtaining high-level security, low latency, and trustworthy information services. With greater security ensured by these emerging methodologies, the future wireless networks will be capable of fortifying their defense against more sophisticated cyber-attacks, thereby making possible consistent and dependable communication services for all the applications (Figure 6.14).

Figure 6.14. *Enhanced security features of digital communication.*

Source: https://www.nist.gov/sites/default/files/images/2020/11/27/Andrey%20Suslov_priacygraphic_SS_640563346_960x600.png.

6.6. CHALLENGES IN DIGITAL MODULATION

The increased application of digital communication systems has brought with it a number of problems associated with the employment of digital modulation techniques. Among the main challenges is electromagnetic compatibility

(EMC). As wireless systems increase and more devices become interconnected, there is a need to make sure that such systems are working in harmony without generating interference. Current EMC standards must be re-visited and revised to accommodate the sophistication of such development. Specifically, traditional compliance methods based on quasi-peak detector levels may not be able to capture the degradation caused by electromagnetic interference satisfactorily. Therefore, the development of new statistical approaches, such as amplitude probability distribution detectors, has become an important objective to make digital communication effective. This evolving environment requires anticipatory action from world standardization bodies to go back on compliance specifications and test procedures aimed at resolving these EMC problems (Azpúrua et al., 2019). The control of optical fiber impairments is another important problem with digital modulation, particularly as the communication networks evolve to more advanced capacity systems. Current optical communication technology is faced with effects from both linear and nonlinear ones, particularly large at high-speed transmission. The introduction of coherent detection systems has provided a means to compensate for some of them through advanced digital signal processing (DSP) algorithms. With networks aiming to move towards 400 Gbps and beyond, effects of fiber nonlinearity such as those caused by the Kerr effect become more pronounced and need to be compensated in order to enhance transmission performance. This growing complexity is inextricably linked to system design; hence, practitioners need to critically examine a variety of fiber nonlinearity compensation (NLC) techniques. The recognition and adoption of appropriate NLC solutions are critical for the dependability and efficiency of modern digital communication systems (Amari et al., 2017). Lastly, the quest for higher spectral efficiency in digital modulation brings with it certain challenges in terms of modulation scheme choice and implementation. As data rates continue to grow in use, using multi-level modulation formats can sometimes be more efficient in making use of available bandwidth; however, advanced technologies are typically required to manage the intrinsic complexity of these systems. The compromise between increasing data throughput while maintaining the integrity of the signal is an onerous task. As modulation methods improve, it becomes essential to make compromises in terms of performance, complexity, and processing power. Developing within the context of coherence and DSP approaches simply increases the intricacy of this dynamic. Therefore, continuous research and development are called for in addressing the sophisticated problems presented by evolving digital modulation techniques so they may effectively cater to the demands of modern communication infrastructure (Azpúrua et al., 2019; Amari et al., 2017) (Figure 6.15).

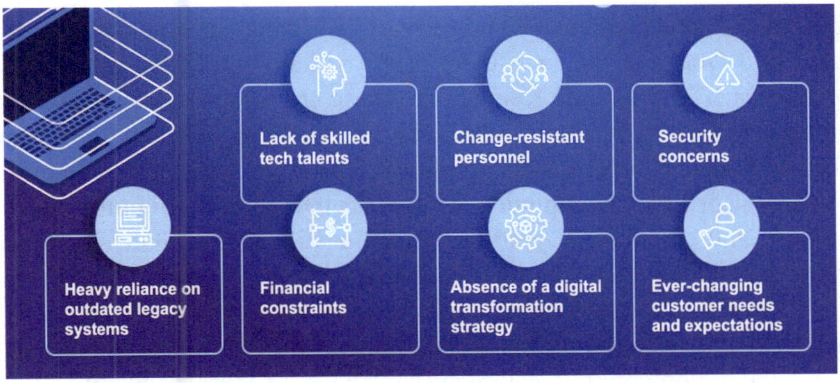

Figure 6.15. *Challenges in digital modulation.*

Source: https://pixelplex.io/wp-content/uploads/2023/04/7-biggest-digital-transformation-challenges-in-2023.jpg.

6.6.1. Interference and Signal Degradation

In reality, growing complexity within digital communication systems has pinpointed major problems of interference and signal degradation. Results from advancements in digital modulation methods have greatly resulted in efficiency in transmission, which is usually interfered with by electromagnetic interference (EMI). In this case, interference can rise from sources both internal and external to the system, which consequently brings complexity in transference reliability of signals. There is a serious focus on why conventional methods of signal integrity analysis, such as those using quasi-peak detector levels, may not necessarily be sufficient anymore in the current situations. Research suggests that adopting new statistical methods using amplitude probability distribution can result in accurate prediction of EMI-induced degradation in digital systems. This emphasis on improved methods points to the contribution of international standardization organizations in placing a high priority for the full-scale renewal of compliance rules with the capability of coping with the dynamic nature of wireless communication and challenges thereof (Azpúrua et al., 2019). In optical communication systems, interference impacts signal integrity immensely, given the high rates of data and complex signal structures involved. As those technologies, for instance, coherent detection become commonplace, impairment compensation like fiber nonlinearity and dispersion emerges to the forefront. Coherent systems, which in current 100 Gbps wavelength-division multiplexing (WDM) technologies are in use, become dependent upon DSP algorithms in order to compensate linear fiber distortions to considerable benefit. In the development of high-capacity systems, such as those working towards 400 Gbps and beyond, there is the added complexity, primarily due to non-linear effects like the Kerr effect. These constitute an essential demand for innovative compensation techniques operating to enhance transmission performance under the threat of non-linear interference generated under high levels of input power, thus depicting the intricate interaction of digital modulation techniques and signal deterioration (Amari et al., 2017). Understanding the dynamic relationship between interference and signal deterioration is essential to advancing digital modulation techniques. By incorporating technologies within DSP and overcoming the shortcomings of current test methodologies, the next-generation communication systems have a superior capability to deal with the exigencies of linear and non-linear impairments. The requirement for an adaptive solution is also enhanced by on-going progress in modulation formats where one needs higher and higher spectral efficiency and interference immunity. This convergence of new methodologies emphasizes an overall cry among researchers and practitioners for some reconsideration of existing standards, and that future trend towards flexibility and precision in compliance testing will be crucial. Ultimately, the complexity of interference will be solved not only to enhance performance but also to lay the groundwork for robust digital communication networks that are equipped to confront the demands of an increasingly interconnected world (Azpúrua et al., 2019; Amari et al., 2017).

6.6.2. Complexity of Implementation

The implementational complexity of sophisticated modulation techniques in digital communication comes as a primary concern when evaluating the efficiency of state-of-the-art communication systems. As technology pushes data rates and spectral efficiencies towards greater limits, the complexity involved with the employment of such modulation techniques increases. For instance, in coherent optical communications, integration of DSP is essential to minimize both linear and nonlinear impairments that predominate over high-speed links. Yet such implementation of DSP technologies has a paradoxical effect where performance enhancement may be overshadowed by increased cost and energy consumption that makes it unavoidably balanced against technological advancement versus economic viability. Therefore, understanding and management of complexity in implementation are critical to ensure the full use of digital modulation techniques with no compromise on system performance and integrity. Computational complexity trade-offs are also illustrated by contrast among different signal processing techniques. An important difference between conventional Orthogonal Frequency Division Multiplexing (OFDM) coherent receivers and filter-bank receivers is highlighted by research.

The latter demonstrates considerable complexity reductions for enabling a more efficient processing architecture in novel communication systems. Such comparative analyses, demarcating performance improvement and computational complexity, are vital to inform the design of future modulation formats. As technology evolves toward ultra-high-speed communications—like 400 Gb/s and beyond—understanding these complexities allows engineers to craft solutions that preserve performance efficiency while minimizing resource commitments, like in research examining DSP in optical transceivers comparing various equalization techniques based on their computational complexity (Bouziane et al., 2016). Furthermore, the consequences of fiber nonlinearity add further complexities in the utilization of digital modulation techniques, particularly in the situation of high-capacity WDM systems. The role of nonlinear effects, such as those resulting from the Kerr effect, assumes greater importance as transmission speeds increase. Compensation schemes for such nonlinear effects require sophisticated DSP solutions not only to reduce the distortion but also to address the attendant implementation complexity. An overall overview of fiber nonlinear compensation methods underlines the imperative to ensure an optimization of performance and processing requirements as well as the incorporation of novel advances to optimize Nyquist and super-Nyquist superchannels (Amari et al., 2017). Ultimately, the solution of such multi-faceted problems represents a compulsory stage in furthering digital modulation methods within high-speed optical systems.

6.6.3. Limitations in Bandwidth

Bandwidth constraints in the area of digital modulation techniques are the determining elements for data transmission effectiveness and efficiency. Bandwidth constraints emerge from regulatory issues, physical channel characteristics, and modulation scheme intrinsic characteristics. In an era where data rates are experiencing further increases with the requirements of quick Internet and streaming, such constraints become increasingly important. More specifically, the frequency band that is available is likely to be saturated, leading to interference as well as reduced quality of service. This necessitates a study of new digital modulation techniques that can make better use of the bandwidth offered, thus minimizing such restrictions and meeting the users' needs in an environment of dynamic communication. One of the methods by which bandwidth restraint might be alleviated can be deduced via the creation of Galois-Division Multiple Access (GDMA). This approach employs finite field transformations to bring bandwidth usage to efficient levels. Employing a binary-to-p-ary mapping approach, the GDMA architecture leverages the opportunistic secondary channel in allocating resources effectively in dense network environments. Performance evaluation of GDMA reveals its potential to compress bandwidth requirements, thereby enabling the handling of more concurrent users without sacrificing transmission quality. With increasing developments in digital systems, new bandwidth-efficient techniques such as these become essential endeavors towards meeting ever-growing data needs while fighting the ailments of limited spectral availability (de Oliveira et al., 2015). Apart from that, developments in transmitter design are also overcoming bandwidth limitations through methods of intricate processes such as balanced encoding and calibration tones. In this, a digital radio communication transmitter can construct a null space through a calibration tone that is synchronized in frequency and phase with the modulated radio frequency output. With such precise synchronization, coherent demodulation is possible at the receiver end and better multipath fading compensation is achieved, ensuring signal integrity under bandwidth restrictions. By exploiting such techniques, digital modulation systems are capable of maximizing the efficiency of transmission, countering the negative impacts of limited bandwidth and maintaining stable means of communication. Thus, the continuous advancement of modulation techniques has seen bandwidth limitations not just identified but also strategically managed in modern systems of communication (Davarian et al., 1987).

SUMMARY

In concluding the subject of digital modulation techniques, it ought to be noted that the methods form the foundation for the efficacy and reliability of modern systems of communication. Digital modulation not only makes the transmission of signals more efficient but also solves issues of noise and interference that are of utmost priority in both wireless and wired systems. Since technology is apt to evolve, these methods of modulation will keep changing, employing more sophisticated algorithms and techniques to further enhance performance. Artificial Neural Network-based control systems, as the one described in Buswig et al. (2011), show how modern methods of computation are being used to maximize signal integrity and provide stability to the output. This is the reason that researchers and engineers need to be mindful of the theoretical advances as well as practical implementations of digital modulation. Additionally, the continuous advancement of digital modulation techniques is seen through the multitude of applications and surroundings in which they are utilized. Under harsher environmental conditions, for example, the effectiveness of modulation techniques significantly varies, as a consequence of which it is essential to modify the transmission methodologies. The findings presented in Lee et al. (2011) support the comparative analysis of several modulation techniques in adverse environments, i.e., turbulence. This creates a deeper understanding of how specific techniques, e.g., polarization shift keying, will surpass traditional methods in maintaining signal integrity. With researchers increasingly in the direction of preventing signal degradation-related problems, ongoing research on these varied techniques remains instrumental in preserving pace in communication technology. Lastly, the future of digital modulation techniques largely rests on studies focused on propelling performance levels simultaneous with addressing the problems already outlined. The integration of new ideas, such as multiple-input and multiple-output (MIMO) systems, could synergize with existing modulation strategies to improve overall system robustness. As digital communication continues to proliferate in sectors ranging from telecommunications to military applications, the ability to adapt modulation techniques will be crucial. With the ever-increasing advancements proved in the form of theoretical modeling and experimental results, for instance, those experienced in Buswig et al. (2011) and Lee et al. (2011), the communication landscape awaits to face fundamental changes. Therefore, an aggressive response towards research and development of digital modulation is not only justifiable but also unavoidable to guarantee the future of effective communication.

The scope of digital modulation techniques is wide and rich, as it reflects the numerous applications to meet the increasing demands of communication systems. The key to this progress is the most popularly deployed Coherent Optical Orthogonal Frequency Division Multiplexing (CO-OFDM), which has drawn significant interest in optical communications through fibers. The architecture of CO-OFDM is emphasized with its simplified digital signal processing (DSP) modules with the high spectral efficiency and flexibility. But it is necessary to note the challenge posed by its high peak-to-average power ratio, in that it renders it prone to fiber-induced nonlinearities. Recent advances in DSP-based machine learning present a promising way forward to surpass the challenges without adding considerable computational complexity, illustrating the interplays dynamic between traditional means and cutting-edge innovation in the field (Aldaya et al., 2018). In addition to CO-OFDM, other architectures have been designed to enhance digital modulation efficiencies within mixed-signal and digital system-on-chip (SoC) environments. Central to the majority of these systems are Phase Locked Loops (PLLs), which are instrumental in clock synthesis and synchronization in multirate digital communications. Of all the various PLL designs, the Charge-Pump (CP) PLL is most prevalent due to its versatility and consistency in generating stable clock signals from lower frequency sources. The uses of CP-PLLs transcend functionality as they ensure the reliability of on-chip system elements such as analog-to-digital converters (ADCs) and digital-to-analog converters (DACs). The working of such elements is irretrievably tied with reliability and accuracy provided through PLLs, highlighting the necessity of comprehensive testing and validation at design and production phases (Burbidge et al., 2008). Thus, use of digital modulation techniques like CO-OFDM and matching PLL structures reflects the necessity for continued innovations in communication technologies. By using machine learning techniques to overcome limitations of the traditional approaches, researchers and engineers are able to enhance system performance at the cost of no efficiency loss. The combination of designing complicated modulation schemes and optimizing accompanying structures is a breathtaking research area in telecommunications. As digital modulation continues to advance, the experience derived from such advancements will be priceless to subsequent generations of communications infrastructure to allow them to meet the more sophisticated demands of international communications and data interchange. Furthermore, with increasing business expansion, the reliance upon deep digital modulation techniques and associated architectures will form the foundation for the ability to grow in the technology-driven economy.

REFERENCES

1. Abouei, J., Brown, J. D., Pasupathy, S., Plataniotis, K. N., et al. (2010). Adaptive demodulation in differentially coherent phase systems: Design and performance analysis. http://arxiv.org/abs/1006.0379 (accessed on 12th September 2025).
2. Abrate, S., Gaudino, R., Nespola, A., Offenbeck, I., et al. (2010). High-speed communications over polymer optical fibers for in-building cabling and home networking. https://core.ac.uk/download/pdf/11420470.pdf (accessed on 12th September 2025).
3. Aldaya, I., Barry, L., Giacoumidis, E., Tsokanos, A., et al. (2018). Harnessing machine learning for fiber-induced nonlinearity mitigation in long-haul coherent optical OFDM. https://core.ac.uk/download/287582088.pdf (accessed on 12th September 2025).
4. Amaral, L., Dris, S., Gill, K., Grabit, R., et al. (2010). Feasibility of using bandwidth efficient modulation to upgrade the CMS tracker readout optical links. http://arxiv.org/abs/1010.1691 (accessed on 12th September 2025).
5. Amari, A., Ciblat, P., Dobre, O. A., Jaouën, Y., et al. (2017). A survey on fiber nonlinearity compensation for 400 Gbps and beyond optical communication systems. http://arxiv.org/abs/1708.06313 (accessed on 12th September 2025).
6. Azpúrua Auyanet, M. A., Pous Solà, M., & Silva Martínez, F. (2019). Improved electromagnetic compatibility standards for the interconnected wireless world. https://core.ac.uk/download/288625020.pdf (accessed on 12th September 2025).
7. Bard, M., Cosmas, J., Song, Y. H., Zhang, Y., et al. (2006). Diversity gain for DVB-H by using transmitter/receiver cyclic delay diversity. https://core.ac.uk/download/334219.pdf (accessed on 12th September 2025).
8. Bertran Albertí, E., Montoro López, G., & Tarrés Ruiz, F. (1998). An experimental course on digital communications. https://core.ac.uk/download/81582988.pdf (accessed on 12th September 2025).
9. Bizon, T. P., Marcopoli, V. R., & Whyte, W. A. (2025). Real-time demonstration hardware for enhanced DPCM video compression algorithm. https://core.ac.uk/download/pdf/42813974.pdf (accessed on 12th September 2025).
10. Bouziane, R., Erkilinc, M. S., Killey, R., Nazarathy, M., et al. (2016). Comparison of linear and nonlinear equalization for ultra-high capacity spectral superchannels. https://core.ac.uk/download/79549921.pdf (accessed on 12th September 2025).

11. Budinger, J., Hoerig, C., Tague, J., Vanderaar, M., et al. (2025). Flexible digital modulation and coding synthesis for satellite communications. https://core.ac.uk/download/pdf/42814834.pdf (accessed on 12th September 2025).

12. Burbidge, M., & Richardson, A. (2008). Phase locked loop test methodology. https://core.ac.uk/download/1558600.pdf (accessed on 12th September 2025).

13. Buswig, Y. M. (2011). Development of a DC-AC power conditioner for wind generator by using neural network. https://core.ac.uk/download/12007862.pdf (accessed on 12th September 2025).

14. Carnegie, D. A., Dorrington, A. A., Drayton, B., Jongenelen, A., et al. (2011). Design and construction of a configurable full-field range imaging system for mobile robotic applications. https://core.ac.uk/download/29198923.pdf (accessed on 12th September 2025).

15. Chen, K. C., Hanzo, L., Jiang, C., Ren, A., et al. (2019). Thirty years of machine learning: The road to Pareto-optimal wireless networks. http://arxiv.org/abs/1902.01946 (accessed on 12th September 2025).

16. Chen, S., Hanzo, L., & Sugiura, S. (2012). A universal space-time architecture for multiple-antenna aided systems.

17. Choi, B. J., & Hanzo, L. (2007). Near-instantaneously adaptive HSDPA-style OFDM versus MC-CDMA transceivers for WiFi, WiMAX, and next-generation cellular systems.

18. Chowdhury, M. Z., Hossan, M. T., Islam, A., Jang, Y. M., et al. (2018). A comparative survey of optical wireless technologies: Architectures and applications. http://arxiv.org/abs/1810.02594 (accessed on 12th September 2025).

19. Davarian, F. (1987). Antimultipath communication by injecting tone into null in signal spectrum. https://core.ac.uk/download/pdf/10396169.pdf (accessed on 12th September 2025).

20. De Leenheer, M., Morea, A., Mukherjee, B., Zhang, Y., et al. (2013). A survey on OFDM-based elastic core optical networking. https://core.ac.uk/download/55762862.pdf (accessed on 12th September 2025).

21. de Oliveira, H. M., & Miranda, J. P. C. L. (2015). On Galois-Division multiple access systems: Figures of merit and performance evaluation. http://arxiv.org/abs/1502.03698 (accessed on 12th September 2025).

22. Escribano, F. J., Evangelista, J. V. C., Kaddoum, G., Wagemakers, A., et al. (2019). Design and performance analysis of an index time frequency modulation scheme for optical communications. http://arxiv.org/

abs/1905.11461 (accessed on 12th September 2025).

23. Fathallah, H., Guesmi, L., & Menif, M. (2018). Modulation format recognition using artificial neural networks for the next generation optical networks. https://core.ac.uk/download/322432956.pdf (accessed on 12th September 2025).
24. Hanzo, L., & Jiang, M. (2007). Multiuser MIMO-OFDM for next-generation wireless systems.
25. Hanzo, L., Liang, D., & Ng, S. (2011). Near-capacity turbo coded soft-decision aided DAPSK/Star-QAM.
26. Karimian, N., & Yazdani, J. (2010). Design and analysis of OFDM system for powerline-based communication. https://core.ac.uk/download/16414585.pdf (accessed on 12th September 2025).
27. Kumbhar, A. (2016). Overview of ISM bands and software-defined radio experimentation. http://arxiv.org/abs/1605.08309 (accessed on 12th September 2025).
28. Lee, C. G., Calvani, S., Rajbhandari, S., Popoola, W. O., Tang, X., & Ghassemlooy, Z. (2011). Coherent optical binary polarization shifts keying heterodyne system in the free-space optical turbulence channel. https://core.ac.uk/download/4149475.pdf (accessed on 12th September 2025).
29. Messerschmitt, D. G. (2011). Interstellar communication: The case for spread spectrum. http://arxiv.org/abs/1111.0547 (accessed on 12th September 2025).
30. Polák, L. (2013). Analysis and simulation of the signal's transmission in the DVB-H/SH standards. https://core.ac.uk/download/30311053.pdf (accessed on 12th September 2025).
31. Pratiarso, A., & Sa'iyanti, N. P. (2025). Pembuatan modul praktikum teknik modulasi digital FSK, BPSK dan QPSK dengan menggunakan software. https://core.ac.uk/download/12342776.pdf (accessed on 12th September 2025).
32. Raoufi, R., & Zinober, A. S. I. (2006). Adaptive sliding mode observers in uncertain chaotic cryptosystems with a relaxed matching condition. https://core.ac.uk/download/49826.pdf (accessed on 12th September 2025).
33. Rummel, K. (2019). The fundamentals of radar with applications to autonomous vehicles. https://core.ac.uk/download/213463684.pdf (accessed on 12th September 2025).
34. Serdean, C. V. (2002). Spread spectrum-based video watermarking algorithms for copyright protection. https://core.ac.uk/download/29816683.pdf (accessed on 12th September 2025).

CHAPTER 7

INFORMATION THEORY AND CODING

Contents

INTRODUCTION

In the arena of digital communication, Information Theory is the basic basis of understanding the process of data transmission and encoding. It revolves around quantifying information first, which is the basis of successful coding and a well-functioning communication system. The entropy and redundancy theory came into existence with the path-breaking work of Claude Shannon, and this illuminated the world regarding quantifying information and achieving optimization in various applications. With the systems it makes possible to effectively encode and compress data so as to minimize the risks of errors when transmitting, Information Theory forms an essential engineering discipline tool. Furthermore, embracing Shannon's concepts can get different fields, from telecommunications to data science, better, paving the way for innovations that are founded on the mathematical foundation of information processing and coding schemes. Moving further into the uses of Information Theory is its combination with engineering principles, primarily in the creation of error-correcting codes. Such codes are absolutely necessary for ensuring the reliability of data, particularly in environments susceptible to noise and interference. Techniques like Hamming codes and Reed-Solomon codes demonstrate the way Information Theory's principles have led to fault-tolerant solutions that restore data loss.

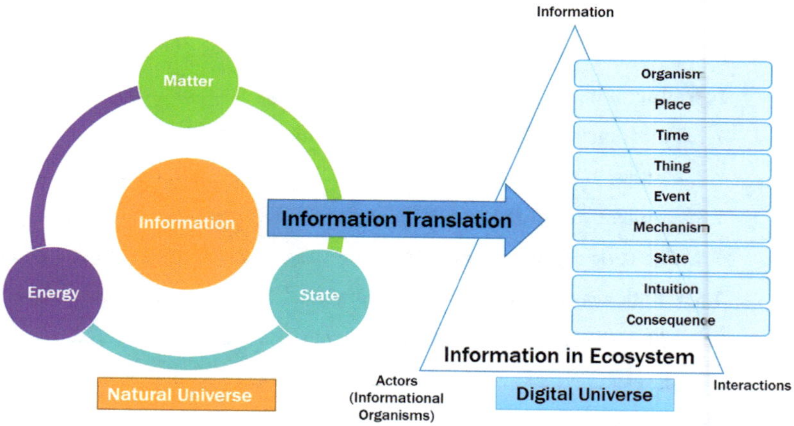

Figure 7.1. *Information theory.*

Source: https://1.bp.blogspot.com/-h-0KDvj--xQ/YOEKmxChZYI/AAAAA AAAAxQ/ aXA4jGJW0oEMQnAMUDD392Gs_avakJEcwCLcBGAsYHQ/s155/ Theory+ of+Information+Trilogy+-+Information+Ecosystem.jpg.

The mathematical theories that provide the foundations for such codes allow reliable communication over noisy channels, and are thus at the heart of digital communications today. This coupling between theoretical abstraction and pragmatic imperatives indicates the importance of Information Theory in assisting to lead on technologies based on optimal data encoding and transmission, that ultimately underpin the paradigm of modern engineering practice. Moreover, the on-going evolution of Information Theory is reflected in its synthesis with emerging and evolving domains like quantum information science. The results which have been established in classical Information Theory are being re-purposed to address the new issues presented by quantum mechanics, as shown in increasing interest in quantum error correction and quantum communication protocols. The marriage of Information Theory and quantum engineering is a step towards a more profound understanding of information in classical and quantum frameworks. As we continue with these developments, it is important that we acknowledge the initial literacy provided by traditional Information Theory that still remains part of forming research and applied development in quantum technologies. Consequently, exchanges of classical information objects and their quantum equivalents are par excellence embodiments of dynamic nature of Information Theory, which supports its significance in researching and applying engineering disciplines (Levi, 2023; Zhu et al., 2017) (Figure 7.1).

Definition and Historical Background

Information theory as a field of study emerged during the mid-20th century and was largely based on the early work of Claude Shannons during 1948. His seminal paper, A Mathematical Theory of Communication, set up a formal theory of communication regarding the transmission and processing of information. Shannon's work formed the basis for the measurement of information, embodied in the concept of entropy, which serves as a uniting supporting pillar in communication systems and data encoding schemes. This theoretical foundation allowed for researchers and engineers to approach the data in a way that was efficient, making it feasible to devise systems that preserved as much information as possible while minimizing redundancy. Consequently, information theory development generated a paradigm shift in the way data was perceived and handled in engineering disciplines that had a far-reaching influence on applications as diverse as telecommunications to data compression. A key aspect of information theory is its emphasis on coding as a procedure for information transmission optimization. Coding theory, one of the branches of information theory, emerged as a solution to sending information efficiently over noisy channels without distorting it. Scientists studied various data encoding methods, leading to the development of error-correcting and error-detecting codes, which are the key to effective communication. The codes enhance the reliability of data transmission for numerous applications, such as computer networks, digital broadcasting, and storage devices. From Hamming to Turbo and LDPC codes, the development of coding methods over the years exemplifies how the field has continued to evolve as the technology grew. The interplay of information theory and coding has given their vital contributions toward technological advancements and developments of communication systems through this perspective. Furthermore, the applications of information theory extended beyond the boundaries of classic engineering applications to interdisciplinary fields. The information processing and transmission principles have been immensely helpful in computer science, biology, and social science, illustrating its broader applicability. Computer science, for instance, uses data compression algorithms and cryptographic mechanisms based on the fundamentals of information theory. The progress toward these applications dates back to Shannons initial work, which effectively bridged the gaps between physics, mathematics, and engineering. Furthermore, contemporary research endeavors, such as the quantification of information in biological systems or the analysis of social network dynamics, also point to how information theory's variety of diversity presents a shared framework for understanding multifaceted phenomena. In positioning information theory at the intersections of several disciplines, its historical and current significance to engineering and elsewhere would be impossible to overstate.

Key Contributors and Their Contributions

The development of information theory and coding has been substantially impacted by several key figures whose work provided the fundamental insights and breakthroughs. Claude Shannon, often referred to as the father of information theory, opened the door with his famous 1948 paper, which presented the concepts of entropy and redundancy. Shannon's theoretical model set the groundwork for effective encoding of data and thorough analysis of communication systems, allowing multiple coding methods to be developed. His work has continued to have an impact, steering later researchers through examining theoretical limits and real-world uses. Shannon's achievements need to be recognized because not only did they assist telecommunications but have also given rise to contemporary uses in data compression and error correction in a range of domains. Another notable contributor is Robert Gallager, whose coding theory further built Shannon's principles into more applied areas. Gallager's invention of low-density parity-check (LDPC) codes and his insight into the error correction processes have revolutionized digital communication, particularly wireless systems. His work illustrates how encoding schemes affect the reliability and efficiency of data transmission, particularly under circumstances of low bandwidth and high error rates. Furthermore, Gallager's work encompasses the contemporary iterative decoding theory, which offers greater flexibility and performance in usage ranging from satellite communications to cellular networks. More and more demand for data transmission reliability has caused Gallager's work to remain at the core of engineering advancement in information systems. Interdisciplinary research has made information theory contributions as well, most notably in computer science and electrical engineering. Recent efforts, primarily with the finite-blocklength information theory problems as the focus, have brought the paramount importance of balancing mathematical rigor with requirements of practicality into the limelight, as witnessed by the highlight on ultra-reliable low-latency communication. These inter-disciplinary efforts established the ineffectiveness of traditional asymptotic methods widely employed in information theory and further indicated the need for methods that have to consider hard realities such as short-packet communication and energy efficiency. By answering these central open questions, researchers today attempt to improve the resilience of next-generation communication networks, enabling innovation as well as future technological trends, thereby creating a sustained trajectory of innovation in information theory and coding engineering.

Importance in Modern Engineering

Information theory's place in modern engineering cannot be overestimated, particularly insofar as it serves as the basis for innovations across various technological areas. As engineering challenges grow in complexity, the methods that find their roots in information theory have also become central to optimizing communication systems and optimizing data transmission efficiency. To illustrate, engineers use coding techniques to minimize errors in data transmission, ensuring the integrity and reliability of the communications. The history of genetic engineering itself continues to show this connection as the focus is on shifting away from traditional sequential gene manipulation to deciphering intricate genomic interactions. This can be observed in recent studies examining how viruses and RNA can edit genetic material with such precision, thus necessitating a reconsideration of the fundamentals in genetics and further solidifying the application of information theory in biological engineering (Witzany et al., 2017). These technologies fully exemplify the contribution of the theory towards the creation of innovative solutions in modern engineering practices. Another convincing aspect of modern engineering is its interface with rapidly evolving digital technologies, driven to a significant degree by the exponential explosion of data in industries such as healthcare. The transition from traditional paper-based systems to digital platforms has generated huge amounts of data that must be processed and analyzed at a high level of efficiency. In this context, information theory principles step in, as they lead engineers to interpret complicated sets of data and create machine learning algorithms that can provide actionable insights. Even with these developments, the significance of a strong grasp of the data is evident in the difficulties of the healthcare field. With increasing diversity of information, engineers need to consider various attributes during preprocessing, model construction and result interpretation, citing that having large volumes of data does not equate to delivering efficient solutions (Arcuri et al., 2018). The combination of information theory and engineering therefore emerges as a critical paradigm for addressing contemporary technology complexities. Furthermore, the applicability of information theory becomes more and more evident when evaluating its contribution to resource management in engineering fields. Engineers have to implement solutions within time, budget, and material constraints while maintaining optimal performance and sustainability.

Progressive coding methods based on information theory enable the better utilization of resources through improved predictive maintenance and minimized waste in different engineering processes. For example, in the manufacturing sector, algorithms analyzing the production data can also predict equipment failures, thus minimizing downtime and optimizing operations. This capability for optimization directly contributes to profitability as well as facilitates environmentally sustainable practices, so it is aligning engineering strategies with overall sustainability goals. With current engineering ever-changing, the information theory principles will inevitably continue to be at the heart of driving innovation and solving critical problems in this ever-dynamic environment, guaranteeing adaptability and resilience for future innovation (Figure 7.2).

Figure 7.2. *Importance in modern engineering.*

Source: https://www.engineeringdesigner.co.uk/wp-content/uploads/2023/05/fostering-innovation-with-computational-engineerin.jpeg.

7.1. FUNDAMENTAL CONCEPTS OF INFORMATION THEORY

One of the very basic principles of Information Theory is that of entropy, which quantifies the uncertainty or information content of a random variable. Entropy indicates unpredictability and has a close association with data compression and coding efficiencies. In engineering terms, an understanding of entropy is crucial as it directs the development of more efficient communication systems with less redundancy and greater information transfer. For instance, Shannon's entropy provides a mathematical foundation for encoding messages in the most efficiently optimal way, thereby saving bandwidth. This fundamental realization is not only theoretical but also applied and has significance in areas such as digital communications and data storage where effective information transfer is paramount in system dependability and performance, showing how Information Theory concepts find application in groundbreaking engineering solutions. The second important concept in Information Theory is that of mutual information, which refers to the amount of information that one random variable has about another. This idea has important use in most engineering operations, such as that of communication channels where an understanding of mutual information can be applied to build improved coding schemes. By quantifying the reduction in uncertainty about one variable when another variable is known, mutual information enables engineers to design signals for improved clarity and less effect of noise. This applicability carries over into data transmission protocols where the assurance of sending maximum information with minimal interference is an essential issue, thus the establishment of the importance of mutual information in the pursuit of reliable and high-fidelity communications systems in existing engineering applications. Other than mutual information and entropy, coding plays a critically significant role in Information Theory because it is required to allow for the reliable transfer of information through channels. Coding techniques such as data compression and error-correction influence the efficiency and robustness of systems directly. For instance, higher-level coding techniques such as convolutional codes and linear block codes utilize Information Theory concepts in counteracting the effects of noise in digital communication. Further, the integration of machine learning algorithms and coding provides more predictive ability in engineering systems, which enables more intelligent communication networks. As the complexity of engineering issues deepens, the confluence between Information Theory and coding principles becomes increasingly essential to providing information that is robust, accurate, and efficient within an ever-evolving technological landscape.

7.1.1. Information Content and Entropy

Content and entropy of information are key concepts in information theory, which guides information understanding and compression in the context of engineering. Content of information is the quantification of the amount of uncertainty that a random variable contains, a quantification of the amazement one experiences upon encountering specific results. In digital parlance, entropy quantifies the average amount of information produced by a stochastic data source and consequently the compressibility bounds. The relationship between these concepts is most critical when designing coding schemes in a most efficient way to achieve the best data storage and transmission. Notably, the concept of an efficient coding scheme identifies the source's entropy and optimizes information transmitted with a decrease in redundancy. The application of the theoretical concepts can be observed by practical applications in the discipline of engineering, particularly image compression.

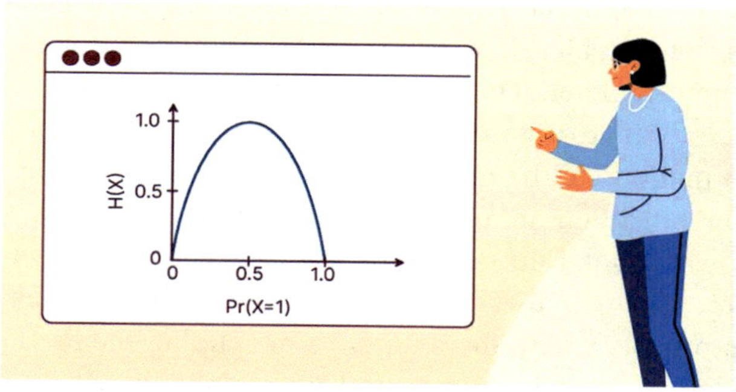

Figure 7.3. *Information content and entropy serve.*

Source: https://images.shiksha.com/mediadata/ugcDocuments/images/wordpress-mages/2023_02_MicrosoftTeams-image-156.jpg.

For instance, recent research has established that entropy coding is capable of measuring the information content of digital images accurately, an implication that applies to the approaches engineers take toward optimal compression strategy. As demonstrated in a study, bound estimation's model proposed for image digital coding capitalizes on the entropy of image data to achieve reversible compression from rate-distortion theory. This approach offers a concession towards increased information inherent in signals and applies the Laplacian probability density function to achieve signal-to-noise ratio gain, thus enhancing predictive coding systems efficacy in noisy conditions. The extensive understanding of information content is thus very important in creating efficient engineering solutions that weigh the demands of quality and efficiency. Using the information theory framework of nonadditivity, exploring other measures of entropy reveals the information content complexity. This contrasts with the conventional models formulated by Shannon, providing perspective into the nonextensive nature of information sources. Nonadditive information content is consistent with a broader understanding of entropy, enabling engineers to study the correlation of the entropy parameter, expressed as q, with codeword length. Examining this connection, researchers can design more general coding designs that support complex data structures without compromising efficient information coding. These advancements in understanding information entropy illustrate the dynamic nature of information theory and its relevance to solving contemporary engineering challenges, paving the way for new methods of coding in the midst of intensifying data environments (Figure 7.3).

7.1.2. Redundancy and Efficiency

In information theory, redundancy and efficiency are most important to understanding how information can be optimized in transmission and storage. Redundancy exists as the repetition of information, a redundancy that while seemingly redundant is vital in error checking and data retrieval. Applied to engineering, redundancy allows for systems to maintain their shape when corrupted or destroyed so communications may stay reliable. For example, data coding redundancy can make it possible to reconstruct lost or corrupted bits, thereby compensating for the impact of channel noise. As outlined in digitized analog data analysis, effective compression schemes try to find a balance between redundancy and efficiency and try to utilize as little wasted bandwidth as possible without compromising on the integrity of the signal being conveyed (Ingels et al., 2025). Thus, redundancy can be controlled effectively in order to enhance the overall system performance through improved data integrity at the cost of relatively little efficiency. The data transmission efficiency is largely dependent on the quality with which redundancy is handled. Engineers often incorporate encoding techniques that reduce redundancy and retain such data as is essential while designing communication systems. This redundancy is necessary because excessive redundancy can cause wasteful bandwidth usage, but insufficient redundancy can compromise data integrity. Utilizing multiple methods of compression allows for the removal of redundant data, which translates to improved throughput. With compressing video signals, for instance, engineers can transmit more content in less time, leading to higher efficiency levels. Further, the concept of redundancy warrants a closer examination of information compression function, not merely as a resource-conserving mechanism but also as a systems response amplifier to errors that are inherent in noisiness. As described, the application of information compression techniques eventually serves to consolidate patterns and optimize communication, thereby enhancing efficiency in all applications of engineering (Wolff et al., 2015). Lastly, the interdependence between redundancy and efficiency in information theory is important to enable optimal data handling in engineering processes. Conscientiousness about the nuances of this interdependence allows engineers to optimize performance and minimize errors in systems by appropriately designing them. Redundancy, while largely viewed as undesirable excess, plays a vital role in ensuring dependable information flow in complex systems. Techniques such as chunking-with-codes, as quoted in information compression studies, show how redundancy can be utilized to provide minimal repetition risk and data accuracy within coded systems. Moreover, the equilibrium that is achieved through redundancy in communication not only helps enhance reliability in systems but also enhances our knowledge of cognitive processes, demonstrating that effective coding patterns are able to reflect our own methods of pattern perception and comprehension (Wolff et al., 2015). As such, redundancy remains a significant factor to support both efficiency and reliability in engineering communications.

7.1.3. Channel Capacity and Its Implications

Channel capacity is central to information theory since it outlines the highest rate at which information may be transmitted with certainty across a communication channel. This idea, spearheaded by Claude Shannon, highlights the importance of measuring the limits imposed by the inherent properties of a channel, such as bandwidth and the amount of noise. When comparing intricate communication models, like the two-user erasure interference channel, one can see that even with local knowledge of channel states, the performance can be highly varying. It has also been demonstrated that though local delayed information might not be beneficial with respect to zero knowledge at all times, it still manages to increase capacity under certain conditions, especially when transmitters use such knowledge strategically to counter interference effects on the received signals (Calderbank et al., 2016). Channel capacity therefore embodies an important framework that influences coding strategies and optimization of performance in contemporary engineering. The channel capacity implications go beyond theoretical discussions, largely shaping real-world applications across different engineering fields. In molecular communication, for example, the notion of channel capacity is realized to show how a single particle or token can be successfully sent and received. This model takes cue from biology when discrete molecules interact to enable signaling processes. By comparing various encoding schemes—e.g., token composition and timing—the work describes capacity limits that define the information transfer efficiency in this novel environment, showing that high rates can be sustained at low energy use (Mian et al., 2016). These observations emphasize the way strong channel capacity understanding is imperative in developing technologies from medical devices to engineering-oriented communication systems, underlining the synergy between theory and application. Furthermore, studying channel capacity forces a rethinking of traditional signaling techniques, enabling engineers to design and allocate resources for systems based on smart decisions. The understanding that marginal distributions may be conserved and yet still effect overall capacity offers a counter-intuitive gain into system behavior. As researchers examine different communication architectures—time-varying channels or multi-user systems, for instance—they can apply these principles to optimize throughput in conditions ranging from bad to good. This is especially important in the age of high-speed telecommunication and low-power molecular communications, where efficient data transfer is increasingly in demand. Thus, a thorough understanding of channel capacity not only satisfies theoretical frameworks but also ensures that engineers are capable of addressing the constantly changing challenges of efficient communications in an increasingly digital age (Figure 7.4).

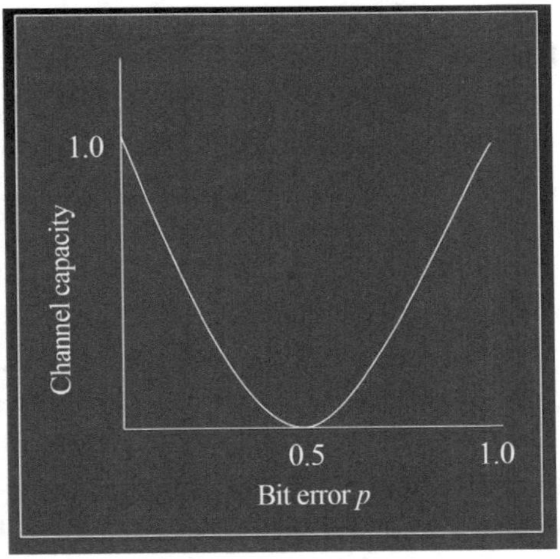

Figure 7.4. *Channel capacity and its implications.*

Source: https://image.slideserve.com/226770/channel-capacity-the-bsc15-l.jpg.

7.2. CODING TECHNIQUES IN INFORMATION THEORY

In the vast realm of information theory, coding techniques are one of the driving forces of maximizing data transmissibility and integrity. The techniques are developed to encode data so that it is resistant to forms of interference and distortion during transmission. Underlying this endeavor is the issue of reducing mean squared error (MSE) with power constraints, particularly in scenarios involving external interference. Via the study of various coding methods, scholars have focused their work on non-linear transmission methods, such as interference concentration (ICO) and one-dimensional lattice (1DL) methods. Notably, the performance of non-uniform quantization has emerged, stating that diminishing quantization intervals with increasing distance from zero can significantly enhance performance under scenarios involving known interference signals (Ertem et al., 2016). These findings highlight the intricate relationship between coding schemes and channel states, demonstrating how innovative methodology preserves data accuracy. With increasingly sophisticated communication systems emerging due to such causes as multi-antenna systems and dispersive channels, the necessity for advanced coding techniques becomes even more critical. In such systems, each receive antenna gathers signals that comprise a linear combination of diverse transmitted symbols, often very frequently tainted with noise. This complexity requires good coding methods to enable accurate symbol detection with interference present. Traditional maximum-likelihood sequence detection is typically deemed impractical due to computational demands; thus, engineers tend to compromise on suboptimal but feasible solutions such as equalizers designed to minimize intersymbol and interuser interference. Such approaches like sphere decoding have been shown to be promising methods, which can provide maximum-likelihood estimates when having reasonable complexity similar to decision-feedback equalizing algorithms (Hassibi et al., 2002). This evolution of coding methods not only enhances performance but also addresses issues of computational complexity brought about by new communication systems. Another significant characteristic of coding techniques in information theory is their adaptability to real field applications, where varying channel conditions dictate the optimal choice of coding mechanisms. Channel conditions, which are subject to environmental variables and technological constraints, require coding mechanisms that will change with conditions to maintain high data transmission rates. Studying different schemes of coding enables engineers to develop adaptive algorithms that respond to changing levels of interference and channel capacity. Adaptability is crucial in real-world applications such as mobile communications, where users frequently find themselves within fluctuating signal environments. As such, research and development remain geared towards optimizing these coding techniques to deliver consistent performance under differing conditions, ultimately maximizing the reliability of data communication in engineering applications. With this continuing inquiry, coding practices are to the forefront of advancing information theory, linking theoretical advances with engineering application.

7.2.1. Source Coding and Its Applications

The source coding principle is core to information theory, or more particularly, the manner in which it is applied to efficient data transmission and representation. By compressing individual bits of information to a more compact form, source coding eliminates redundancy while maintaining the original message's inherent content. The practice is core to numerous applications such as telecommunications and data compression, where bandwidth and storage capacity are limited. For instance, algorithms derived from source coding principles enable compression of multimedia files, making storage and transmission of vast volumes of information possible with minimal bandwidth use. Additionally, innovation in lossless and lossy compression is a reflection of balance between fidelity and efficiency, which can be used across many fields from streaming to digital archiving (Gao et al., 2025). These developments underscore the significance of source coding in optimizing data flow and enabling end-to-end communication in contemporary technological scenarios. For new wireless communication systems, the restrictions of traditional asymptotic information theory become increasingly evident. Ultra-reliable low latency communication and massive machine-type communication services in the future necessitate source coding techniques to be re-evaluated to address their unique challenges. Traditional models that are infinite blocklength- and vanishing error probability-based are slow in catching up to the demands of practical applications. Research that justifies finite-blocklength information theory brings insight into such pressing demands, necessitating optimized source coding approaches that compromise between short-packet communication and low energy consumption at the cost of no performance loss (Gao et al., 2025). This paradigm is not only pushing the boundaries of theoretical models but also guiding the system engineering that would be able to support the greater needs of contemporary wireless networks and thus bridge the gap between practice and theory. The cross connection between source coding and channel coding is equally critical in optimizing overall efficiency of communication. Joint source and channel coding has received interest given its ability to surpass performance by individual coding techniques, particularly in non-asymptotic regimes where performance limits are significantly influenced by finite constraints. By integrating both the source and channel coding procedures, the technique allows more extensive dealing with data transmission through optimization of error correction and still maintaining data integrity. The analysis of the correlation is most important as communication technology evolves to cope with the increasing volume of data generated in global networks. Finally, the evolution of joint coding schemes and their applications provides engineers new methods to tackle the nature-imposed challenges of modern communication systems. Through constant research in the field, the options of devising strong and effective codes continue to grow and ultimately result in the formulation of engineering practices pertaining to information theory.

7.2.2. Channel Coding and Error Correction

Channel coding and error correction are required to ensure error-free communication via various systems, particularly in noisy and interference-rich environments. The primary requirement of channel coding is to add redundancy to the transmitting information so that the receiver can detect and correct errors when they arise during transmission. This redundancy is encapsulated in coding protocols, i.e., block codes and convolutional codes, for enhancing the reliability of the received signals. In information theory, the decision of being capable of designing efficient coding methods is based on having a thorough knowledge of the channels capacity, which characterizes the maximum rate of information transfer without error under specified conditions of noise. Advances in coding techniques have far-reaching implications for a wide variety of applications, ranging from digital communications to storage devices, highlighting the necessity of dependable channel coding in contemporary engineering designs. As communication systems rely increasingly on digital signal methods, their capacity for error correction must be understood. A majority of the contemporary systems apply mapping methods such as phase shift keying (M-PSK) to transmit data through various states.

However, such systems are confronted by a number of challenges, such as thermal noise and harmonic interference, that can compromise the integrity and quality of signals. The implementation of error correction methods, such as convolutional encoding with Viterbi decoding, and Hamming codes, has been proved to enhance noise immunity and therefore general transmission quality in the presence of such interference. Studies indicate that the use of such codes facilitates improved performance under light interference but may break down under higher intensity noise levels, necessitating ongoing research in coding methods (Nguyen, 2024). Ongoing research on the topic is thus important for the advancement of error correction technology and achieving increased information transmission reliability. In addition to reducing interference and noise, the research of error-correcting codes is relevant to novel fields such as quantum information science. Quantum states' unique requirements, including their intrinsic sensitivity to noise, call for new coding methodologies that extend classical error correction into the quantum realm. Techniques such as quantum error correction and fault-tolerant architecture play a crucial role in realizing the potential of quantum computers and quantum communication networks. Such methods are critical to preserve the integrity of quantum information by removing errors while performing computation and transmission. Since this field continues to evolve, the intersection of quantum and classical error correction codes will make profound breakthroughs, shaping the destiny of secure communication and advanced computation (Pasupuleti, 2025). By linking fundamental theories with real-world problems, scientists and engineers can better harness the power of coding to advance industries (Figure 7.5).

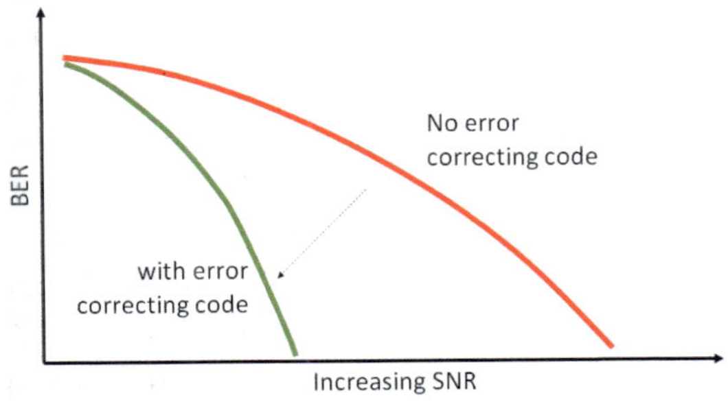

Figure 7.5. *Channel coding and error correction various systems.*

Source: https://www.wasyresearch.com/content/images/size/w2000/2022/09/COV-ER_channel_coding.png.

7.2.3. Lossless vs. Lossy Compression

In data storage and transmission, compression algorithms have a crucial role in managing the quantity of information that is generated in engineering and computer science and other fields. Lossless compression allows for perfect reconstruction of original data from its compressed state and, therefore, is invaluable for applications where data integrity is of particular concern, i.e., text files or executable code. By means of algorithms that effectively reduce file size without loss of quality, the average compression ratios ranging from 1.35:1 to 4.69:1 are achieved by lossless compression, as indicated by existing research on astronomical data with the GERLUMPH project, where the average ratios of 1.35:1 to 4.69:1 were observed in working with big datasets (Fluke et al., 2015). This is particularly useful when data fidelity must be maintained, so minor differences could result in catastrophic analysis failure. Conversely, lossy compression discards some information to achieve significantly higher rates of compression and is thus the optimal application for file formats like multimedia, where detail loss is acceptable. For instance, audio and video media routinely employ lossy approaches—JPEG2000 being a notable example—to maximize quality against file size. While lossy compression can provide enormous savings in storage and bandwidth use, it requires thoughtful consideration of the distortion tolerance level against data usage when employing such media in applications like astronomical imaging or numerical simulations. Researchers have to weigh convenience against degradation of data, so that the resulting file is still of a suitable quality to analyze, as indicated by findings from various tests of lossy formats against observational data (Fluke et al., 2015). The choice of whether to use lossy or lossless compression will depend significantly on data purpose and what trade-offs are acceptable between size and quality. In information theory, a similar decision can be represented through the ideas of rate-distortion theory, which really traces out the relationship between the compression ratio and the distortion achieved with lossy methods. This has implications for a scientific assessment of how compression will impact usability of data, especially in applications involving accurate measurements or prediction, like engineering applications. Using sophisticated coding methods and algorithms, including those under research in source coding with fixed lag side information, engineers can optimize data transmission and minimize loss, thereby achieving more effective and efficient information processing systems (Figure 7.6).

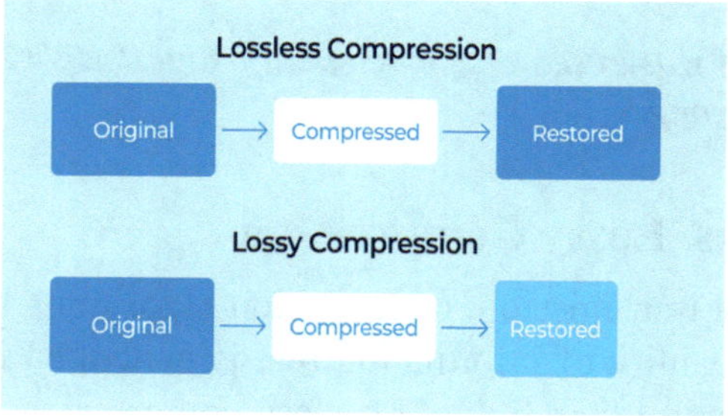

Figure 7.6. *Lossless vs. lossy compression.*

Source: https://mlfk3cv5yvnx.i.optimole.com/cb:HA53.300ea/w:930/h:485/q:mauto/f:best/https://www.ninjaone.com/wp-content/uploads/elementor/thumbs/N1-0921-Lossy-vs-Lossless-Compression-blog-image-1-2-qvjtaq0fhxhz-kuuel3xr3mwst5thvoor3jjybjtboy.jpg.

7.3. APPLICATIONS OF INFORMATION THEORY IN ENGINEERING

Information theory, first conceived by Claude Shannon in the mid-20th century, has had a profound impact on engineering disciplines by providing a mathematical framework for communication and data transmission. The use of information theory in the development of coding techniques is arguably the most fundamental application of information theory since effective data transfer in any channel depends on coding techniques. Shannons work focused on the significance of data encoding in order to optimize bandwidth usage and reduce errors, a factor that has become only more crucial with the advent of modern digital communications. With a focus on improving the performance of communication systems, investigating novel coding techniques and how they can be combined with new technologies is now of major importance in ensuring data consistency and reliability in our rapidly globalizing world (Such et al., 2020). Signal processing is another fundamental use of information theory in engineering, particularly in fields such as biomedical imaging and multimedia communication. Signal manipulation and analysis require a knowledge of the data underneath and also of efficient encoding that can be applied. The interdisciplinary nature of signal processing comprises mathematical modeling, probability, and statistics, thereby creating a robust foundation for applications in a variety of areas including audio, video, and medical imaging. Recent advances here have spawned new algorithms and approaches that assist in dealing with massive amounts of complex data, delivering timely and accurate interpretation of signals. With the growth in demand for processing and communicating real-time data, further study of information theory's role in advancing signal processing methods is becoming increasingly important to engineering advancement (Ruiz-Alzola et al., 2013). Secondly, its use in engineering fields like artificial intelligence and machine learning is a testimony to its far-reaching applications in engineering. With engineers being confronted more and more with complex problems, information theory provides important measures and paradigms to design algorithms that can efficiently handle and interpret data. For instance, in machine learning, information-theoretic principles are utilized to improve model performance, particularly classification and data compression. With data-driven decision-making becoming more prevalent, understanding how information theory can be applied to support engineering solutions is necessary. Such collaborative partnership puts emphasis on the importance of interdisciplinary collaborations, whereby engineers utilize knowledge from various disciplines to produce innovation and enhance engineering practice within an era of rapid technological transformation (Such et al., 2020).

7.3.1. Telecommunications and Data Transmission

In the world of telecommunications and data communications, the evolution of advanced technologies has significantly improved the efficiency and efficacy of communication networks globally. A key innovation is the Cloud Radio Access Network (C-RAN) that has surfaced as a groundbreaking paradigm for fifth-generation (5G) wireless communications. C-RANs are configured to reduce both capital and operational costs while, at the same time, enhancing spectral and energy efficiency. One of the key elements in this architecture is the fronthaul, connecting the baseband unit and the remote radio head. With the growing complexity of C-RANs, the fronthaul capacity can be the bottleneck that requests new solutions for optimization such as compression and quantization. Such measures are important in maintaining service quality in situations of constrained resources, hence representing the crossroads of advanced telecommunication theory and practical engineering challenges (Lau et al., 2015). Telecommunication data transmission is finally based on information theory concepts brought about by it, which addresses quantification and maximization of information transfer. Security and reliability are fundamental aspects within this scope, particularly when addressing challenges posed by eavesdropping.

Figure 7.7. *Telecommunications and data transmission.*

Source: https://questsys.com/wp-content/uploads/2023/07/Blog-Quest-1920-x-1080-px-600x338.jpg.

In the case of a real communication channel with a potential external eavesdropper attack, secure coding and transmission protocols play an important role in avoiding interception. Physical-layer security is an evolution of concepts by considering trade-offs in reliability of communication vs. security under adversarial settings. Techniques such as opportunistic relay selection have been proposed to enhance the security-reliability tradeoff, enhancing the data transmission resilience with reduced opportunities for successful eavesdropping. As such, the integration of security into data transmission protocols is a core evolution in telecommunications engineering (Hanzo et al., 2013). Also, advances in telecommunication and data transmission equally align with the vision of the Internet of Things (IoT) where humongous numbers of inter-connected devices need efficient and dependable communication technologies. The increasing proliferation of IoT applications require higher capabilities of data exchange, which demands advancements in coding and modulation technologies that can handle increased data rates and enhanced reliability. Technologies such as resource optimization and software-defined networking provide the flexibility required to adapt to shifting traffic patterns and user demands. The creation of such technologies is evidence of the dynamic character of telecommunications, whereby engineers constantly seek means of addressing emerging problems associated with the transmission of data. This relentless pursuit reflects an inherent tie between information theory and engineering processes, which testifies to the central role of coding and modulation in the relentless development of complex communication systems (Lau et al., 2015) (Figure 7.7).

7.3.2. Computer Science and Data Storage

The intersection of computer science and data storage has evolved completely by virtue of advances in information theory and coding. Data storage not only includes the physical placement of data but also ensuring its integrity and usability, and this poses significant challenges in today's computing environment. Efficient coding protocols play a crucial role in this case to facilitate error detection and correction and data recovery functions. With the sheer amount of data generated daily, effective storage and management systems are important. Mathematical modeling and emerging coding technologies improve reliable systems that can effectively tackle data storage problems. The performance of storage systems can be enhanced by using such techniques, storing data securely and making it accessible, thus expanding computer science usage in the real world (Adams et al., 2019). Moreover, the efficiency of data storage systems is also demonstrated through the advent of distributed systems, with the distributed data files across locations. In such systems, the role of coded computation techniques comes into play to mitigate issues of noise and crashes of the system. Coded computation presents new findings on reducing operations such as matrix multiplication, which is a common operation used in machine learning and data analysis. Through the application of codes to shatter the power of straggler nodes, distributed matrix multiplication can be significantly enhanced, with speedup gains directly proportional to the number of workers employed. Ultimately, this interconnection between coding theory and actual storage technology reveals the path toward robust systems that can manage gargantuan databases with reliability and efficiency (Lam et al., 2018). As data storage grows, computer science, information theory, and coding converge further. The use of innovative coding methods, such as coded shuffling, leads to tremendous reductions in communication costs, making it possible to transfer the data rate among many users at higher levels of efficiency. Cost-effectiveness becomes increasingly important in large-scale systems where traditional methods break down due to bottlenecks in communication infrastructure. Through the utilization of excess storage capacity, these new coding techniques maximize system performance and user satisfaction. Theoretic advances in coding reflect their value of application to future data storage technologies with the guarantee that systems will continue to be robust and responsive to the nature of today's computing environments. Thus, the intersection of these disciplines not only advances academic research but also paves the way for future innovations in the realm of engineering (Adams et al., 2019; Lam et al., 2018) (Figure 7.8).

Figure 7.8. *Computer science and data storage.*

*Source: https://miro.medium.com/v2/resize:fit:1400/1*mcS5PUKZsDieTW1H5zo rQ.jpeg.*

7.3.3. Signal Processing and Multimedia

Signal processing became an increasingly focal area in multimedia applications to facilitate the encoding, transmission, and decoding of the audio-visual content. As digital media got more pervasive, demand for effective coding techniques improving quality while minimizing data consumption also increased. At the center of this pursuit is the understanding of how multiple video coding architectures, and in this instance, Distributed Video Coding (DVC), operate to achieve optimum performance in resource-limited conditions, such as wireless video sensor networks (WVSNs). Recent advancements in DVC have shown that it can leverage the Slepian-Wolf and Wyner-Ziv theorems to efficiently compress video information, and therefore it offers an interesting prospect for future multimedia applications. This new body of work highlights the compelling need to explore the current gaps and potential future directions of DVC architectures, notably their working conditions under practical uses (Fong et al., 2015). In addition, innovations in coding schemes in multimedia processing have also been a matter of growing impact of various user demands and technical advancements. A fascinating development is Costa coding, which has recently gained new attention based on its usage in numerous fields, including data-hiding and cooperative transmission in wireless systems. Algebraic binning is used by this method of coding to maximize source-channel coding interaction and thus improve performance. By employing nested lattice codes and proposing nested turbo codes for efficient encoding, Costa coding shows how new coding structures can have a direct impact on the efficiency of signal processing functions in multimedia applications. The tradeoff optimization between turbo-like trellis-coded quantization (TCQ) and turbo trellis-coded modulation (TTCM) is essential to achieving improved reliability and quality in signal transmission, and thus further reinforcing the application of coding theory in engineering practice (Cheng et al., 2008). The interplay between information theory and multimedia signal processing hence finds expression both in theoretical contributions and practical applications. In this ever-evolving environment, the coding methodologies established not only advance the effectiveness and quality of multimedia systems but also serve the wider discipline of engineering. With researchers constantly developing new ways of doing things, they add to both the theoretical body of work and actual applications that inform the directions in which multimedia technologies evolve. With expanding demands for greater fidelity, reduced latency, and efficient bandwidth utilization, the ongoing innovation in signal processing methods and coding frameworks will remain at the forefront of engineering research. This ongoing innovation is an indicator of interdisciplinary approaches since information theory knowledge is applied to resolve complex problems in multimedia systems and hence drive development in the field (Fong et al., 2015; Cheng et al., 2008).

7.4. THE ROLE OF ALGORITHMS IN CODING

Algorithm use in programming is basic in the sense that they assist in defining and optimizing the operations that are tasked with encoding and transferring data. Algorithms guide information to be transformed mathematically into a compatible form that can carry out effective communication, ensuring that data is decoded, transmitted, and compressed with minimal loss and maximum integrity. This is emphasized in various applications, particularly in coding theory, where finite fields are used heavily. Finite fields provide the mathematical structure that serves as the basis for most coding algorithms, allowing efficient error detection and data integrity checking. With the increased research, books such as the Handbook of Finite Fields bring together colossal works of worldwide researchers highlighting avant-garde advancements which further consolidate the theoretical characteristics and uses of coding algorithms (Gary et al., 2013). The efficiency of coding algorithms not only affects computational effectiveness but further enhances data transfer security. In a world where cyber-attacks grow more sophisticated every day, strong algorithms keep sensitive data safe during its passage through even insecure channels.

Powerful encryption algorithms are a prime example, as they employ sophisticated mathematical models to transform plaintext into a form intelligible only to the supposed owner. They range from the application of algorithms such as those found in computer algebra system encyclopedic volumes to demonstrate their pivotal relevance in secure cryptographic protocol development. In comprehensive studies that span across a wide range of applications in coding theory, cryptography, and other related areas, the importance of these algorithmic structures can be observed by illustrating how they underpin theoretical and applied aspects of computer science and engineering (Joachim et al., 2013). With advancements in the world of technology, the application of algorithms in programming remains at the core of solving emerging problems in information technology. Advances in machine learning and artificial intelligence remain a driving force behind the design of new algorithms that can adapt to encode and process data adaptively. These advances cut across various fields, including telecommunications, where algorithms at peak levels enhance signal processing and data compression algorithms to ensure efficient communication under various conditions. Besides, the interface between algorithmic design and actual application in areas such as quantum information theory constitutes a crossover towards interdisciplinarity, expanding the boundaries of coding efficiency. Consequently, understanding the sophistication of algorithms and their applications not only promotes engineering but also enhances the appreciation for information theory principles that govern them. This convergence is a key area of focus as scholars and practitioners grapple with the complexities of modern engineering (Gary et al., 2013).

7.4.1. Huffman Coding and Its Efficiency

Huffman coding is a landmark technique in data compression in that it dramatically changed the process of encoding information and its transmission efficiently. It is developed from the fundamental components of information theory and uses a variable-length code scheme to minimize the average length of the resulting binary string. By assigning shorter codes to more frequent symbols and longer codes to less frequent symbols, Huffman coding also shrinks the overall space that is required for transmission and storage. This is not an abstract concept; empirical evidence has been shown to prove that Huffman coding can provide far superior compression ratios over fixed-length encoding methods, making it a real-world application in uses as varied as file compression through to data transmission via telecommunications links. In addition, the algorithm design has employed a binary tree data structure that supports encoding as well as decoding operations and therefore became a significant part of information system engineering education and coding practices. Quantitatively, Huffman coding efficiency can be measured based on some parameters such as average code length and the corresponding compression ratio, which has a direct influence on the performance of data processing systems. In constructing a Huffman tree, the algorithm employs a greedy strategy that is in such a way that the best symbol ordering based on frequency counts is attained. This property not only assists in efficiency but also renders the algorithm extremely easy to implement.

The computational complexity analysis reveals that the construction of the Huffman tree is of $O(n \log n)$ time, and the encoding and decoding processes usually take linear time. Such performance metrics entrench Huffmans position in adaptive computing environments requiring data to be processed in real time. The effectiveness of Huffman coding is also emphasized in environments characterized by non-uniform symbol distributions, which offer a wide range of benefits compared to traditional, fixed-length encoding techniques. Here, Huffman coding serves as a timeless example of the scenario of engineering, where demands for effective data management and transmission feature prominently.

Moreover, the emergence of machine learning and artificial intelligence paradigms has hastened the evolution of Huffman coding, integrating it with contemporary technologies to extend its field of application. With industries demanding increasing amounts of huge datasets, innovations such as the integration of Huffman coding and neural networks evidence a syncretic combination of ancient coding methods with cutting-edge computational methodologies. These hybrid approaches improve the likelihood of optimal representation of data, with the promise of dramatic reduction in storage requirements while maintaining information fidelity in retrieval and processing. By applying frameworks founded on Huffman coding principles, educators can prepare engineering students to meet such higher-order paradigms of data compression and manipulation. This holistic understanding emphasizes not only the historical importance of Huffman coding but also its relevance in shaping future technological evolution and confronting challenges posed by rising volumes of data in the evolving information technology scenario (Figure 7.9).

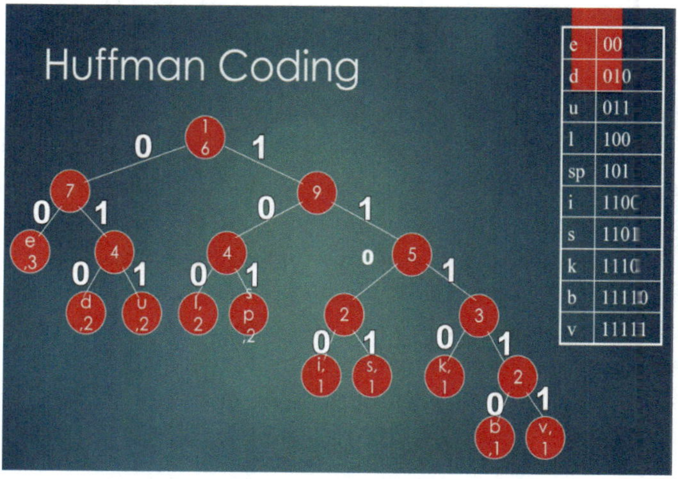

Figure 7.9. *Huffman coding stands as a pivotal technique.*

Source: https://image.slidesharecdn.com/huffman-coding-170911184817/75/Huffman-coding-14-2048.jpg.

7.4.2. Reed-Solomon Codes for Error Correction

Reed-Solomon codes are now a central technique in the field of error-correction, a pillar of protection of data integrity in most communication systems. Block codes, through their strength found using polynomial interpolation in finite fields, work very well correcting burst errors—where several consecutive bits get lost or displaced. The structure of the Reed-Solomon code provides effective error detection and correction through encoding data as sequences of symbols, typically the form of an element within a finite field. The unique encoding is accomplished in such a way as to be able to recover original information despite corruptions, which has been invaluable in applications such as CDs, QR codes, and data communication networks. The stability of Reed-Solomon codes under varied environments justifies their importance in contemporary information theory and coding. Mathematical background on which Reed-Solomon codes are established is well based on algebraic coding theory, whose mathematical foundation supports a theoretical framework of encoding and decoding strategies. Notably, these codes can correct errors by exploiting redundancy; a Reed-Solomon code over an algebraic field is able to recover data if the number of corrupt symbols falls within some predetermined limit. This versatility is not theoretical; it has been experimentally verified on a large scale. Combinatorics Coding and Security Group (CCSG) is concerned with the actual implementation of Reed-Solomon and other forward error correction (FEC) codes of similar nature to real-world applications, demonstrating better performance in communication error-prone channels (Capdevila et al., 2015). The capacity to solve both theoretical and practical issues turns Reed-Solomon codes into a cornerstone of modern digital communication techniques. Follow-up research on Reed-Solomon codes illustrates ongoing progress and improvements aimed at making them more efficient and reliable. Research endeavors in groupings such as CCSG reflect a number of initiatives addressing improving encoding algorithms and testing their application with emerging technologies such as quantum computing and wireless communications. These endeavors seek to counteract the growing demand for reliable methods of data transmission within emergent complexities in technological environments. Understanding the various structural characteristics and fault correction abilities of Reed-Solomon codes not only anchors their position in engineering application but also spurs innovation in coding theory. Continued research in this field is in line with broader trends throughout information theory, prioritizing work on advancing classical codes to provide solutions to contemporary data integrity and transmission performance issues (Capdevila et al., 2017).

7.4.3. Turbo Codes and Their Impact on Communication

The emergence of turbo codes revolutionized the error correction landscape, making communications far more reliable. Turbo codes, which were invented in the 1990s, use a parallel concatenation of convolutional codes and iterative decoding. It enables the codes to achieve performance levels on the cusp of the Shannon limit, making them of great value when strong transmission of data is required. For example, in deep space communications, where bit-error rates must be kept as low as possible, turbo codes effectively reduce errors without adding the unnecessary complexity of standard coding methods. Their flexibility makes possible effective use of resources, essential with the severe hardware constraints typically encountered in these applications, as evidenced in the developments in hardware structure allowing for high resource reuse in turbo code designs (Condo et al., 2013). The use of turbo codes goes beyond the traditional application, with impacts being felt in state-of-the-art technologies such as wireless communication and data-hiding techniques. Costa coding is one of the fields that have gained significant attention due to its role in enhancing data protection and communication channel effectiveness. Researchers have considered the integration of turbo codes and nested lattice codes and developed new approaches in source-channel coding. These developments indicate that turbo codes are not only capable of enhancing error correction ability but also support complex network environments, including multiple-input multiple-output (MIMO) systems and cooperation from the transmitter. As researchers continue to refine these coding methods, fundamental principles of turbo codes enable improved and safer information flow across different platforms, observing their flexibility in modern communication technologies (Cheng et al., 2008). In evaluating the broader implications of turbo codes, it is evident that their applications to other areas of engineering spur continued advancements in information theory. Turbo codes encourage a collaborative approach to coding design in which researchers benefit from the respective merits of turbo-like trellis-coded quantization and turbo trellis-coded modulation. This synergy maximizes not only performance trade-offs, but also broadens the scope of possibility for the deployment of turbo codes into real-time applications like high-speed data transfer and streaming media. In expanding the boundaries of coding theory, the useful insight developed by turbo code deployments contributes to an overarching understanding of how data integrity is maintained through the vicissitudes of mounting data pressures within our interconnected world. The effect of turbo codes is evident, giving way to novel communication solutions (Condo et al., 2013; Cheng et al., 2008).

7.5. CHALLENGES IN INFORMATION THEORY

In exploring the issue inherent in information theory, there is the need to address the constraints being placed by existing communication systems, particularly as they relate to cloud radio access networks (C-RANs). Fronthaul, or the link for data transfer between the remote radio head and baseband unit, demands high capacity for efficient data transfer. But the majority of C-RAN deployments are marred by fronthaul capacity limitations, which can significantly erode spectral efficiency and energy efficiency. To work against these challenges, researchers are turning to advanced techniques such as compression and quantization which aim to counteract the adverse impacts of limited fronthaul on users quality of service. This exchange of technological innovation and existing limitation recognizes key challenges in the delivery of effective and efficient communication on a range of applications in engineering and highlights the need for continuous research in optimizing network performance (Lau et al., 2015). While problems in information theory emerge in various fields, the health sector illustrates the complexity introduced by the transition from traditional paper-based information systems to digital and data-based systems. The burst of electronic health records and other computable data types has created a historic quantity of information with the potential to be mined by integrated machine learning solutions. However, this trend is also reflected in catastrophic amounts of data diversity and complexity. It is not how much information, but rather the qualities of such information that must be well-picked and honored via the machine learning process to derive valuable understanding. Consequently, the practitioner is confronted with challenges at every stage in the pipeline—from preprocessing to model interpretation—demonstrated by the multiple dimensions of the challenges involved in successfully navigating the availability of digital information in healthcare (Arcuri et al., 2018). The application of information theory into engineering disciplines also poses challenges concerning the theoretical foundations of data transmission and error correction. Established principles in information theory, although strong, encounter fresh challenges as systems become increasingly complicated and the demands for speed and reliability increase. For instance, in coding theory, the task of designing efficient algorithms for correcting errors with little overhead continues to be an insurmountable challenge. Besides, the widespread use of heterogeneous communication channels complicates the construction of a universal general-error-correcting code since it needs to be accommodated in various types of conditions and requirements. With further technological progress, the theoretical and practical issues should be resolved to make the information transmission system more efficient and reliable, which also testifies to the fact that information theory is a dynamic science that requires constant academic effort and creativity (Lau et al., 2015).

7.5.1. Limitations of Current Coding Techniques

In the fast-evolving world of information engineering and theory, current coding schemes exhibit fundamental limitations that undermine the efficiency and reliability of data transmission. Primarily, wireless communication is self-limiting because of inherent degradations of the transmission channel that have a tendency to degrade signal quality and restrict efficient data transmission. This is readily evident in wireless channels, where it is still difficult to ensure reliable, high-data-rate transmission even with advances in coding design. While utilizing multiple antennas is capable of enhancing performance through spatial diversity and power gains, ad-hoc networks are typically hindered by such applications because of size, cost, and hardware complexity constraints. Thus, alternative approaches such as cooperative relaying techniques are being explored to counteract such constraints, even though implementation issues remain to hamper widespread applications and efficiency (Quan et al., 2010). Further, dependence on traditional coding schemes, i.e., amplify-and-forward (AF) and decode-and-forward (DF) strategies, proves the performance limitations of current approaches in real-world applications. Although these methods have been building blocks in the arena of information theory, they do not account for the advanced complexities of contemporary multi-stage relay networks. Recent advances highlight the advantage of integrating symbol-level modulation with space-time coding to steer clear of such issues, suggesting innovation to overcome the staticity of present schemes. Advances hold the promise of significant performance improvements, though the challenge remains to integrate such innovations as well as possible within present frameworks without disproportionate increase in system complexity or cost (Quan et al., 2010). The introduction of machine learning (ML) techniques in the realm of wireless communications renders the coding technique scenario even more daunting. The transition to 6G networks, with ubiquitous, reliable connectivity demanded, will necessitate a realignment of traditional coding techniques with regard to what MLs can deliver. Whereas ML algorithms can optimize transmission protocols and optimize network performance, their successful application is highly reliant on robust datasets, hardware resources, and new methodologies that redefine the existing coding paradigms' inherent limitations. The interaction of ML with information theory raises a number of questions regarding the trade-off between innovation and complexity, especially because coding methods have to adapt to handle the increasing need for efficient and reliable wireless communication systems. As such, comprehending and solving the limitations of current coding methods will be crucial in determining the direction of future overall network efficiency towards high levels of reliability and capacity (Figure 7.10).

Figure 7.10. *Limitations of current coding techniques.*

Source: https://woz-u.com/wp-content/uploads/2022/02/Coding-Tips-Featured.jpg.

7.5.2. The Impact of Noise on Data Transmission

The role of noise in data communications is important, as it affects the accuracy and reliability of information being communicated. Noise, which can be caused by a wide variety of sources such as thermal movement and electromagnetic interference, causes errors in the signal being communicated. These kinds of interference pose significant challenges to communication systems, especially in regions with high interference. This demands the development of good coding techniques that can reduce the undesirable effect brought about by noise. In Information Theory, the foundation of understanding the interaction of noise with data transmission rate is central to engineers and scientists in an effort to enhance error-correcting codes enabling clean data transmission via noisy channels. Thus, management of this complex interaction is crucial for optimizing communication systems and achieving successful data transmission, thereby underscoring the inherent importance of noise consideration in engineering. Coding schemes are pivotal in combating noise-driven errors in data transmission. Forward error correction and automatic repeat request methods are among the techniques employed, with the objective of upholding the integrity of the transmitted message regardless of noise occurrence. Significant advancements have also been witnessed in Hybrid Automatic Repeat request (HARQ) schemes, which provide dynamic data control transmission in noisy conditions. Kang et al. (2015) assert that buffer management in HARQ has also emerged as a critical area of research that impacts modem design and performance in contemporary wireless standards such as LTE and LTE-A. These developments reveal a refined interplay between compression methods, buffer control, and the basic coding schemes employed in communication systems, driving the intelligence of effective noise management techniques. Secondly, accessible capacity theory becomes crucial for understanding the impact of noise in data communication systems. By problem-formulating using Gaussian interference channels, scientists try to gain insights into the ability of secondary users to communicate correctly notwithstanding the noise generated due to primary user transmissions. As explicated in Bai et al. (2013), this involves characterizing the interference and achieving bounds on transmission rates with desired error performance levels. This discussion establishes a more detailed understanding of how noise limits available bandwidth and informs the development of encoding methods that ensure effective communication. Ultimately, these findings emphasize the importance of noise control, leading to more enhanced transmission efficiency and robustness in modern applications in engineering, required for developing next-generation information systems (Figure 7.11).

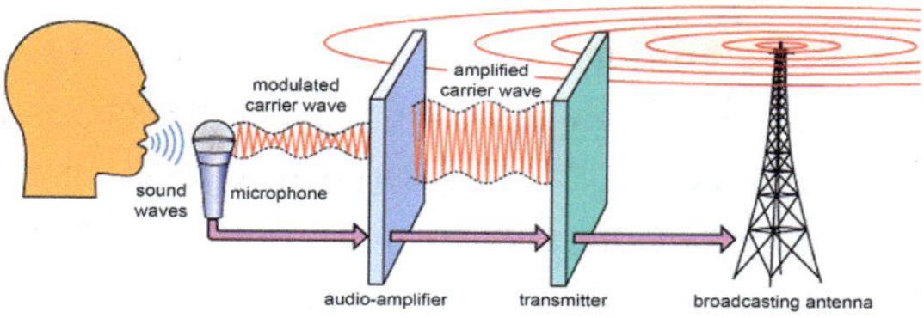

Figure 7.11. *The impact of noise on data transmission.*

Source: https://cdn.britannica.com/97/62897-004-51785FB1.gif.

7.5.3. Security Concerns in Information Transmission

The integrity of the digital environment comes with a wide range of security issues that challenge individual privacy as well as the confidentiality of the organization. Data transmission involves inherent vulnerabilities with the increasing use of the Internet and reliance on digital communication. Hackers and adversaries use ever more advanced schemes to steal or alter data, requiring strong measures of security. The applicability of information theory in this context is brought out by the fact that it can both model and analyze the effectiveness of communication systems as well as address potential threats at the same time. By considering such threats from the point of information theory, engineers could develop more effective coding techniques for safeguarding data when transmitting it. Among the new techniques for enhancing security in the transmission of information is random linear network coding with redundancy. By adding hash symbols as a polynomial function of data symbols, networks can be designed to implement protection against malicious tampering. This is based on the hypothesis that the attacker is unaware of the random coding coefficients, thereby making it difficult for them to design attacks against the conveyed information. This attack is simple to implement as a function of the size of the field employed in coding and the overhead caused by the inclusion of hash symbols, as explained in Effros et al. 2008. These advancements not only add strength to data communication but also create trust in the communication networks essential to the interconnectivity of the world. Furthermore, the introduction of autonomous software agents on social networks creates additional layers of complexity in terms of security. With the agents processing huge volumes of user data, ethical considerations of privacy and surveillance come to the forefront. Stakeholders in these environments must balance the own imperative of enhancing user experience with personalization and safeguarding personal data. Recent studies emphasize the necessity of better knowledge of users' attitudes toward privacy and the roles of software agents within social networks. Focus on user-centered designs and scouting algorithms, as indicated by Arazy et al. (2014), is able to facilitate the development of agents that respect user privacy but are still operational in their use. All these issues of security are required not only to foster trust with users but also to maintain ethical deployment of technology in social interactions.

7.6. FUTURE TRENDS IN INFORMATION THEORY

As we enter the age of information theory's future, integration with machine learning is a top trend, particularly in healthcare. The accelerating rate of digital data from electronic health records, imaging, and public health data necessitates sophisticated methods to process efficiently. This change places emphasis on the ability to understand not just the volume of information but also its complexity and variability, particularly since health care data are obtained from multiple sources. Consequently, machine learning programs will have to be modified to analyze such complex data so that conclusions drawn from them are not just correct but also insightful. This appeal to the unified approach indicates the need for future development of information theory with an emphasis on the unique challenge from the data of healthcare, affirming its central position in creating good healthcare practice and administration. As an illustration, being aware of the data properties during preprocessing phases can significantly enhance later analyses in accuracy and dependability of models (Arcuri et al., 2018). A second significant trend in designing the future of information theory is the convergence of synthetic biology and bioinformatics, namely through the design of synthetic cells. Such cells, composed of biomolecules in lipid vesicles, offer a new field where information theory can extend further into bioengineering. Synthetic cells' programmability allows them to perform specific biomolecular functions, and future potential for advanced applications in biotechnology is possible. This dynamic field demands strong communication protocols between synthetic and biological systems, enabling a new level of information exchange that needs to be well-modeled and researched. When researchers build bio/chemical-based Information and Communication Technologies (bio-chem-ICTs), information theory will be crucial in enabling effective synthetic communication. This convergence brings great hope for the future of information theory because it now begins to transcend traditional technological domains to significantly impact biological systems (D'Angelo et al., 2013). The future of information theory will also likely be centered on resilience due to rising concerns regarding data privacy and cybersecurity. As digital infrastructures expand, the volume and sensitivity of data continue to increase, and thus robust encryption and coding methods are highly sought after. Subsequent innovations will be focused on the development of future-generation cryptographic mechanisms that can withstand smart cyber-attacks, drawing from information theory principles to reinforce security protocols. In addition, prioritizing resilience will also involve reconsidering current coding methodologies in order to address vulnerabilities exposed through emerging technologies like quantum computing. As these developments unfold, information theory concepts will be the basis for quantifying and improving the resilience of data storage and transmission systems. This is in concert with a broader cultural appreciation of the requirement for secure and reliable information systems in an increasingly interconnected world. Information theorys adaptability will thus be key to coping with the sophisticated challenges in the pipeline (Figure 7.12).

Figure 7.12. *Future trends in information theory with machine learning.*

Source: https://mvjce.edu.in/wp-content/uploads/2024/09/The-Future-of-Machine-Learning.jpg.

7.6.1. Quantum Information Theory

The ground of quantum information theory is rooted firmly in the quantum mechanics of physics, which itself fundamentally transforms information processing from a classical point of view. At its core is the concept of qubits, the quantum equivalent to the classical bits. Unlike traditional bits that are in a state of 0 or 1, qubits exist in multiple states simultaneously due to the superposition property. This allows quantum systems to handle sophisticated calculations at velocities even the fastest traditional computers can't reach. Furthermore, entanglement—a condition where states of qubits get entwined—allows for more advanced communication protocols and error correction. As technologies quantum continue to grow and mature, an understanding of these core principles is not just theoretically challenging but also practical engineering concepts for secure communication and computation in the context of quantum computing (Pasupuleti, 2025). In the midst of quantum information theory development is the maturity of robust error-correcting codes designed to offset the inner fragility of quantum states. Unlike classical error correction, where one is generally allowed to use redundancy to recover lost or corrupted information, quantum error correction must account for the no-cloning theorem and susceptibility of quantum information. This necessitates new methods, such as surface codes and stabilizer codes, that permit the fidelity of quantum information over many channels without seriously diminishing the impact of noise. Additionally, the implementation of such codes in fault-tolerant architectures is crucial towards achieving operational quantum computers that can run intricate algorithms in a reliable manner. Research over the years has shown great advancements in this direction, not only highlighting theoretical foundations but also actual implementations necessary for the scaling-up of quantum computer technologies (Pasupuleti, 2025). The application of quantum software design patterns is an indication of the collaborative work involved in taking theoretical underpinnings to practice in the area of quantum information theory. As quantum algorithms become increasingly sophisticated and there is a demand for efficient algorithmic organization, design patterns such as initialization, uniform superposition, oracle mechanism, entanglement, and uncomputation provide developers with standardized approaches to building quantum circuits. Recent quantum code repository trends like Qiskit and OpenQASM show that specific patterns, specifically initialization and uniform superposition, predominate recent quantum software implementation. Such a trend reflects standardization in practices of quantum software development that researchers and engineers can follow to develop more efficient and accessible quantum programs. Through strict scrutiny of these design patterns, the field can expect to stimulate continuous innovation and improvement of the efficiency of quantum computing systems.

7.6.2. Advances in Machine Learning and AI

Recent developments in machine learning (ML) and artificial intelligence (AI) have impacted the engineering fields, such as coding and information theory, by a great margin. These developments have enabled highly efficient algorithms for handling massive amounts of data, thereby enhancing information encoding and decoding techniques' accuracy and rate of processing. The advent of ML models has enabled engineers to develop systems that can learn and adapt to their environment, optimizing performance for a variety of applications. With developments in technologies such as neural networks, they are being applied on a large scale in applications ranging from error correction, data compression, to predictive modeling. Such computational complexity, however, raises pressing questions about the moral meaning of excessive use of AI-based decision-making in engineering work, and there has been a demand for more inclusive frameworks ensuring transparency and accountability to AI use in this field. Machine learning also intersects with emerging technologies that build immersive environments in virtual worlds, loosely aggregated as the metaverse. With the use of augmented and virtual reality technologies, engineers can test out ideas in information theory in such virtual worlds, paving the way for data representation and transmission innovation.

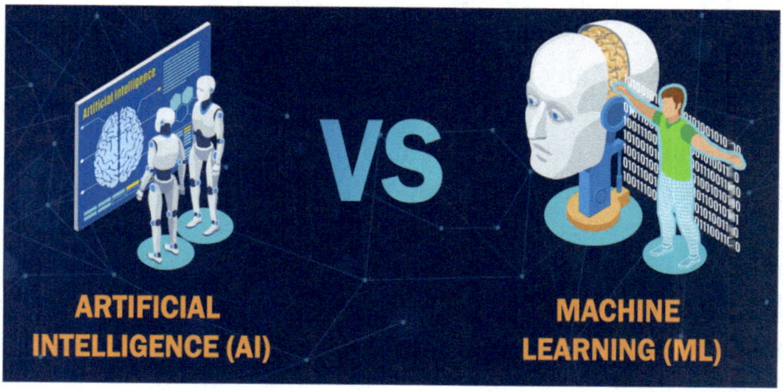

Figure 7.13. *Machine learning (ML) and artificial intelligence (AI).*

Source: https://futureskillsacademy.com/wp-content/uploads/2024/03/Artificial-intelligence-AI-vs.-machine-learning-ML-1.png.

The ability of the metaverse to blur the boundaries between physical and virtual interactions redefines the methods used in data coding and transmission since information systems must provide room for instant input from dissimilar sources and contexts. This convergence of AI, ML, and the metaverse signifies the necessity of cross-disciplinary research in engineering, as behavioral science contributions can help the development of adaptive systems to become more immersive and informative for users, as reflected in the growing number of papers on the topic. While there are potential benefits, swift development in AI is accompanied by significant challenges and ethics. The increasing complexity of ML systems has a tendency to obscure the decision-making process behind, leading to further exacerbation of problems such as bias, disinformation, and privacy violations. Engineers and stakeholders must address these challenges while maintaining powerful frameworks for accountability and ethics on the front burner. Also, the intersection of AI application and social impact calls for the analysis of trust and transparency across engineering fields on a number of levels. These are, as seen in existing research, critical to appropriately actualizing the potential of ML and AI technologies while mitigating risks and addressing the legal issues at play in their applications in terms of such fields as information theory and coding (Figure 7.13).

7.6.3. The Role of Big Data in Information Processing

Big Data has transformed the landscape of information processing dramatically by enabling the collection, storage, and analysis of big data from diverse sources. From an information theory perspective, Big Data enhances the ability to derive useful information from complicated data sets, thereby making data processing systems more efficient and effective. For instance, the introduction of graph signal processing (GSP) technology has witnessed improved ways of processing and analyzing inherently imbalanced data, e.g., data that are gathered using sensor networks and biological systems. Such advances emphasize the synergy between traditional digital signal processing and Big Data, paving the way for groundbreaking applications in many fields where data interpretation and learning algorithms can best shine, rendering Big Data a key component in current information processing paradigms (Frossard et al., 2018). Besides, integrating Big Data into information processing systems facilitates massive innovations in communication systems, especially fifth-generation (5G) wireless systems.

Using cloud radio access networks (C-RANs) is a case where Big Data can be applied to improve resource allocation and spectral efficiency, leading to an improved user experience. With fronthaul constraints, usage of Big Data-inspired techniques like compression, quantization, and joint processing is essential to circumvent the issue due to bandwidth constraints. Thus, real-time processing of vast amounts of data becomes crucial to attain quality of service with maximum energy efficiency. This convergence of Big Data and communication technologies highlights the indispensable role of robust information processing techniques to deal with today's demands (Lau et al., 2015). Impacts of Big Data reach beyond real-time processing constraints to longer-term strategic advantage in engineering disciplines. Its use in predictive analytics, for instance, allows businesses to forecast trends and make data-driven decisions, enhancing operational efficiency and innovation. In information theory, the vast reservoirs of data make it possible to refine existing methods of coding and design new algorithms that can effectively process and interpret data streams. This key mechanism of feedback guarantees a continuing process of improvement in techniques of processing information, and this, in turn, has impact on engineering applications in a wide variety of fields from smart cities to medical applications. As the size and intensity of data keep growing, Big Data integration of information processing will keep playing a pivotal role in advancing the frontiers of engineering and technology, thereby shaping the future of data (Figure 7.14).

Figure 7.14. *Role of big data in information processing.*

Source: https://external-content.duckduckgo.com/iu/?u=https%3A%2F%2Fdata-flair.training%2Fblogs%2Fwp-content%2Fuploads%2Fsites%2F2%2F2019%2F12%2FBig-Data-Job-Roles.jpg&f=1&nofb=1&ipt=1c9a82eac3d7e563fcdf05e02bf725cd4676b95030a96993503e64b2bbad8f14.

SUMMARY

The study of information theory and coding within engineering reveals the inherently transmutational nature of such fields within much of the technological industry. With a summary of what was discovered throughout this research, we find that information theory applications greatly exceed theoretical frameworks; information transmission is improved by them in terms of effectiveness and efficiency. Increased data complexity and numbers call for advanced coding methodologies that can keep pace with evolving digital environments. In particular, the novel solutions to data representation, as witnessed through the rise of deep learning frameworks, mark a paradigm shift in program analysis and the attainment of more robust information systems. Hence, it is imperative to determine how such advances can influence future work on engineering applications. Besides, the meeting point of information theory and working coding practices indicates continued need for advancing techniques and technologies. As seen, the suggested coding criterion for program representation not only improved classification accuracy but fortified theoretical bases for deep learning application to program analysis (Jin et al., 2014). This innovation depicts the strength that emerges when traditional concepts of information theory are paired with modern machine learning algorithms to deliver improved outcomes. Different methodologies compared here highlight the merits and challenges pertaining to different coding schemes, making it evident that more studies are needed to enhance these approaches. Thus, the outcomes of this study strongly favor a synergy-based approach to coding and information theory that fosters innovation. Lastly, application of coding techniques to digital images most clearly demonstrates the broader relevance of these theoretical ideas to engineering practice. By leveraging entropy coding techniques and utilizing rate-distortion theory, advances in compression of digital images have been realized with improvements in data integrity and representation that are well into orders of magnitude (Samcovic et al., 2015). The need for robust solutions that account for variability in signal quality and the effect of noise remains crucial in ensuring efficiency in the transmission of data. The implications obtained through the performance analysis of the proposed model with seabed images justify the applicability of information theory with various datasets. This crossing over of theory and practice not only shows the enduring relevance of coding methods to engineering but also promises the beginnings of further studies that could employ such ideas to resolve some of the most intractable issues of information management.

Research in coding in engineering and information theory is fundamentally concerned with the transmission, processing, and protection of data in communication systems. One of the key concerns in this area is the study of bandwidth efficiency and error correction schemes, which significantly enhance the integrity of data transfer. One significant development in this area is the development of coding schemes that minimize redundancy and maximize data integrity. In the case of next-generation wireless communications technologies, such as the scheduled rollout of 5G networks, these strategies are becoming increasingly indispensable. The unification of advanced coding methods allows more data rates and more channel capacity, hence addressing the growing bandwidth requirements of various applications, such as mobile and Internet-of-Things systems. Another leading concept in information theory is that of physical layer security, which has received extensive research focus in recent times. This approach depends on the randomness inherent in transmission channels to utilize robust security mechanisms at the physical layer of communication systems. Physical layer security is highlighted by recent research as a new technique for maintaining data confidentiality, especially by the emergence of future threats brought about by technologies such as enormous multiple-input multiple-output (MIMO) and millimeter wave-based communications for 5G networks. The transformation these technologies can contribute to security models is tremendous since they possess built-in capabilities in counteracting eavesdropping at high rates. Consequently, ongoing research in this area is necessary to resolve unaddressed technical problems and building security frameworks suitable for upcoming communication standards (Caire et al., 2018). Research on studying the exploration of user experience in information systems has also determined the need for improved access to knowledge, particularly in schools. A study conducted at Imperial College Londons Library analyzed the behavior of students in seeking information and hence untangled significant findings regarding their preferences and challenges when making use of library resources. An understanding of such user behavior helps make interfaces like library catalogs more effective, thereby making accessibility to academic literature in general better. As information theory and coding continue to make advancements, impacting engineering practice, the same principles can be utilized to be applied to information systems such that tool design and resources efficiently meet the user needs. Bridging the gap between theoretical advances in information transmission and practical application in information retrieval will ultimately yield improved learning outcomes and user engagement (Barron et al., 2017).

REFERENCES

1. Abbas, R., Ahmad, I., Ali, S., Amuru, S., et al. (2020). 6G white paper on machine learning in wireless communication networks. http://arxiv.org/abs/2004.13875 (accessed on 12th September 2025).
2. Adams, R., Amafabia, D., Barker, T., Beka, A., et al. (2019). Proceedings of Abstracts Engineering and Computer Science Research Conference 2019. https://core.ac.uk/download/287582059.pdf (accessed on 12th September 2025).
3. Arazy, O., Baran, P., Baym, N., Bazeley, P., Bignell, K., Boyd, D., Bryman, A., et al. (2014). A qualitative study of stakeholders' perspectives on the social network service environment. https://core.ac.uk/download/29139553.pdf (accessed on 12th September 2025).
4. Arcuri, A., Rector, A. L., Wood, A. M., Glas, A. S., Kulis, B., Cortes, C., Sammut, C., et al. (2018). Beyond volume: The impact of complex healthcare data on the machine learning pipeline. http://arxiv.org/abs/1706.01513 (accessed on 12th September 2025).
5. Bai, B., Huang, X., Lin, L., Ma, X., et al. (2013). Accessible capacity of secondary users. http://arxiv.org/abs/1012.5197 (accessed on 12th September 2025).
6. Barron, S., Larose, K., Mackenzie, S., Preater, R., et al. (2017). Library search UX report summer 2016. https://core.ac.uk/download/78075187.pdf (accessed on 12th September 2025).
7. Caire, G., Gao, X., Khisti, A., Wong, K. K., et al. (2018). A survey of physical layer security techniques for 5G wireless networks and challenges ahead. https://core.ac.uk/download/158170707.pdf (accessed on 12th September 2025).
8. Calderbank, R., & Vahid, A. (2016). Two-user erasure interference channels with local delayed CSIT. http://arxiv.org/abs/1503.03449 (accessed on 12th September 2025).
9. Chang, C. L. (2021). Development of physics-based transition models for unstructured-mesh CFD codes using deep learning models. AIAA AVIATION 2021 Forum. https://www.semanticscholar.org/paper/818257d9c0ae9596a28769d0cfbc09fa9b5ee930 (accessed on 12th September 2025).
10. Cheng, S., Liveris, A. D., Stankovic, V., Sun, Y., et al. (2008). Nested turbo codes for the Costa problem. https://core.ac.uk/download/9022774.pdf (accessed on 12th September 2025).

11. Condo, C. (2013). Concatenated Turbo/LDPC codes for deep space communications: Performance and implementation. https://core.ac.uk/download/pdf/11432814.pdf (accessed on 12th September 2025).

12. D'Angelo, F., Damiano, L., Leoni, L., et al. (2013). Chemical communication between synthetic and natural cells: A possible experimental design. https://core.ac.uk/download/589959131.pdf (accessed on 12th September 2025).

13. Dwivedi, Y. K., Hughes, L., Baabdullah, A. M., Ribeiro-Navarrete, S., Giannakis, M., Al-Debei, M. M., Dennehy, D., et al. (2022). Metaverse beyond the hype: Multidisciplinary perspectives on emerging challenges, opportunities, and agenda for research, practice and policy. International Journal of Information Management, 66, 102542–102542. https://doi.org/10.1016/j.ijinfomgt.2022.102542 (accessed on 12th September 2025).

14. Dwivedi, Y. K., Kshetri, N., Hughes, L., Slade, E., Jeyaraj, A., Kar, A. K., Baabdullah, A. M., et al. (2023). Opinion paper: "So what if ChatGPT wrote it?" Multidisciplinary perspectives on opportunities, challenges and implications of generative conversational AI for research, practice and policy. International Journal of Information Management, 71, 102642–102642. https://doi.org/10.1016/j.ijinfomgt.2023.102642 (accessed on 12th September 2025).

15. Effros, M., Ho, T., Karger, D. R., et al. (2008). Byzantine modification detection in multicast networks with random network coding. https://core.ac.uk/download/4878953.pdf (accessed on 12th September 2025).

16. Ertem, T., Gunduz, D., & Varasteh, M. (2016). Zero-delay joint source-channel coding in the presence of interference known at the encoder. https://core.ac.uk/download/77016207.pdf (accessed on 12th September 2025).

17. Fluke, C. J., Vernardos, G., & Vohl, D. (2015). Data compression in the Petascale Astronomy Era: A GERLUMPH case study. http://arxiv.org/abs/1505.05617 (accessed on 12th September 2025).

18. Fong, A. C. M., Imran, N., & Seet, B. C. (2015). Distributed video coding for wireless video sensor networks: A review of the state-of-the-art architectures. https://core.ac.uk/download/46558677.pdf (accessed on 12th September 2025).

19. Frossard, P., Kovačević, J., Moura, J. M. F., et al. (2018). Graph signal processing: Overview, challenges and applications. http://arxiv.org/abs/1712.00468 (accessed on 12th September 2025).

20. Gao, J., Chen, S., Wu, Y., Liu, L., Caire, G., Poor, H. V., & Zhang, W. (2025). Finite-blocklength information theory. https://www.semanticscholar.org/

paper/ad36d4666e042ff8da75df90d8fa4b9f0b85380b (accessed on 12th September 2025).

21. Hanzo, L., Shen, W., Wang, X., & Zou, et al. (2013). Security versus reliability analysis of opportunistic relaying. http://arxiv.org/abs/1311.6020 (accessed on 12th September 2025).
22. Hassibi, B., & Vikalo, H. (2002). Maximum-likelihood sequence detection of multiple antenna systems over dispersive channels via sphere decoding. https://core.ac.uk/download/4876842.pdf (accessed on 12th September 2025).
23. Ingels, F. M., & Shahsavari, M. M. (2025). A Study of Data Coding Technology Developments in the 1980–1985-Time Frame, Volume 2. https://core.ac.uk/download/pdf/42871670.pdf (accessed on 12th September 2025).
24. Jin, Z., Li, G., Liu, Y., et al. (2014). Building program vector representations for deep learning. http://arxiv.org/abs/1409.3358 (accessed on 12th September 2025).
25. Kalita, K., Ramesh, J. V. N., Čepová, L., Pandya, S. B., Jangir, P., & Abualigah, L. (2024). Multi-objective exponential distribution optimizer (MOEDO): A novel math-inspired multi-objective algorithm for global optimization and real-world engineering design problems. Scientific Reports, 14. https://www.semanticscholar.org/paper/c9b37d8d20d05ac32fd588339a35c902fd0862e5 (accessed on 12th September 2025).
26. Kang, J., Lee, W., Popovski, P., et al. (2015). HARQ buffer management: An information-theoretic view. http://arxiv.org/abs/1501.06166 (accessed on 12th September 2025).
27. Katz, J., & Lindell, Y. (2020). Introduction to Modern Cryptography. https://www.semanticscholar.org/paper/ca8de6c6b6ac63c882909a56b38c7f3ed99c4de0 (accessed on 12th September 2025).
28. Khan, B., Aslam, S., & Fatima, R. (2024). Exploring the themes of horrors of the war, power and hegemony in the manga "Barefoot Gen Vol. 1" in the light of dual coding theory. Academy of Education and Social Sciences Review. https://www.semanticscholar.org/paper/34de8a3f40ef4b9da7695a2d53571c7ab44f9285 (accessed on 12th September 2025).
29. Lam, M., Lee, K., Papailiopoulos, D., et al. (2018). Speeding up distributed machine learning using codes. http://arxiv.org/abs/1512.02673 (accessed on 12th September 2025).

30. Lau, V., Peng, M., Poor, H. V., et al. (2015). Fronthaul-constrained cloud radio access networks: Insights and challenges. http://arxiv.org/abs/1503.01187 (accessed on 12th September 2025).

31. Levi, F. J. (2023). Applied Quantum Mechanics. https://doi.org/10.1017/9781009308083 (accessed on 12th September 2025).

32. Martinian, E., & Wornell, G. W. (2004). Source coding with fixed lag side information. http://arxiv.org/abs/cs/0410008 (accessed on 12th September 2025).

33. Mian, I. S., & Rose, C. (2016). Inscribed matter communication: Part I. http://arxiv.org/abs/1606.05023 (accessed on 12th September 2025).

34. Mo, T., Li, W., & Zhang, L. (2025). Teaching practice and innovation in the professional master's degree program in electronic information at Peking University in the era of artificial intelligence. 2025 IEEE International Conference on Software Services Engineering (SSE). https://www.semanticscholar.org/paper/07957154d36a030ddb92cf88361362aa0c[illegible]d[illegible]e2c (accessed on 12th September 2025).

35. Moreira, P. H. de L. R., de Siqueira, R. N. C., & Vilani, C. (2024). A simple chemical equilibrium algorithm applied for single and multiple reaction systems. Computer Applications in Engineering Education, 32. https://www.semanticscholar.org/paper/0032f8252d12e54a[illegible]80d6d44[illegible]135fa51936cf00 (accessed on 12th September 2025).

36. Narváez, C. A., & Solarte Solarte, C. M. (2024). Analysis of the organizational culture of the companies of the massive cargo transportation sector in the city of Pasto. Management (Montevideo). https://www.semanticscholar.org/paper/261e661bf190e10534aef37cb13f256eaf463436 (accessed on 12th September 2025).

37. Nguyen, V. D. (2024). Studying the influence of correction codes on coherent reception of M-PSK signals in the presence of noise and harmonic interference. Russian Technological Journal. https://www.semanticscholar.org/paper/3919c799f525769712c3bb375ee6baff6ebe1aee (accessed on 12th September 2025).

38. Pasupuleti, M. K. (2025). Quantum systems for the 21st century: Theory, technology, and applications. International Journal of Academic and Industrial Research Innovations (IJAIRI). https://www.semanticscholar.org/paper/81070fe4ef5a19ccbd2b800f225c4e7359590faa (accessed on 12th September 2025).

39. Pérez-Castillo, R., Fernandez-Osuna, M., Cruz-Lemus, J., & Piattini, M. (2024). A preliminary study of the usage of design patterns in quantum software. 2024 IEEE/ACM 5th International Workshop on Quantum

Software Engineering (Q-SE). https://www.semanticscholar.org/paper/12e5f48d4e97bda9708ba943e729e3390d0f6364 (accessed on 12th September 2025).

40. Pujol Capdevila, J., & Villanueva, M. (2015). Binary codes: Binary codes databases. https://core.ac.uk/download/154765973.pdf (accessed on 12th September 2025).

41. Pujol Capdevila, J., & Villanueva, M. (2017). Q-ary codes. https://core.ac.uk/download/154765975.pdf (accessed on 12th September 2025).

42. Quan, Y. (2010). Cooperative symbol-based signaling for networks with multiple relays. https://core.ac.uk/download/70419449.pdf (accessed on 12th September 2025).

43. Rahat, K., & Sharma, H. (2025). Research trends in visual impairment: A bibliometric analysis of Indian contributions. International Journal for Multidisciplinary Research. https://www.semanticscholar.org/paper/d5d43f9b51c84d407ad0e7db4864370abb379014 (accessed on 12th September 2025).

44. Rajagopal, K., Renyi, A., Shannon, C. E., Tsallis, C., et al. (2001). Information theory based on non-additive information content. http://arxiv.org/abs/cond-mat/0010074 (accessed on 12th September 2025).

45. Ruiz-Alzola, C., Alberola-López, C., & Westin, C. (2013). Advanced signal processing methods for biomedical imaging. International Journal of Biomedical Imaging, 2013. https://www.semanticscholar.org/paper/32df9e61d2090603dbed9bf6ba9e815de0564b27 (accessed on 12th September 2025).

46. Samcovic, A. (2015). Model for estimation of bounds in digital coding of seabed images. https://core.ac.uk/download/30312879.pdf (accessed on 12th September 2025).

47. Such, O., Fabricius, R., Klimo, M., & Juhár, J. (2020). Innovating instruction of communication theory with machine learning and speech analysis. 2020 18th International Conference on Emerging eLearning Technologies and Applications (ICETA). https://www.semanticscholar.org/paper/12a8389db1e5a2d71a26d89249127e7ed10845e3 (accessed on 12th September 2025).

48. Von Zur Gathen, J., & Gerhard, J. (2013). Modern Computer Algebra. https://doi.org/10.1017/cbo9781139856065 (accessed on 12th September 2025).

49. Witzany, G. (2017). Artificial and natural genetic information processing. https://core.ac.uk/download/186332662.pdf (accessed on 12th September 2025).

50. Wolff, J. G. (2015). Information compression, intelligence, computing and mathematics. http://arxiv.org/abs/1310.8599 (accessed on 12th September 2025).

51. Zhu, J. Y., Park, T., Isola, P., & Efros, A. A. (2017). Unpaired image-to-image translation using cycle-consistent adversarial networks. https://doi.org/10.1109/iccv.2017.244 (accessed on 12th September 2025).

CHAPTER 8

TRANSMISSION MEDIA AND CHANNELS

Contents

INTRODUCTION

The advent of new communication technology has introduced an era where transmission channels and media are prerequisites for effective information exchange. The various transmission media, i.e., wired, wireless, and optical, are major elements of communication systems that allow data transfer across different environments. As connectivity is the foundation of our modern society, an understanding of these mediums offers us a chance to see how communication structures function, from their inherent limitations to their functional implementations. As complicated as it gets, a detailed study of transmission media is the basis needed for comprehending the broader entity of communication systems, which eventually decides how organizations and people are connected in an increasingly interconnected world. Furthermore, interoperability between different transmission channels is also essential in making the communication barrier-free between platforms.

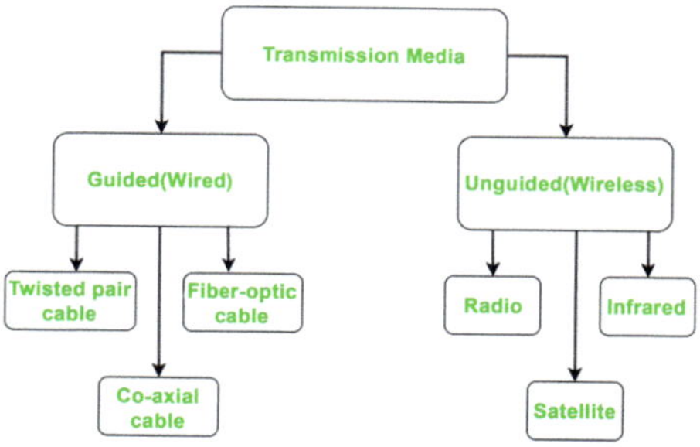

Figure 8.1. *Transmission media.*

Source: https://media.geeksforgeeks.org/wp-content/uploads/20210924165[illegible]1/ transmission.jpg.

With advancements in smart technologies and the Internet of Things (IoT), protocols like ZigBee demonstrate that there is potential for communication protocols to act as the intermediary between heterogeneous devices and networks. ZigBee, in its definition by adherence to IEEE 802.15.4, talks about the applicability of low-power, short-range communication solutions in enhancing network efficiency. This suggests the fundamental importance of selecting the correct transmission channels, as compatibility and performance are direct predictors of the effectiveness of communication systems. These become particularly important in advanced environments, where aggregating the application of numerous disparate technologies requires a discerning appreciation for the strength and weakness of each channel. Beyond this, safety remains a top consideration in the design and deployment of communication systems, particularly in safety-critical infrastructural applications like rail transit. Employment of proprietary safety procedures, such as RSSP-1 and RSSP-2, highlights the importance of being certain about reliability in signal transference to protect against potential risk. As highlighted in recent trends, the adoption of technologies such as FPGAs has the capacity to revolutionize the transmission of safety information by greatly reducing expenses and enhancing the reliability of communication channels. A procedure-based approach to safety information transfer not only enhances the effectiveness of communication systems but also addresses the critical need for personnel and asset protection in various industry sectors. Hence, effective knowledge regarding transmission media, its associated channels, and safety procedures is required for designing effective communication systems (Figure 8.1).

Definition of Transmission Media

Transmission medium is the physical path along which signals are propagated in a communication system. It can take numerous forms that consist of cables, optical fibers, and wireless channels, each being intended for specific needs in terms of bandwidth, reach, and environmental factors. Twisted-pair cable, coaxial cable, and fiber optics each possesses distinct characteristics that dictate data transmission speeds and susceptibility to electromagnetic interference. The efficiency of these forms of transmission is critical in the determination of the efficiency of communication systems overall, and with fiber optics, to some degree, there is a lot of bandwidth capacity due to less attenuation of light waves (Renzo et al., 2020). Identification of the forms of transmission media is crucial in the determination of their uses and limitations in modern communication frameworks. The transmission media type impacts the reliability of the systems and integrity of the data significantly. Even though wired media such as copper cables are typically more vulnerable to environmental factors and physical interference, fiber-optic cables provide advantages of signal constancy over lengths and velocity. Nevertheless, wireless transmission media has unique challenges like signal attenuation, interference, and security vulnerability. These factors necessitate strategic selection of transmission medium to maximize communication efficacy. Additionally, technological advancements, such as the development of reconfigurable intelligent surfaces (RISs), are transforming traditional paradigms by offering state-of-the-art solutions for maximizing transmission capacity in wireless environments (Renzo et al., 2020). Communications engineers and technologists need to have a clear understanding of these kinds of media. In the background of evolving technological surroundings, functionality and definition of the transmission media also need to adapt to emerging needs and uses, particularly under increasing wireless communication dependency. Bandwidth, latency, and cost determination drive continued innovations and advances in transmission technologies. As emphasized in the guidelines for the optimal communication practices, medium selection must be consistent with the target application to support both reliability and efficiency. As such, the continuous research that has surrounded transmission media not only assesses the current technologies but also seeks emerging methods, such as hybrid systems that may guarantee data delivery in complicated environments. Lastly, a detailed knowledge of transmission media and what defines it is the foundation for building additional communication systems and technology.

Importance of Communication Channels

Effective communication channels are essential in the transmission of information, particularly in times of public health crisis such as the COVID-19 pandemic. Communication channels are the means by which essential messages are relayed to the general public and affect their behavior and attitudes. The importance of utilizing available and credible means of communication cannot be overestimated, as evidenced by the WHOs assertion that when messages are perceived to be consistent, competent, and empathetic, individuals are better placed to make informed decisions regarding their health (Betsch et al., 2020). Misinformation and lack of clarity in communication can lead to confusion, mistrust, and non-compliance that can worsen the situation. Therefore, having well-framed communication structures is crucial to channel the public towards healthy behaviors, particularly during a crisis where concise and clear information can be a matter of life and death. Further, what social media and technology can contribute in terms of communication cannot be ignored. Today, in the modern landscape, these platforms have become a vital tool to reach out to people in huge numbers within short spans of time. But the dual characteristics of these tools bring both strengths and vulnerabilities. While they facilitate rapid transmission of vital health information, they also pose the risk of rapid dissemination of disinformation that can dilute public health interventions (James et al., 2020). This necessitates health authorities not only to leverage such platforms but also to carefully monitor and counteract disinformation that will arise. The ability to communicate two-ways through social media also offers a platform for health practitioners to directly address public concerns and establish trust, which is key in the provision of compliance among communities with health recommendations. Lastly, integration of communication strategies among various sectors is still a necessary component towards the optimization of public health messages. As witnessed with the pandemic, lack of unified messaging from institutions can lead to disjointed awareness of health policies therefore affecting compliance (Betsch et al., 2020). There needs to be concerted effort from the government agencies, health practitioners, schools, and the media for there to be a single message front. If these institutions collaborate in unifying the message, the message becomes clearer and genuine, and the public is able to participate better while preventing anxiety during uncertain times. In this regard, examination of the convergence of communication channels not only determines their applicability for health information dissemination, but also ability to make contributions to a collaborative public health program response (Figure 8.2).

Figure 8.2. *Importance of communication channels.*

Source: https://www.consultationmanager.com/wp-content/uploads/2023/08/Tailor-your-messaging-to-specific-stakeholder-groups-2000x1415.png.

8.1. TYPES OF TRANSMISSION MEDIA

The types of transmission media are crucial components of communication systems because they define the way data is transmitted between devices. Transmission media, on a general note, can be categorized into guided and unguided media. Guided media, such as twisted pair cables, coaxial cables, and fiber-optic cables, involve a physical medium to transport signals. Twisted pair cables, which are frequently used in telephonic communications, are composed of pairs of wires that are twisted with each other for the specific reason of reducing interference. Coaxial cables are utilized for carrying television signals and internet data, while fiber-optic cables possess huge bandwidths and are increasingly utilized for high-speed internet and long-distance communication. Unguided media like radio waves, microwaves, and infrared signals all rely on electromagnetic wave propagation in air. Understanding these types is helpful in selecting the suitable mediums for fulfilling specific communications needs in various situations. Recent advances in transmission technologies have also created new systems, such as reconfigurable intelligent surfaces (RISs). RISs, due to their capacity to include massive arrays of cost-effective antennas, can customize wireless communication conditions using fewer resources than traditional technologies. In enabling smart radio environments (SREs), they enhance signal quality, optimize efficiency, and assist with the minimization of wireless communication system costs. The revolutionary design of RISs makes it possible to implement very powerful communication networks with the least possible power and space requirement. This advancement in transmission media demonstrates a major tendency towards cost-effective and green communication solutions that can change the present scenario of wireless technology to a large extent (Renzo et al., 2020). Moreover, it cannot be ignored that social media usage in the realm of transmission media cannot be overlooked, particularly during global crises, such as the COVID-19 pandemic. The study said that social media platforms, such as Twitter, are necessary in order to facilitate rapid information dissemination. The study on the information that is spread regarding COVID-19 via social media led to the discovery of healthy practices of communication. Specifically, the user network sharing news of a health nature signaled a preference for medically framed content and highlighted the critical function served by transmission media within public information campaigns (Park et al., 2020). The ability of these sites to facilitate connectivity and allow for real-time information exchange highlights their function within the broader transmission media environment and makes them indispensable in modern communication systems (Figure 8.3).

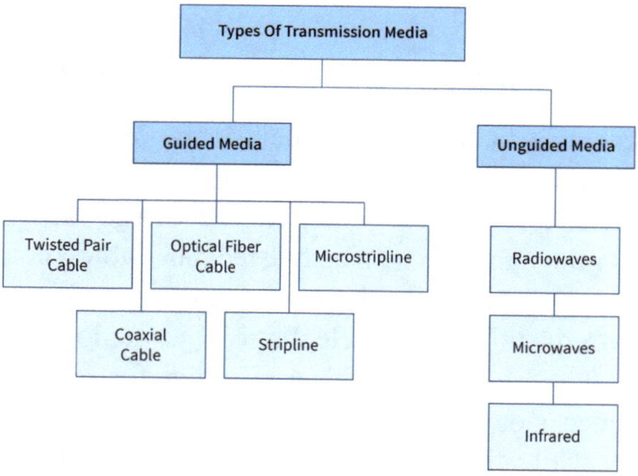

Figure 8.3. *Types of transmission media.*

Source: https://external-content.duckduckgo.com/iu/?u=https%3A%2F%2Fscaler.com%2Ftopics%2Fimages%2Ftypes-of-transmission-media.webp&f=1&nofb=1&ipt=4c0e1414cdda7f68e621a5a5015f0bc6b5730e18e37fd0784003031c55c9ee97.

8.1.1. Guided Media

Guided media is a significant medium of transmission of information in various networks of communication, providing a controlled channel where signals can travel. As compared to unguided media, i.e., radio waves, guided media consists of physical channels like cables and fiber optic cables. This arrangement provides signals fewer interferences and attenuations, thus ensuring better quality of communication. The individuality of guided media allows it to be possible to offer a wide range of applications, including computer networks, telephone networks, and television broadcasting. Each medium, such as copper cable or fiber-optic cable, has certain qualities that influence bandwidth, transmission rate, and overall capacity. The importance of guided media assumes special significance during high-demand applications, when speed and data integrity are the prime concerns, adding to its position as a pillar of contemporary communication schemes. Guided media has undergone tremendous development with innovation in the arena of transmission technologies, with the field of fiber optics being the most prominent area of development. Fiber-optic technology employs extremely thin strands of glass or plastic for conveying light signals, with nominal loss of signal over long distances. Unlike traditional copper cables, fiber optics have immensely superior bandwidth and transmission rates and enable high-definition video and high-speed internet communications. Quantum communication networks, using fiber optics for the secure transmission of data, are one such area that research has put under the limelight. With clear advantages, fiber-optic systems can meet the greater data transmission and processing requirement in a rapidly digitalizing world. The possibility of change of guided media with fiber optics proves the necessity to make continuous investment in research and technology to be able to harness its optimum extent and pliability. As guided media advances, greater emphasis is laid on integrating quantum communication techniques within mainstream transmission systems. For instance, recent studies indicate a quantum communication system capable of transmitting compressed video information in an effective way over noisy channels. It employs quantum superposition and error correction codes like three-qubit and five-qubit models for optimal performance in noisy environments. The outcome indicates that three-qubit error correction method offers significant improvements over traditional methods like 1/3 rate polar codes with significant improvement in peak signal-to-noise ratio (PSNR) and structural similarity index measure (SSIM) (Jayasinghe et al., 2025). Furthermore, learning about these new models seeks to highlight the importance of guided media in addressing future communication needs, representing a turning point towards more powerful and efficient data transmission systems (Jayasinghe et al., 2025) (Figure 8.4).

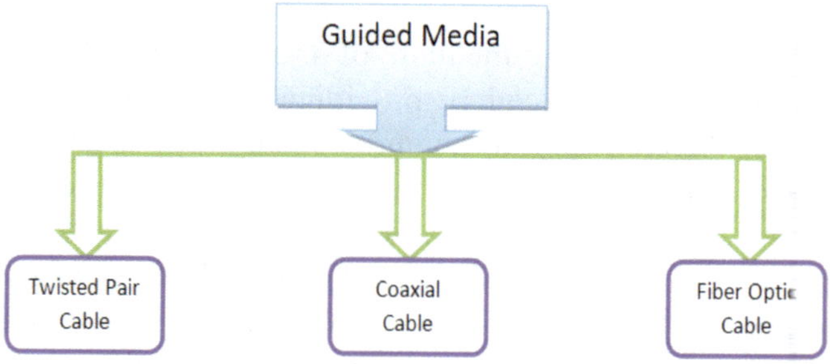

Figure 8.4. *Guided media.*

Source: https://studyopedia.com/wp-content/uploads/2017/08/Guided-Transmission-Media.png.

8.1.2. Unguided Media

Unrestricted space, or the other term used for it being unguided media or wireless communication, is an important component of the transmission medium and communication system channel. In contrast with guided media, which relies on physical mediums such as cables or fibers for its operations, unguided media employs unbounded variables like vacuum and air to facilitate the transmission of the signal data. This type of communication utilizes various forms of electromagnetic waves like radio frequency and visible light to send information over distances without the need for physical connections. With the evolution of technology, unguided media has been taking a more important role in meeting the growing demand for mobility and flexibility in communication networks, particularly in urban areas where the deployment of traditional wired facilities is burdensome and costly. The arrival of wireless optical communication (WOC) is a major improvement in the technology of unguided media, providing a novel and novel option for the data transmission technology. WOC employs light as the carrier but air as the unguided medium of propagation, differentiating itself from optical fiber communication, which depends on direct physical connections. This technology offers significant benefits such as ease of installation and the minimal amount of licensing necessary, which make it a highly appealing option for various uses. Specifically, WOC has expanded to encompass a vast array of applications from indoor communications to advanced uses in deep-space satellite systems, reflecting its versatility and capacity to cope with modern communications issues (Fate et al., 2021). Visible Light Communication (VLC), being a type of unguided media, once again offers the potential of wireless transmission systems based on light sources for illumination and data transmission. The multifunctional nature of such provision not only supports high data rates but also minimizes electromagnetic interference typical of traditional radio frequency systems (Abuella et al., 2021). The broad bandwidth offered by VLC is particularly beneficial in applications that require reliable, low-latency communication, such as vehicle safety communications and indoor networking. However, it is still a challenge, with the requirement for line-of-sight connectivity putting limitations on the performance of the systems in certain cases. To counteract these limitations, hybrid RF/VLC systems have been created, capitalizing on the benefits of both technologies to enhance connectivity as well as maximize network performance. This unifying vision captures the ongoing evolution of unguided media and its focal role in communication technology innovation (Figure 8.5).

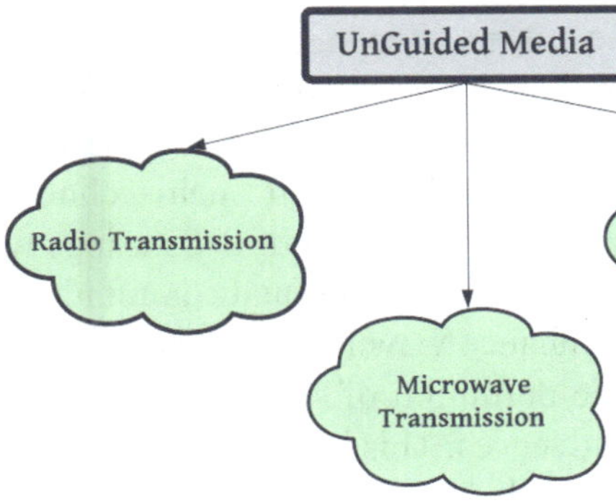

Figure 8.5. *Unguided media.*

Source: https://files.codingninjas.in/article_images/unguided-media-3-1677354173.webp.

8.1.3. Comparison of Different Types

Among media of transmission, various types possess distinctive characteristics that significantly impact communication systems. Guided media such as fiber optics and coaxial cables are such a type, with characteristics that are particularly notable for the high-bandwidth and low-loss transmission of signals. Fiber optics further make use of light signals to transmit data over considerable distances with minimal attenuation, making them well suited for backbone installations in telecommunication. On the other hand, unguided media like microwaves and radio waves are mobile and flexible in the sense that signals are transmitted through the air. Such transmission modes are usually susceptible to interference and obstruction by objects around them, degrading communication quality. Hence, an awareness of the advantages and disadvantages of various transmission media is needed to ensure high performance in networks as well as to support the requirements of new communication systems. Reconfigurable intelligent surfaces (RISs) are a groundbreaking extension of wireless communication with immense capability versus traditional methods. Unlike traditional methods that are mostly based on the employment of transmit antennas and amplifiers, RISs utilize many low-cost scatterers for management of incoming signals in a passive way. This distinctive architecture not only lowers the complexity and expense of network infrastructure, but also better-quality signals by managing the scattering, reflection, and refraction of radio waves. Such technology's effects on existing networks, as well as networks to be developed in the future, particularly in terms of 6G, are significant. Rather than only taking advantage of the existing propagation characteristics, RISs enable operators to actively control and mold the environment of the transmission medium, effectively transforming wireless network design and optimization towards better performance (Renzo et al., 2020). Furthermore, the technological innovations in transmissions highlight the need for an in-depth evaluation of such technologies against existing communication frameworks. The arrival of RISs brings with it several issues that need to be rethought in terms of traditional communication-theoretic frameworks. As these surfaces enhance wavefront property management, e.g., amplitude and phase, the classic definition of signal propagation needs to be adapted to accommodate these advances. This is particularly so in 6G networks, which will need more sophisticated solutions to handle the anticipated data traffic surge and connectivity requirements. By looking into the open research questions that challenge RIS technologies, we become familiar not only with their functionality but also with the manner in which mathematics and optimization strategies might transform to meet prospective wireless networks, and most importantly, it becomes imperative to harmonize these legacy theories with new evidence (Ertuğrul et al., 2019).

8.2. GUIDED TRANSMISSION MEDIA

Guided transmission media are integral components of communication systems, as they provide a physical medium for signal conveyance. They consist of cables in various forms such as twisted pair, coaxial cable, and fiber optic cables. They possess distinct characteristics influencing their behavior in different applications. Twisted pair cabling, for example, is typically employed in local area networks (LANs) and telecommunication networks, where its configuration works to reduce electromagnetic interference. Coaxial cables which have an insulated conductor covered with a shield layer, are usually used for cable TV and internet due to the fact that they are good at maintaining high-frequency signals over a long distance. Fiber optic cables, on the other hand, transmit data using light, so there is greater bandwidth and less attenuation, which is ideal for long-distance and high-speed communications. Guided transmission media choice affects the overall efficiency and capacity of a communication system significantly. Fiber optic technology has revolutionized the transmission media arena in guided transmission. It enables data to be transmitted at speeds and over distances not achievable with traditional metallic cables. Fiber optics utilize light instead of electrical signals, minimizing loss and susceptibility to electromagnetic interference. The fundamental working of fiber optics relies on total internal reflection of the signal in the glass or plastic core, which allows for the signal to propagate with less loss. Additionally, advances in fiber optic technology have also expanded the potential for applications in communication systems with the help of metamaterials.

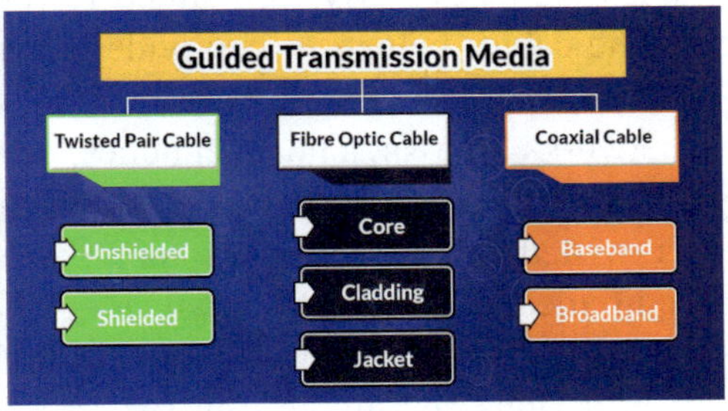

Figure 8.6. *Guided transmission media.*

Source: https://external-content.duckduckgo.com/iu/?u=https%3A%2F%2Fdata-flair.training%2Fblogs%2Fwp-content%2Fuploads%2Fsites%2F2%2F2021%2F09%2FGuided-Transmission-Media.jpg&f=1&nofb=1&ipt=f8968c64168d760467059a234ee67699dc4db410c7adffae54fadfba51a5a687.

These metamaterials can manipulate electromagnetic waves in a manner not achievable with conventional materials, providing beyond-normal-material capabilities, thus enhancing important performance metrics such as data rate and distance efficiency in communication channels (Hou-Chen et al., 2016). Efficiency in guided transmission media also depends on the design and deployment of the communication network. As businesses and organizations increasingly rely on fast internet for their operations, the utilization of the correct transmission media gains importance. For instance, where space is limited in cities, installations of fiber optic cable are prioritized at the highest level since they provide higher bandwidth capacities without the spatial requirements of copper cables. Besides, guided transmission technologies must adapt to enhanced demands for safe and fast communication based on the advancements in the Internet of Things (IoT) and smart devices. By providing robust channel infrastructure through proper selection and integration of guided media, organizations can make their communication systems more dependable while meeting growing demands for data transmission speed and security (James et al., 2020) (Figure 8.6).

8.2.1. Twisted Pair Cables

Twisted pair cables are one of the fundamental building blocks of modern communication networks, playing an important role in data transmission across distances. Insulated copper wires twisted into pairs constitute the twisted pair cables, which are meant to neutralize electromagnetic interference and enhance signal transmission quality. Inexpensive and easy to install versus other cable types, twisted pair cables are particularly well-suited for both household and business applications. The cables belong to two wider categories: unshielded twisted pair (UTP) and shielded twisted pair (STP). UTP cables are usually found in local area networks (LANs) and telephones, whereas STP cables are employed in systems where a great deal of electrical interference exists. Such flexibility allows twisted pair cables to accommodate a very wide scope of communication requirements, earning them an integral pillar in the case of transmission media and communication channels. In looking at the utility of twisted pair cables in various communication systems, it is of paramount importance to understand their bandwidth capabilities. While twisted pair cable does allow for high-speed data transfer—especially in newer versions optimized for greater effectiveness—their effectiveness can be significantly reduced over extended distances. As a rule, UTP cables are limited to a maximum distance of approximately 100 meters at maximum performance. This limitation makes it necessary to think about other cable types, like fiber optic, which has a higher degree of speed and distance, and also immunity to environmental factors. Yet their widespread use in twisted pair cables speaks volumes about their cost-effectiveness and suitability in the majority of circumstances, making them a widely accepted alternative in the majority of networking environments whose deployment size does not necessitate the performance offered by more advanced technology. Future advancements in twisted pair cable technology may well consolidate their position within the shifts in communication system dynamics. Material and structural developments, such as the application of sophisticated shielding technologies or designing hybrid models, could manage to overcome the limitation while maintaining the benefits of low-cost deployment. As increasingly more societies worldwide are relying on good-quality communication infrastructures, twisted pair cables can supplement next-generation technologies, e.g., the Internet of Things (IoT) and intelligent environments. For instance, advancements in communication system-related sensor technologies will need reliable media for transmission that can effectively deal with increased data loads and interaction complexity among devices (Hassani et al., 2023). Thus, twisted pair cables remain one of the leading fields to research and develop, reflecting their ongoing relevance to the discourse relating to transmission media and communication system channels (Jahanbakht et al., 2021) (Figure 8.7).

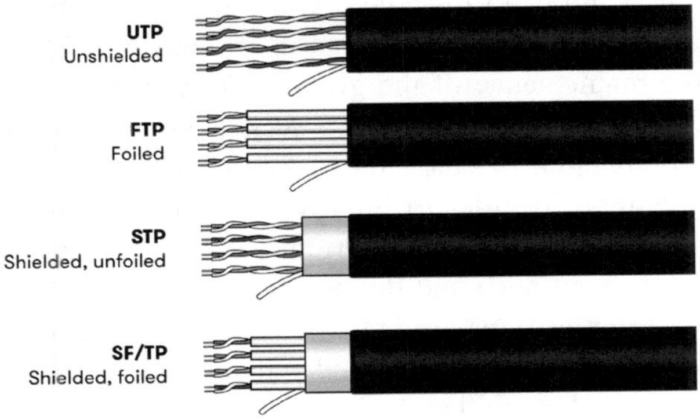

Figure 8.7. *Twisted pair cables.*

Source: https://res.cloudinary.com/rspoc/image/upload/f_auto/q_auto/v1726664302/twisted_pair_screening.png.

8.2.2. Coaxial Cables

Coaxial cables with their dedicated structure of internal conductor, insulator, outside conductor, and protection cover are a fundamental transmission medium for most communication systems. The structure protects the internal conductor from external electromagnetic interference, hence ensuring a stable signal transfer over long distances. Coaxial cables are particularly prevalent within television broadcasting, telecommunications, and broadband internet connectivity due to their ability to provide high-frequency signals with minimal loss. Their inherent structure provides more bandwidth than traditional twisted pair cables, making them a perfect option for supporting numerous data-hungry applications. Coaxial cables remain pertinent even with evolving technology since they have learned to accommodate in accordance with increasingly complex communication needs. The performance of coaxial cables is significantly influenced by parameters like material properties, cable geometry, and installation environment. For instance, the use of high-quality materials for the inner conductor, i.e., copper or aluminum, will enhance conductivity while reducing signal attenuation. Finite element analysis-based studies in recent years have reported the central importance of cable structure in determining the effectiveness of transmission. Specifically, the development of coaxial cables with an optimum geometric configuration will highly minimize signal loss at high frequency levels, a central element for applications in today's smart grids. Understanding the aforementioned factors is central towards optimizing the performance of coaxial cables in other communication systems to enable them to continue well positioned to handle the demands of emerging technologies (Mahjoob et al., 2024). Moreover, advances in coaxial cable technology are essential for the maintenance of the growing infrastructure of networks and communications. The enhanced designs—such as bridge joints that have been applied to the production of coaxial superconducting cables—exhibit steady technological progress in order to improve existing transmission efficiency. Bridge joints enable a symmetrical current flow, reducing risk factors such as concentration of resistance and thermal loads, which are the utmost when high-performance applications are being considered (Guo et al., 2025). With industries moving toward more integrated and complex systems, the evolution of coaxial cable technology will undoubtedly play an important role in ensuring communication is efficient and reliable. Lastly, the flexibility of coaxial cables shows their enduring position in the contemporary transmission media and channels of communication systems.

8.2.3. Fiber Optic Cables

The evolution of communication technology has seen a significant shift with the utilization of fiber optic cables, which have numerous advantages over traditional metallic conductors. These cables, composed of glass or plastic fibers, transport light signals along great distances with minimal loss of quality. In contrast to copper wires that are subject to electromagnetic interference and signal degradation, fiber optic cables are not susceptible to these issues, providing improved signal integrity. This characteristic makes them particularly beneficial in applications with heavy electronic noise in the environment. The amplification of optical communication highlights the growing utility of numerous applications ranging from telecommunication to data centers where data capacity is a primary concern. The advantages of fiber optic technology are not theoretical; they have also been validated by universal research comparing conventional electrical systems and the optical systems, underlining the many benefits that fiber optics offers to modern communications networks (St. Denis, 2025). The most astounding feature of fiber optic technology is that it has the ability to enable high-bandwidth applications, which is vital in today's data age world. For instance, the Fiber Distributed Data Interface (FDDI) provides an incredibly high 100 Mbit/s network interface for real-time data, voice, and video transmission. Such high-speed, interrupt-driven hardware topology to support such features results from efficient handling of stream and packet data transfer. Such functionalities are essential in the handling of on-demand video streaming and huge data transfer, activities that are increasingly widespread in business and academic circles. Additionally, modular single-mode laser-diode links enhance node separation, allowing for the vast flexibility when designing networks.

This flexibility, coupled with technological innovations like optical bypass switches, results in improved fault tolerance and network reliability, to the extent that high levels of performance are not only realized but also maintained in the long term (Bergman et al., 2025). As businesses come to appreciate increasingly the value of trustworthy systems of communication, fiber-optic cable integration becomes not just advantageous but essential. These cables not just facilitate faster data transmission but also more effective use of resources, whether physical or operational. For applications that require continuous connectivity—like remote system monitoring or high-speed trading—fiber optics are the solution, since they provide the appropriate degree of reliability and speed. In addition, greater deployment of fiber optic networks is just one aspect of a broader trend where industries are investing in future-proof technologies. With increasing demand for bandwidth and increasing network complexity, fiber optic cable steps into the limelight, enabling the systems to carry increasing loads of data while remaining immune to threats that have accrued with electrical systems. Their role in the transmission media cannot, therefore, be overstated (St. Denis, 2025) (Figure 8.8).

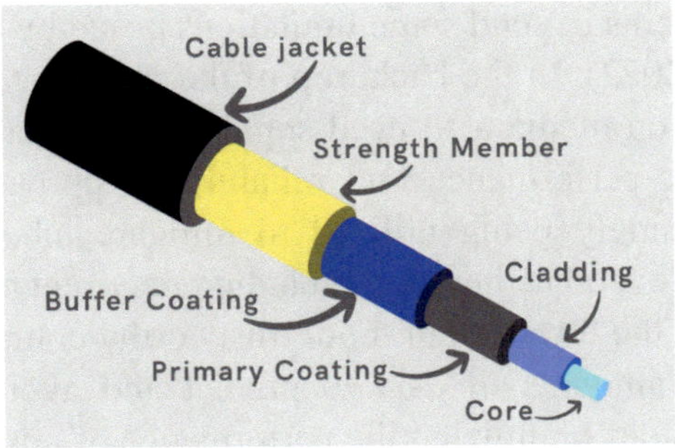

Figure 8.8. *Fiber optic cables.*

Source: https://cdn.fiberopticx.com/wp-content/uploads/2024/01/Components-of-optical-fiber-cable.webp.

8.3. UNGUIDED TRANSMISSION MEDIA

In communication systems, unguided media of transmission play a significant role in allowing electromagnetic waves to pass freely in space or air without any physical conduit. This medium comprises a number of wireless means of communication such as radio waves, microwaves, and infrared signals. The advantages of unguided transmission are aplenty, particularly in their capacity to offer information over long distances with relatively low installation costs. They are also susceptible to interference by environmental factors like weather and physical obstacles, which induce signal loss. With increasing requirements for interconnectivity, particularly with the wanton increase in the Internet of Things (IoT), the relevance of unguided transmission media is even greater. By providing device-to-device connectivity without the constraints of wired transmission, unguided media facilitate seamless integration into various applications like smart homes and industries. The emergence of optical wireless communication (OWC) technology is a fascinating trend in unguided transmission media. Unlike traditional radio frequency systems, which are plagued by shortage of spectra and interference with a high number of devices, OWC employs light waves for data transmission. This approach offers much bandwidth capacity and minimum propagation loss, thus improving overall network performance. Under the scenario of IoT, OWC presents unique opportunities for enhancing connectivity within other fields, including underwater and biomedical applications. Combining OWC with traditional wireless technology indicates its potential as a complement and alternative to radio-based systems. While research goes on to reveal the actual applications of OWC in the real world, the integration of these advanced technologies is poised to transform the realm of unguided transmission media and expand the functionality of modern communication systems beyond some limitations posed by traditional methods (Abdulkadir et al., 2022). In the backdrop of these advancements, issues with unguided transmission media also need sophisticated data-driven approaches to enhance network performance and reliability. Approaches such as deep learning are increasingly being utilized to mitigate inherent limitations of wireless networks. Several techniques, including recurrent neural networks and autoencoders, are at the forefront of more trustworthy communication through learning from vast amounts of data to predict and avoid potential system malfunctioning. Besides optimizing the performance of unguided transmission media, these intelligent systems also dynamically modify themselves as per varying network environments. Moreover, with wireless communication advancing very quickly, integrating these learning techniques will be essential in the development of flexible communication systems that can accommodate rising connectivity demands in different applications. Hence, the research on unguided transmission media as well as machine learning is an emerging field for the development of communication technologies (Ajay et al., 2022).

8.3.1. Radio Waves

Radio waves, being a part of the electromagnetic spectrum, are employed in the heart of modern communication systems. Radio waves have the characteristic of propagating through several media to increase the information transmission in long distances without employing physical paths. In transmission media, radio waves are employed to facilitate communication to occur in various types of systems such as radio broadcasting, television, and mobile phones. Their capacity for propagation across different environments, including urban and rural terrains, is indicative of their importance in ensuring broad access to communication technology. Moreover, technological innovations have enhanced the efficiency and capability of radio wave communication, especially the establishment of digital modulation techniques, to enable improved signal quality and data rates within contemporary networks. Reconfigurable intelligent surfaces (RISs) are the technology revolution for the future that will optimize the transmission of radio waves in wireless networks. RISs, designed to smartly change the wireless propagation context, can be utilized to improve coverage and capacity for any communication system.

The innovation is particularly well timed as the industry transitions to sixth-generation (6G) networks, which promise to address the increasing demand for data and connectivity. With the integration of RIS and machine learning technologies, researchers are seeking answers in resolving issues related to wireless dynamic channels and mobility of users. The development of RIS applications hence is a dynamic field of research that holds the promise of enhancing radio wave transmission performance such that channels not only become robust but also dynamic, adapting to shifting user and environmental needs (Liu et al., 2021). Also, the applicability of satellite communications also demonstrates the applicability of radio waves in facilitating global connectivity. With advances in technology propelling the new advancements in satellite systems, there is a greater focus on tapping radio wave frequencies for various applications ranging from 5G integration to Earth observation. The unique attributes of satellite communications, including broad coverage areas and the ability to cover distant regions, highlight the pivotal role played by radio waves in today's communication architectures. As a paradigm shift takes place toward advanced systems with on-board processing and non-terrestrial networks, researchers have a plethora of challenges and opportunities ahead. The exploration of these innovations will not only demand the refinement of the existing models of satellite communication but also enhance the overall efficacy of transmission of radio waves from different platforms (Kodheli et al., 2020) (Figure 8.9).

Figure 8.9. *Radio waves.*

Source: https://www.thoughtco.com/thmb/1paorklh4MK4_2R_wqH5yIM0X7s=/1200x808/filters:fill(auto,1)/VLA730B_med-58b846845f9b5880809c6d6d.jpg.

8.3.2. Microwaves

Microwaves, a specific family of electromagnetic waves, play a vital role in modern communication networks. They have frequencies ranging from approximately 1 GHz to 300 GHz and offer terrestrial and satellite communications with the unique strengths of high bandwidth and the ability to penetrate various atmospheric media. Their relatively shorter wavelengths enable the development of more miniature antennas, leading to the development of advanced technologies such as microwave antennas and phased array systems. As a result of this, uses of microwaves have spread to numerous other areas like cellular networks, Wi-Fi, and satellite communications, which form the backbone of the connectivity that is enjoyed today. Microwave technology's continued advancement underscores its contribution to enabling higher transmission capacity, making it an essential component of contemporary communications infrastructure. Microwave propagation in various media offers opportunities as well as challenges in communication systems. The atmosphere is among the most important microwave transmission factors since weather conditions such as rain, fog, and humidity result in attenuation and scattering of signals. This impact is particularly significant in long-distance ground microwave links. New solutions are being developed to reverse these issues, like adaptive modulation techniques and diversity configurations that boost signal strength. Notable studies indicate advancements in the control of microwaves over propagation in dynamic scattering media, ensuring the highest stability of signal transmission under varying conditions.

Through the use of such methods, communication systems are able to effectively manage the limitations incurred by atmospheric interference in order to improve overall performance and reliability for end-users under varying scenarios and environments. The addition of microwaves into wireless communication systems has made new applications possible within varying fields such as medical imaging and sensor networks. As an example, new technology in the form of giant magnetoimpedance (GMI)-type receivers has demonstrated promising performance in unconventional environments, such as underwater media, where traditional communication paradigms are not able to function efficiently. Magnetic communication using a GMI receiver signifies the versatility and robustness of microwaves at adverse conditions, highlighting their adaptability beyond typical earth-based applications. With more emphasis on optimizing microwave channels to have higher data transmission rates and lower susceptibility to noise from the environment, the effect on communications is huge. Advances in this regard guarantee stronger digital communication links that could eventually transform the means of information transmission across various communication platforms, including across media like those addressed in Kim et al. (2023) (Figure 8.10).

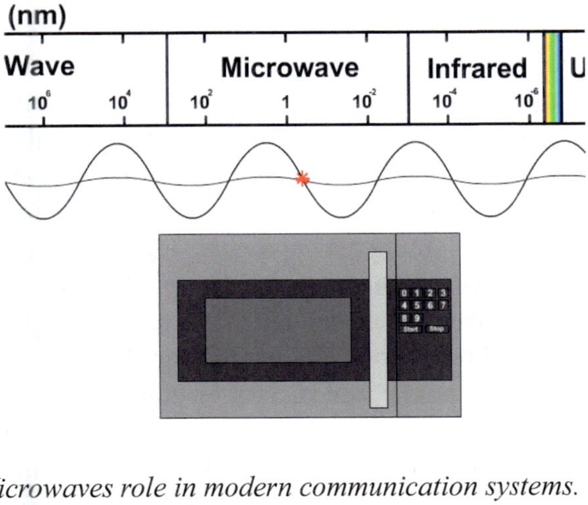

Figure 8.10. *Microwaves role in modern communication systems.*

Source: https://stickmanphysics.com/wp-content/uploads/2020/03/EM-Waves-Microwave2.gif.

8.3.3. Infrared Communication

Infrared communication is a fine aspect of today's transmission media, providing an extensive and effective system of short-range wireless communication. Due to the infrared frequency range, slightly above visible light, infrared communication allows data transfer by the use of infrared light waves. Infrared technology is used extensively in remote controls, computer data links, and wireless personal area networks (WPANs). The intrinsic ability of infrared communication to be immune to electromagnetic interference and simple also enhances its applicability in most environments. Additionally, the line-of-sight transmission employed reduces the eavesdropping attack, hence a level of security necessary in current communication systems. As a result, infrared communication has carved out a niche within the larger set of data transmission technologies, and its value is emergent both for industrial and consumer applications.

One of the considerations in assessing infrared communication is its physical limitations and environmental sensitivities. In contrast to radio frequency signals which may travel through walls and extend over vast distances, infrared signals require a clear line of sight between the transmitter and receiver. This restriction may keep usability out of advanced settings or obstructed environments. Additionally, external environmental factors, like ambient light and weather, can affect the effectiveness of infrared communication. For instance, direct sunlight may add noise, disrupting communication, while atmospheric factors may affect signal quality.

These limitations necessitate appropriate planning for deployment conditions to ensure optimal performance of infrared communication systems. As highlighted by recent advances in technology, studies still attempt to mitigate these issues so that infrared communication will still be included in an evolving environment of communications. The future of infrared communication is full of great potential for innovations which may further increase its range of uses along with forthcoming technologies. For instance, innovations in secure device pairing (SDP) schemes (Fomichev et al., 2017) can provide new security functionalities that take advantage of the unique properties of infrared for personal devices and Internet of Things (IoT) applications. Simultaneously, innovations such as free-space optical (FSO) communications identify the potential of infrared technology in addressing the requirement of high-speed communications in mobility applications, e.g., high-speed trains (Ghassemlooy et al., 2011). With these advancements, infrared communication will advance with the capability of delivering higher data rates and higher resistance to environmental conditions. In general, with some trends toward research and development, infrared communication will continue to advance and integrate into the overall broadband of transmission media and channels of communications system in order to retain its role as a future technology contender.

8.4. CHARACTERISTICS OF TRANSMISSION MEDIA

Properties of the transmission medium are of utmost significance in determining the effectiveness and reliability of communications systems. Transmission media can be grouped generally into guided and unguided mediums, and each of these has some features influencing signal propagation. Guided transmission media, for instance, such as fiber optic cable and coaxial cable, enclose signals in a specified path, hence restricting interference and enhancing data integrity. On the other hand, unguided media such as radio waves rely on the atmosphere medium to transmit signals and can suffer from high attenuation and multipath fading. The specific characteristics of such media directly relate to the suitability of such media for particular communication purposes and indicate the need for applying the right medium to the appropriate purpose. Consequently, knowledge of transmission media properties enables the maximum utilization of communication systems to efficiently transport data over different environments. One of the most important transmission media properties is bandwidth, which is a measure of the amount of data that can be conveyed through a medium over a period of time. Bandwidth is what dictates the rate at which information is transmitted and significantly affects the overall system performance. For instance, the fiber optic cables offer enormous bandwidth capacity, which allows for large volumes of data to be transmitted at minimal delay. Copper cables, however, offer smaller bandwidth that can limit their use in high-demand scenarios. Furthermore, the nature of the medium directly correlates with the latency or delay in transmitting data. Good communication systems should therefore consider bandwidth as well as latency to meet the needs of modern applications ranging from high-definition video streaming to real-time video conferencing. Additionally, forces of the environment also impact transmission media's performance characteristics. Physical parameters such as temperature fluctuations, humidity, and extraneous electromagnetic interference can all negatively impact signal quality, especially in unguided transmission mediums. For example, wireless communication networks are very prone to obstacles that may disrupt signals and hence make connectivity more difficult, resulting in suboptimal performance. Mitigating such characteristics often calls for using sophisticated methodologies, such as error-correcting codes and adaptive modulation, to enhance communication reliability. In addition, the integration of new technologies in transmission media, such as adaptive beamforming and multiple-input and multiple-output (MIMO) systems, aims to compensate for the impact of environmental challenges. As noted by new studies, these technologies emphasize the continuous development of transmission media to increase the robustness and flexibility of communication systems (Qian et al., 2024; Griber, 2024).

8.4.1. Bandwidth

In the context of channels and transmission media of communication systems, bandwidth is arguably the most significant parameter with a direct influence on data transmission capacity. Bandwidth refers to the speed of data transmission along a particular route in a network, typically measured in hertz (Hz) or bits per second (bps). The significance of bandwidth is evident from the fact that it not only decides data communication speed but also the quality of service for the users. While the need for communication increases, particularly with the growing amount of multimedia content and HD video streaming, higher bandwidth is now acutely required. Higher bandwidth is required for concurrent data transmission of higher volumes of data to facilitate applications demanding real-time data transfer. This offers a critical requirement for system designers and network managers to understand and optimize bandwidth. Recent technological developments have particularly highlighted the importance of bandwidth in various communication systems. Visible light laser communication (VLLC), for instance, has been shown to be an effective strategy for deploying high bandwidth solutions, particularly in underwater communications. Experiments indicate that GaN-based laser diodes possess the potential to operate in the water's blue-green transmission window, providing excellent bandwidth potential. For example, experiments indicate that the devices possess the potential to provide a −3dB E-E bandwidth of 5.4 GHz, thereby enabling a full-duplex system capable of supporting data rates of up to 21.79 Gbps for downlink and 1[illegible].40 Gbps for uplink utilizing advanced modulation formats. These advancements demonstrate the promise of future generations of communications systems, showing how bandwidth optimization can yield substantial gains in terms of overall throughput and system efficiency (Li et al., 2025). In addition, effective bandwidth management also calls for stable media access protocols, which lie at the heart of making shared media efficient in communications systems. These protocols dictate data transfer to minimize collisions and manage access for high channel throughput and low transmission delay. For wireless networks, where bandwidth is often a limited resource, the use of effective media access control (MAC) protocols becomes all the more necessary. Through the examination and selection of proper protocols using a structured method and matching service requirements, engineers can maximize network performance and enhance user experience. Current literature is an exhaustive summary of the majority of MAC protocols, highlighting their classification, impact on network parameters, and their generalizability to other uses. This is a reflection of the vital interdependence between bandwidth and efficient protocol development in making modern communication systems efficient and reliable (Saini et al., 2025).

8.4.2. Attenuation

In communications systems and transmission media, attenuation is a significant factor in impacting signal quality and system performance. Attenuation refers to the weakening of signal power along a transmission media distance, and it can significantly influence data rates and distances of transmission. Understanding the principles of attenuation is required to maximize communication channels, especially with the growing needs for high-speed internet and reliable connectivity. Attenuation factors consist of the type of medium, operating frequency, and environmental parameters. Attenuation in optical fibers, for instance, can be caused by scattering, absorption, and bending losses that all lower signal quality with distance and have technologies to mitigate such losses. Emergent advances in modulation techniques, such as discrete multitone (DMT) modulation, have been highly promising against the challenges of attenuation in optical communication systems. Specifically, when used for traditional 1-mm core-diameter step-index plastic optical fiber (SI-POF), DMT can significantly accelerate transmissions by efficiently addressing the interaction between light signals and the fiber medium. This improvement is essential to achieve effective communication over longer distances, previously limited to around 100 meters with data rates around 100 Mbit/s. By applying digital signal processing to maximize the signal in attenuation-distorted conditions, researchers have demonstrated extensive capacity improvement in SI-POF systems (Bluschke et al., 2009). These developments bring into focus the importance of studying various techniques to overcome attenuation as the telecommunication environment advances. Besides, attenuation is not only a feature of fiber-optic systems; it is an issue in other transmission media such as free-space optics (FSO) and wireless communications. FSO signal propagation can be greatly hindered by atmospheric conditions such as turbulence and fog. Attenuation in such atmospheres can cause signal quality reduction, hence the research into emergent solutions. Recent studies have investigated hybrid systems that involve combining different transmission means and utilized some structures of modulation and coding in order to maximize performance when the conditions are poor (Awan et al., 2009). These cross-disciplinary approaches reflect the need for ongoing innovation and research towards reducing loss, thereby enhancing the reliability and efficiency of modern communication systems for channels and media (Figure 8.11).

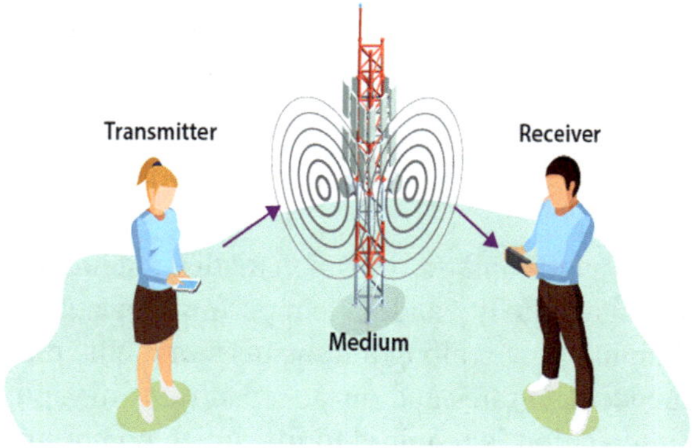

Figure 8.11. *Attenuation of transmission media and communication systems.*

Source: https://cdn1.byjus.com/wp-content/uploads/2021/04/Communication-Systems-01.png.

8.4.3. Interference

Interference is a critical issue in the transmission medium as well as communication system channels, with significant effects on data transmission dependability and efficiency. Interference can be caused by various reasons, varying from natural causes to man-made devices, leading to distortion of the signal being transmitted. Assessment and mitigation of interference are required for the quality-of-service assurance of communication networks. In vehicle communications, where numerous devices are transmitting information at the same time, interference is far greater. The issue is particularly relevant with increasing numbers of connected cars on the highways, highlighting the need for advanced techniques to amplify communication channels and minimize disruptions, hence more seamless and safer transportation networks. The application of Cognitive Radio (CR) technology provides a new solution in reducing interference problems experienced in high vehicular communication environments. By enabling vehicles to exploit unused frequency bands in an opportunistic manner, CR technology enables more dynamic utilization of the available spectrum, which is of significance considering the limitations of the existing assigned frequencies. Experiments demonstrate that the CR Assisted Vehicular NETwork (CRAVNET) system can significantly enhance vehicular communication by virtue of vehicles' ability to autonomously sense and utilize other spectrum resources without being deleterious to licensed users. As a result, such dynamic spectrum management not only minimizes interference but also achieves the highest transmission capacities in increasingly crowded communication channels, paving the way to more efficient intelligent transportation systems (Eze et al., 2018). Quantitative models to quantify the performance of networks in cellular mobile communications systems often play up the significance of interference in both spatial spectrum and energy efficiencies. The development of a spatial spectrum and energy efficiency model for random cellular networks provides important insights into the effect of interference on service quality. The use of a Markov chain model to illustrate user access to the network allows researchers to analyze parameters such as outage and blocking probabilities during simultaneous transmissions. These commonly used parameters are essential in measuring how interference can result in increased call drop rates and service degradation. Lastly, by employing such models, stronger communication systems are developed that can withstand interference challenges, hence guaranteeing efficient and stable network operations (Ge et al., 2015).

8.5. COMMUNICATION CHANNELS

One of the most distinctive characteristics of modern systems of communication is the diversity of channels for data transmission. These channels define paths along which information travels and both the data rate and quality of communications. Among conventional approaches, wired channels such as fiber optic and coaxial cables have prevailed due to their high bandwidth capacity and reliability over long distances. With advancements in technology, the use of wireless communication mediums has grown, e.g., radio frequencies and infrared systems, which offer mobility and flexibility. The availability of such wireless systems facilitates easy communication in various situations where physical connections may not be feasible. As Nefed'ev et al. (2023) points out, greater utilization of wireless systems for the transmission of data points to a shift toward more sophisticated communication systems that leverage the characteristics of various media in an effort to maximize performance. Further, the ability of communication channels is generally subject to the constraints of the medium employed to transmit data. The quality of a channel is not only an issue of its shape but also the characteristics of the transmit medium. Ultrasonic channels, for example, have proven to be good potential candidates for data transmission in elastic media such as liquids and certain solids. Empirical evidence indicates that effective operation within some frequency ranges drastically enhances the performance of these channels, particularly in remote sensing where precision is everything. Such discoveries have critical applications because they open the door to employing ultrasonic communication systems in many industrial and scientific applications, essentially broadening the possibility of using communication channels beyond the expected (Nefed'ev et al., 2023). To complement structural issues, channels of communication will have to provide room for the intricacies involved in data communication of large quantities. The runaway growth of geographic information systems (GIS) has highlighted the necessity of proper data handling and communication. The necessity of large datasets to analyze spatial data implies that there is a demand for innovative methods of data compression and communication. Techniques such as fractal image compression have come to symbolize key methods for increasing the efficiency of data transmission on communication links, especially for satellite remote sensing. While the encoding algorithms may prove to be computationally intensive, the results of significant data compression are necessary to manage bandwidth and storage constraints. This dual focus on effective channel communication and new ways of processing information reflects the increasingly sophisticated nature of communication systems in handling modern informational loads (Figure 8.12).

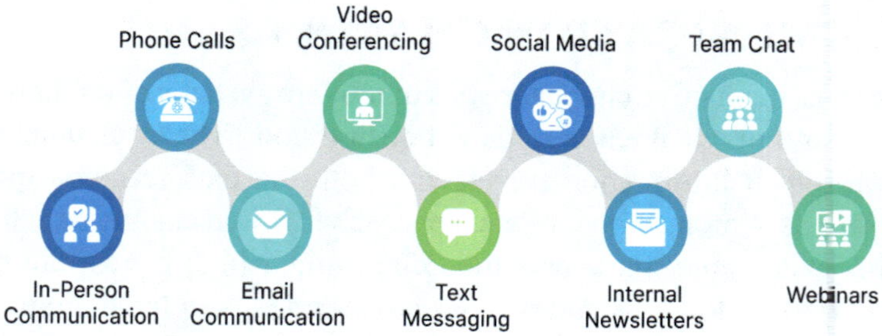

Figure 8.12. *Communication channels.*

Source: https://blogimages.dialerhq.com/blog/wp-content/uploads/2024/05/21112108/9- essential-business-communication-channels.jpg.

8.5.1. Definition and Role of Communication Channels

Channels of communication are the avenue through which data is conveyed from a sender to a receiver, and they are extremely crucial in interpretation and transmission of messages. Channels are categorized into various types, from physical (such as cables and fiber optic cables) to virtual (such as radio waves and the world wide web). The effectiveness of a communication channel also depends greatly on the ability to minimize distortion and maximize clarity, and by extension, it impacts the quality of the received message. Communication channels keep evolving, especially with digital technologies, and therefore organizations are required to adapt their methods on how to utilize such channels to their benefit. For instance, during the COVID-19 pandemic, telehealth was an indispensable channel of communication that exemplified its potential in facilitating healthcare delivery with less physical contact, thereby affirming the imperative nature of communication channels in whatever circumstances, including public health initiatives (Monaghesh et al., 2020). Furthermore, the planning and construction of communication channels are also essential for successful transmission media. Emerging technologies such as reconfigurable intelligent surfaces (RISs) illustrate the realm of innovation in communication systems, which offers a means to enhance wireless communications. RISs achieve this through the manipulation of wireless signals by strategically positioned reflecting surfaces that are capable of making overall data transmission efficiency better. Through this technology, the employment of expensive traditional components is avoided and allows more flexibility and applicability within communication networks. The RIS unification represents a paradigm shift in how communication channels can be established and optimized and thereby facilitating connectivity and throughput in the applications. Therefore, as more research is conducted in this area, the vision for balancing early-on communication theories and cutting-edge technology becomes even more crucial to advancing the development of communication systems (Renzo et al., 2020). Beyond technology development, understanding the inherent characteristics of different channels of communication is also required to facilitate effective information transmission. Bandwidth, delay, and noise level significantly influence the performance of a channel. High-bandwidth channels, for instance, can support more information and at higher speeds, a factor that is essential in situations that call for real-time communication, like telehealth consultations. However, channels with excessive latency or high noise have the ability to break efficient communication and lead to miscommunication. It is hence the responsibility of organizations not just to select the appropriate channels but also to monitor them continuously so that the evolving demands of communication in a globalized era may be met. Monitoring in this manner is essential in optimizing communication strategies and facilitating proper and effective message transmission.

8.5.2. Types of Communication Channels

Channels of communication are the media upon which information is conveyed from one party to another, and communication channels can be broadly classified as wired and wireless channels. Wired communication channels, such as coaxial cables, fiber optics, and twisted-pair cables, carry electrical signals over physical media. These channels are appreciated for their reliability and wide bandwidth that they can provide, enabling a lot of data to be transported far away with minimal interference. Wireless communication channels, i.e., radio waves, microwaves, and infrared signals, instead, possess mobility and ease since customers are able to communicate without physical connection. Each of these types of channels plays a basic function in the communication systems of today, even though they are appropriate for various types of user application in varying situations, from single communications to large telecommunication networks.

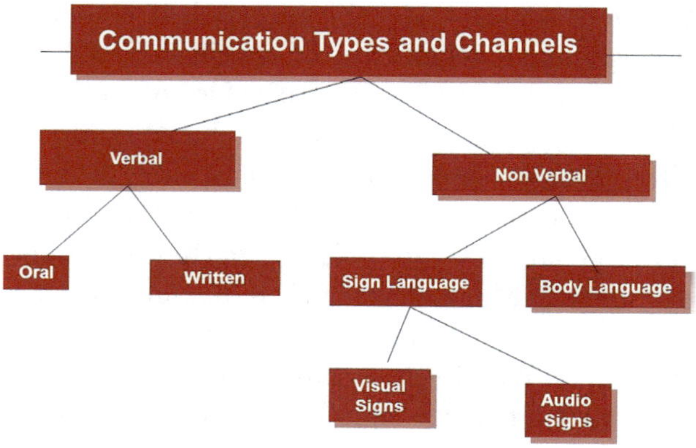

Figure 8.13. *Types of communication channels.*

Source: https://i.pinimg.com/originals/d8/70/4a/d8704ab065a90f0bb1fa3157c-fae0f4d.jpg.

One such significant wireless communication channel is the satellite communication channel, which is founded upon the use of satellite technology to offer facilities for long-distance transmission of signals. Satellite communication is particularly valuable where traditional wire-based systems are unavailable or would be impractical. Satellites, for instance, facilitate telecommunications broadcasting, and data communication across geographical barriers to provide essential connectivity in dispersed regions as well as during emergency situations when ground networks fail. As discussed, communications satellites provide routes for sending and receiving all manner of media services, with applications ranging from global positioning systems (GPS) to meteorological pattern information gathering. The central components of these systems are ground stations and the satellites, which allow the reception and transmission of signals effectively around the world, in turn serving different industries like military, telecommunications, and broadcasting media (N/A, 2024). Cellular networks are another strong example of wireless communication channels, characterized by their ability to offer extensive coverage and higher user capacity. The emergence of cellular technology has surmounted overall challenges such as wavelength crowding, enabling high-capacity transmissions without extreme technological advancements in the existing infrastructure. Cellular systems employ complex models of propagation to predict signal strength, thus ensuring effective communication despite challenging topographies. The models are crucial since they factor in environmental issues influencing the signal quality that can be fundamentally different in urban and suburban areas. As noted, radio signals may be subject to a variety of types of path loss due to interactions with the environment, and such information is quite critical to network planners and researchers to maximize communication channels' efficiency (Yahia et al., 2021). Hence, knowing different types of communication channels is necessary to assist in maximizing transmission media and enhancing systems in numerous applications (Figure 8.13).

8.5.3. Factors Affecting Channel Performance

The performance and dependability of communication channels are influenced by a wide range of factors beyond mere technology. Physical circumstances such as weather conditions and geographical constraints play a significant role in the efficiency of channels. For instance, the propagation of radio waves can be severely influenced by physical barriers, e.g., buildings and hills, and atmospheric conditions, e.g., rain and fog. These factors can lower signal strength and clarity, effectively impairing the functionality of transmission media. In an increasingly dependent world where quick communication is paramount, realizing the relationship between these extrinsic influences and transmission quality is crucial. With next-generation communications system advancement to come in the future, like 5G and beyond, improved knowledge regarding geographical and environmental conditions will be critical to maintaining the performance standards aligned with the requirements of modern society as discussed in the present context of communication evolution in Tataria et al. (2021). Apart from environmental conditions, technological factors play a crucial role in channel performance. System design, for instance, the choice of modulation and frequency bands, makes an important contribution. Different techniques of modulation may have varying degrees of efficiency for transmitting data across specific media. Greater-frequency bands are capable of handling greater data rates but may also suffer greater attenuation. Also, the use of newer technologies such as multiple-input and multiple-output (MIMO) systems enhances channel capacity and reliability through transmission of data multiple times over multiple paths simultaneously. However, the application of these technologies presents numerous challenges related to hardware capabilities and network design. As the technology continues to evolve with the transition to 6G systems, the assessment and optimization of such technological variables will determine the effectiveness of communication channels as a whole, in coordination with the strategic implications arising from Tatania et al. (2021). Lastly, human factors for communication channels are often ignored but can contribute significantly to performance outcomes. Inability to provide education and training equality for digital literacy can prevent efficient use of communication technologies. Like in the COVID-19 pandemic, numerous people suffered from not having access to technology or assured internet connectivity, as presented in. These pedagogical and technological impediments ultimately hinder the ability of the users to use the channels efficiently, resulting in communication efficiency gaps. Further, social dynamics in communication such as the lack of human interaction in rural areas, can affect motivational levels and engagement, complicating channel use further. Resolution of these human-oriented issues is essential to ensure that complex communication systems are successfully implemented to benefit various communities effectively while improving channel performance overall.

8.6. APPLICATIONS OF TRANSMISSION MEDIA

The applications of transmission media are a functional feature of the operational technologies of the modern communication systems, serving to facilitate the exchange of information between different media. Different forms of media, including wired (fiber optic cables and coaxial cables) and wireless (radio waves and microwaves), can meet the multiple requirements of users of the modern era. All media possess unique physical properties and bandwidth characteristics, which guide their suitability for various applications. For instance, fiber optic cable carries information at high speeds through long distances with minimal signal loss and is therefore used in telecommunication network backbone connections. Wireless media, however, allow for mobility and convenience, essential when employed in personal communication devices. The choice of the media has a direct impact on the efficiency, reliability, and performance of a transmission system and thus on the communication technology. Artificial intelligence (AI) is rapidly transforming the application of transmission media, particularly in maximizing data communications effectiveness. AI-based solutions enhance the effectiveness of electronic communications through the analysis of humongous volumes of data in an attempt to predict system performance and optimize connectivity. The application of AI-driven algorithms, according to various studies, allows for better network optimization and resource allocation, eliminating congestion and latency in real-time. For example, machine learning-driven models used in communication networks can control bandwidth in real time to ensure un interrupted data flow even during peak usage times. These proactive measures not only improve the customer experience but also reduce service providers' costs of operation. Therefore, the symbiotic relationship between transmission media and AI represents a quantum leap in the evolution of communication systems to enable future breakthroughs in connectivity and user experience (Vetrivel et al., 2024). Moreover, emerging transmission methods such as rate-splitting multiple access (RSMA) are gaining prominence for facilitating the efficient management of communication channels. RSMA is a significant departure from traditional multiple access methods in that it facilitates greater spectral and energy efficiency and improved user fairness and reliability. With the use of sophisticated mathematical frameworks, RSMA enables non-orthogonal transmission, thus allowing multiple users to share a common bandwidth more effectively. This technique not only mitigates interference in high-density network environments but also yields reliable communication under imperfect channel state information. The versatility of RSMA in handling different network loads and channel conditions underscores its potential for future wireless communications, particularly as growing demands for higher data rates continue to escalate. Thus, technology development of this nature attests to the ongoing innovation of transmission media in meeting the demands of communication systems in today's world (Mao et al., 2022).

8.6.1. Telecommunications

Telecommunications is a key part of contemporary communication structures, making it possible to move data through media and channels. The building blocks of telecommunications are built upon sending and receiving information, which can range from text data to multimedia. The process often consists of the transmission of binary code, where data points become bits, enabling perfect communication between devices. After describing the topology of trees in data communication networks, the link between more than a single computer system demonstrates how complex structures can efficiently manage data exchange (Octaviani et al., 2024). With the advent of digitization in the modern era, these networks facilitate a plethora of applications that enable individuals and business organizations to communicate and share information and resources at unprecedented velocities. One key aspect of telecommunication is the application of satellite technology, which transformed information transmission over long distances. Satellites provide connection channels by serving as intermediary between two points on the ground, transmitting data, voice, and broadcast signals. This technology is particularly vital in areas with inadequate ground infrastructure, providing life-saving communication services where other systems are not feasible. Communications satellites operate in different orbits, e.g., geostationary and low-earth orbits, each performing specific functions in telecommunication systems (N/A, 2024). Furthermore, the satellites enable critical services such as GPS and disaster relief communications, affecting different sectors, e.g., agriculture, military, and weather forecasting, ultimately enhancing global connectivity. The development of telecommunications using multiple transmission media has expanded the line of communication channels available. These include fiber optic, coaxial, and wireless communications, which all contribute towards the construction of communication systems to evolve increasingly. Fiber optics, for instance, utilize light to transmit messages, allowing unprecedented data rates and bandwidths compared to the traditional metallic wires. Moreover, innovations in wireless technologies have paved the way for mobile communications, which allow the user to communicate over distances without physical constraint. The combination of these various media has created a platform on which real-time sharing and communications are made possible and reachable to larger segments of the population. With the continued improvement of telecommunications, knowledge of the internal workings of transmission media is essential in fulfilling the complicated communications requirements of the 21st century, as well as building effective and efficient systems for generations to come (Figure 8.14).

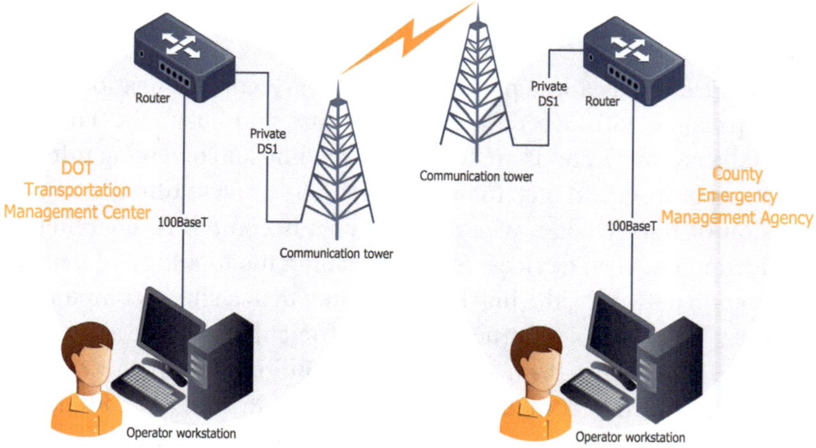

Figure 8.14. *Telecommunications is an essential component of modern communication systems.*

Source: https://www.conceptdraw.com/solution-park/resource/images/solutions/telecommunication-network-diagrams/Computer-and-Networks-Telecommunication-Network-Diagrams-TMC-to-EMA-Comm-Link-Rural-Setting.png.

8.6.2. Networking

Improvement in networking technologies has substantially enhanced the reliability and effectiveness of communication networks. By interconnecting varied transmission media, networks are able to offer simultaneous sharing of information across a variety of platforms, thus optimizing bandwidth use. This is particularly pronounced in advanced underwater communication systems, where the need for guaranteed data delivery is critical. Existing research highlights the importance of compact multichannel devices that can transmit and receive multiple signals at the same time, as explained in Abdi et al. (2024). This development enables improved communication performance without the need for many spatially separated devices, which can be cumbersome when being set up and limit functionality. The ability to minimize physical space needs without loss of signal quality captures advances in networking in existing communication infrastructures.

In addition to enhancing the physical components of networking, there is also a critical need for sophisticated methods for effectively managing data transmission. Deep learning techniques are increasingly being applied to channel estimation problems in future networking systems, such as wireless-powered sensor networks (WPSNs). The issue is effectively managing the uplink and downlink channels without compromising performance from cascading signal paths. The new technique introduced in Jung et al. (2023) demonstrates the method of constructing full-duplex communication systems with the assistance of channel estimation made easy via deep learning. Employing advanced algorithms like feedforward neural networks and convolutional neural networks, these systems are capable of achieving the maximum efficiency of transmission as well as overcoming nonlinearity and non-convexity issues inherent in intricate networking. The convergence of new transmission media and new networking methods is an indication of a profound paradigm shift in communication infrastructures. Multichannel devices and deep learning are opening new paradigms for data sharing with low overhead, particularly in the adverse environments of underwater and IoT. With an optimal utilization of hardware and software entities, modern networking solutions are more flexible and reliable than ever before. This coincidence of transmission technology and intelligent data management not only enhances performance but also lays the foundation for new possibilities for future application in communication systems. As study continues along these lines, it's likely that the changing developments in networking will produce even more robust and effective systems that can handle the increasingly massive demands of digital communication (Figure 8.15).

Figure 8.15. *Networking technologies.*

Source: https://quadrant-solutions.com/wp-content/uploads/2022/04/IT-Infra-structure-and-Networking-1-2-scaled.jpeg.

8.6.3. Broadcasting

Broadcasting forms an important component in the media of transmission and communication channels of communications systems used to distribute information and entertainment to a mass of people. Broadcasting channels utilize both satellite and terrestrial technologies to provide content to multiple recipients simultaneously, hence enhancing access to important content. Broadcasting is performed in various formats like radio and television broadcasts and newer ones like web streaming and podcasting. In the last few years, digital broadcasting has revolutionized classic paradigms by bringing forth high definition and interactivity that promote a more immersive viewing experience. Broadcast's ongoing adjustment to evolving technologies notwithstanding, it is still essential for conveying information to heterogeneous populations, playing its role as a fundamental conduit in contemporary communication systems. At the heart of broadcasting's functionality lie the technologies that facilitate efficient message transmission. Satellite communication is a critical element in this context. Relay stations are employed using artificial satellites through broadcasting to reach broad geographical regions, sending signals to remote locations where the media may not extend. As pointed out, communications satellites are applied in different uses such as telecommunication and military applications, further showing their versatility and necessity in the communications sector. The use of satellite technology in broadcasting allows for a wide-reaching coverage, being one of its key roles in media access, especially in rural or distant areas (N/A, 2024). Furthermore, as satellite networks continue to evolve, the potential for improved service delivery increases, hence guaranteeing that broadcasting keeps pace with the needs of users. As the digital age matures, the requirement for efficient transmission of data is exemplified by the development of broadcast technologies. The transmission of high-definition video and images necessitates efficient compression techniques for proper use of bandwidth. The innovation in new techniques, particularly in image and video data compression, presents the supreme importance of this aspect within broadcasting technology. As the dimensionality of multimedia information continues to increase, these compression methods are key to maximizing storage space and facilitating real-time processing on various communications networks. Building on advances in artificial intelligence and deep learning, the broadcasting sector will maximize audience experience through optimizing the delivery of content (Meitei et al., 2024). Thus, the interaction among broadcasting and data efficiency does reflect further development of transmission media and is a reflection of the dynamic nature of communication systems in today's world.

SUMMARY

In summarizing the discussion of transmission media and channels in communications systems, there is evident proof that the evolution of these technologies has had significant effects in various sectors, most notably education and emergency communication. With the addition of social media platforms, it has been able to assist organizations, particularly in higher education institutions, to achieve instant engagement and facilitate information sharing. As commented by Андрей et al. (2024), the necessity to avoid repetition of channels of information transfer is significant in ensuring effective communication, hence enhancing decision-making processes within organizational systems. This observation highlights the necessity to use tailored communication devices that align with users' likes and socio-technological trends. Secondly, advances in communications technology continue to impact our understanding of connectivity in challenging environments. The new applications of magnetic communication systems, particularly employing GMI-based receivers, provide unparalleled prospects for information exchange with unparalleled robustness against aggressive challenges, including noise in various media. As the findings presented in Kim et al. (2023) illustrate, these developments affirm the importance of adapting communication strategies to meet specific environmental requirements. Consequently, the efficient functioning of such systems not only widens the horizons of traditional channels of communication but also becomes integral to improving operational effectiveness in technologies like underwater communications. Ultimately, the destiny of channels of communication systems and transmission media is linked with the continuing innovativeness of current forms to accommodate place for emerging technologies. The implications derived from the application of social networks for communication within organizations and trying new channels of communication, such as the application of GMI receivers, confirm that constant innovation is crucial. With advancement, it is the same crucial takeaway that still stands in the guise of the identification of the technology-communication need dynamic interplay; thus, businesses should be cautious to evaluate and embrace channels as driven by user demand alongside possible technology. Through such vision, communication system development will keep driving social, educational, and business advancements in different fields

The study of transmission media and channels in communication systems is greatly enhanced by the development of Reconfigurable Intelligent Surfaces (RISs). The systems have garnered significant research attention due to their promise of enhancing the performance capabilities of wireless networks through the dynamic reconfiguration of the wireless environment. RISs possess the potential to optimize signal propagation, which is a key factor in achieving efficient and reliable communication, especially in consideration of the growing demand for higher data rates. RIS integration is reflective of changing communication medium trends, indicating a migration towards the utilization of capabilities of high-tech materials to provide better performance at lower cost. This focus on RIS is a paradigm shift in communication infrastructure design and deployment potential, and this has placed increasing pressures on innovative solutions in the next generation network (Liu et al., 2021). Furthermore, RIS is unique from traditional transmission technologies in both architecture and operation. Unlike phased arrays or multi-antenna transmitters, which usually incorporate complex elements, RISs are founded on the premise of exploiting extensive arrays of affordable scattering elements. This paradigm is enabled by the ability to use numerous basic elements at a reduced cost, size, weight, and power consumption—attributes included in the conceptual model known as C-SWaP design. The absence of heavy power amplifiers is consistent with current trends towards efficiency and minimization in communications systems. Moreover, RISs make the concept of Smart Radio Environments (SREs) and consequently adaptive communication methods that adapt to the requirements of the users and the condition of the environment possible, a critical consideration for next-generation wireless networks (Renzo et al., 2020). In general, the consequences of merging RIS technology with the transmission medium are mind-boggling. By extensive performance analysis and machine learning-based applications for managing resources, RISs can efficiently enhance efficiency in communications. The future of communication channels now demands the integration of classical theories of communications with recent advances in electromagnetism to further offer a complete framework for further investigation. Most chief areas of research remain to be explored, namely, how to optimize RIS deployment towards practical applications in various environments. With these technologies set to converge, the vision of enriching communication networks is increasingly tangible, opening up to new applications that will redefine the rulebook for connectivity in the years to come (Liu et al., 2021; Renzo et al., 2020).

REFERENCES

1. Abdi, A., Zhang, E., & Rashid, R. (2024). Multichannel signal transmission and reception using compact multichannel underwater communication devices: A unified theory and experimental results from different underwater media. *The Journal of the Acoustical Society of America*. https://www.semanticscholar.org/paper/1f76cdad2d8696e246f82db6_85e718491faaec8 (accessed on 12th September 2025).
2. Abuella, H., Elamassie, M., Uysal, M., Xu, Z., Serpedin, E., Qaraqe, K., & Ekin, S. (2021). Hybrid RF/VLC systems: A comprehensive survey on network topologies, performance analyses, applications, and future directions. IEEE Access, 9, 160402–160436. https://doi.org/10.1109/access.2021.3129154 (accessed on 12th September 2025).
3. Ajay, P., Nagaraj, B., & Huang, R. (2022). Deep learning techniques for peer-to-peer physical systems based on communication networks. Journal of Control Science and Engineering, 2022, 1–12. https://doi.org/10.1155/2022/8013640 (accessed on 12th September 2025).
4. Awan, M., Brandl, P., Capsoni, C., et al. (2009). Current optical technologies for wireless access. https://core.ac.uk/download/4149_43.pdf (accessed on 12th September 2025).
5. Başar, E., Di Renzo, M., de Rosny, J., Debbah, M., Alouini, M. S., & Zhang, R. (2019). Wireless communications through reconfigurable intelligent surfaces. *IEEE Access, 7*, 116753–116773. https://doi.org/10.1109/access.2019.2935192 (accessed on 12th September 2025).
6. Bergman, L. A., Cassell, P., Edgar, G., et al. (2025). Fiber optic tactical local network (FOTLAN). https://core.ac.uk/download/pdf/42818096.pdf (accessed on 12th September 2025).
7. Betsch, C., Wieler, L. H., & Habersaat, K. B. (2020). Monitoring behavioral insights related to COVID-19. *The Lancet, 395*, 1255–1256. https://doi.org/10.1016/s0140-6736(20)30729-7 (accessed on 12th September 2025).
8. Bluschke, A., Antonius, M. J., Bosco, G., Breyer, F., et al. (2009). Discrete multitone modulation for maximizing transmission rate in step-index plastic optical fibres. https://core.ac.uk/download/pdf/11406250.pdf (accessed on 12th September 2025).
9. Çelik, A., Romdhane, I., Kaddoum, G., & Eltawil, A. M. (2022). A top-down survey on optical wireless communications for the Internet of Things. *IEEE Communications Surveys & Tutorials, 25*, 1–45. https://doi.org/10.1109/comst.2022.3220504 (accessed on 12th September 2025).
10. Chen, H. T., Taylor, A. J., & Yu, N. (2016). A review of metasurfaces: Physics and applications. Reports on Progress in Physics, 79, 076401–

076401. https://doi.org/10.1088/0034-4885/79/7/076401 (accessed on 12th September 2025).

11. Di Renzo, M., Zappone, A., Debbah, M., Alouini, M. S., Yuen, C., de Rosny, J., & Tretyakov, S. (2020). Smart radio environments empowered by reconfigurable intelligent surfaces: How it works, state of research, and the road ahead. IEEE Journal on Selected Areas in Communications, 38, 2450–2525. https://doi.org/10.1109/jsac.2020.3007211 (accessed on 12th September 2025).
12. Eze, E. C., Eze, J. C., Liu, E., et al. (2018). Cognitive radio-enabled Internet of Vehicles (IoVs): A cooperative spectrum sensing and allocation for vehicular communication. https://core.ac.uk/download/156784622.pdf (accessed on 12th September 2025).
13. Ferri, F., Grifoni, P., & Guzzo, T. (2020). Online learning and emergency remote teaching: Opportunities and challenges in emergency situations. *Societies, 10*, 86–86. https://doi.org/10.3390/soc10040086 (accessed on 12th September 2025).
14. Fomichev, M., Gardner-Stephen, P., Hollick, M., et al. (2017). Survey and systematization of secure device pairing. http://arxiv.org/abs/1709.02690 (accessed on 12th September 2025).
15. Ge, X., Han, T., Mao, G., et al. (2015). Spatial spectrum and energy efficiency of random cellular networks. https://core.ac.uk/download/29177868.pdf (accessed on 12th September 2025).
16. Ghassemlooy, Z., Le Minh, H., Livingstone, B., et al. (2011). Investigation of turbulence effect on the free space optical link for ground-to-train communications. https://core.ac.uk/download/4148057.pdf (accessed on 12th September 2025).
17. Griber, Y. (2024). Peach fuzz: Conceptual engineering of color categories. Nauchnyi Dialog. https://www.semanticscholar.org/paper/812847a80d7a2beb2a6d8e3058dfd5a6659fc19d (accessed on 12th September 2025).
18. Guo, R., Dai, S., Dong, Q., Hu, L., Cheng, J., Yu, P., & Wang, B. (2025). Current transmission and resistance characteristics of bridge joints in three-phase coaxial superconducting cables. IEEE Transactions on Applied Superconductivity, 35, 1–12. https://www.semanticscholar.org/paper/18807870f22bf06db4d8398f06df99177ff6c753 (accessed on 12th September 2025).
19. Hassani, S., & Dackermann, U. (2023). A systematic review of advanced sensor technologies for non-destructive testing and structural health monitoring. Sensors, 23, 2204–2204. https://doi.org/10.3390/s23042204 (accessed on 12th September 2025).

20. Jahanbakht, M., Xiang, W., Hanzo, L., & Rahimi Azghadi, M. (2021). Internet of underwater things and big marine data analytics—A comprehensive survey. IEEE Communications Surveys & Tutorials, 23, 904–956. https://doi.org/10.1109/comst.2021.3053118 (accessed on 12th September 2025).

21. Jayasinghe, U., Samarathunga, P., Fernando, T., Ganearachchi, Y., & Fernando, A. (2025). Image transmission over quantum communication systems with three-qubit error correction. Electronics Letters. https://www.semanticscholar.org/paper/3096784d1ce483cca97fad33c467[illegible]a[illegible]7b644550 (accessed on 12th September 2025).

22. Jayasinghe, U., Samarathunga, P., Pollwaththage, N., Ganearachchi, Y., Fernando, T., & Fernando, A. (2025). Quantum communication for video transmission over error-prone channels. IEEE Transactions on Consumer Electronics, 71, 1148–1155. https://www.semanticscholar.org/paper/[illegible]4[illegible]6 2dbb2e114a5cff791bd532f14e1bd3bd644d (accessed on 12th September 2025).

23. Jung, C. Y., Kang, J., & Kim, D. I. (2023). Deep learning-based channel estimation for full-duplex backscatter communication systems. *2023 International Conference on Artificial Intelligence in Information and Communication (ICAIIC)*. https://www.semanticscholar.org/paper/[illegible]d[illegible]8 414e431d039dadee9d9f82df8e607e79ffac (accessed on 12th September 2025).

24. Kim, J. Y., Lee, H. J., Oh, J., Lee, J. H., & Cho, I. K. (2023). Experimental results of magnetic communication using the giant magnetoimpedance receiver in underwater environments. Journal of Electromagnetic Engineering and Science. https://www.semanticscholar.org/paper/dbee132ab7bbf9b2dee89a045cbda40d7d4d05ff (accessed on 12th September 2025).

25. Kodheli, O., Lagunas, E., Maturo, N., Sharma, S. K., Shankar, B., Mendoza Montoya, J. F., Merlano Duncán, J. C., et al. (2020). Satellite communications in the new space era: A survey and future challenges. IEEE Communications Surveys & Tutorials, 23, 70–109. https://doi.org/10.1109/comst.2020.3028247 (accessed on 12th September 2025).

26. Li, Z., Xu, Z., Li, S., Zhang, X., Zhou, Y., Wang, J., Hao, Z., et al. (2025). 21.79/17.49 Gbps full-duplex visible light communication enabled by short-cavity blue and green InGaN/GaN laser diodes. Journal of Lightwave Technology, 43, 3348–3357. https://www.semanticscholar.org/paper/b0f6a32c75cb272b7dcea6a506ba6dc234c14180 (accessed on 12th September 2025).

27. Liu, Y., Liu, X., Mu, X., Hou, T., Xu, J., Di Renzo, M., & Al-Dhahir, N. (2021). Reconfigurable intelligent surfaces: Principles and opportunities. IEEE Communications Surveys & Tutorials, 23, 1546–1577. https://doi.org/10.1109/comst.2021.3077737 (accessed on 12th September 2025).
28. Mahjoob, A., Liu, Y., Bouchkati, S., & Ulbig, A. (2024). Investigation of characteristics of cables in high frequency for power line communication using finite element analysis. *2024 IEEE PES Innovative Smart Grid Technologies Europe (ISGT EUROPE)*. https://www.semanticscholar.org/paper/2d7268eddcadae328ad36585eddfdfd1def6c70c (accessed on 12th September 2025).
29. Mao, Y., Dizdar, O., Clerckx, B., Schober, R., Popovski, P., & Poor, H. V. (2022). Rate-splitting multiple access: Fundamentals, survey, and future research trends. IEEE Communications Surveys & Tutorials, 24, 2073–2126. https://doi.org/10.1109/comst.2022.3191937 (accessed on 12th September 2025).
30. Meitei, C. M. M., Dabas, S., & Kumar, A. (2024). Image compression for communication: Topic modelling of Scopus abstracts using BERTopic. *2024 27th International Symposium on Wireless Personal Multimedia Communications (WPMC)*. https://www.semanticscholar.org/paper/e0069c5e9d2d47ce710715a2212552163d24abf0 (accessed on 12th September 2025).
31. Mididoddi, C. K., Kilpatrick, R. J., Sharp, C., del Hougne, P., Horsley, S., & Phillips, D. (2023). Threading light through dynamic complex media. *Nature Photonics, 19*, 434–440. https://www.semanticscholar.org/paper/cf6832d8d501a071cad65a7f59c535eb240a0327 (accessed on 12th September 2025).
32. Monaghesh, E., & Hajizadeh, A. (2020). The role of telehealth during COVID-19 outbreaks: A systematic review based on current evidence. *BMC Public Health, 20*. https://doi.org/10.1186/s12889-020-09301-4 (accessed on 12th September 2025).
33. Nefed'ev, I., & Isaeva, L. A. (2023). Study of the operating modes of the ultrasonic receiving/transmitting channel. *2023 International Russian Automation Conference (RusAutoCon)*. https://www.semanticscholar.org/paper/700d8c4d5dad1fadef9a61e464653e495047bdab (accessed on 12th September 2025).
34. Octaviani, S. N., Raihan, R., Andrianto, M. I., Arrosyid, M. R., & Ariwibowo, D. (2024). Komunikasi data menggunakan topologhy tree dalam pembelajaran sistem jaringan dan telekomunikasi. Jurnal Teknik Mesin, Industri, Elektro dan Informatika. https://www.semanticscholar.

org/paper/a1c447c093121b0427151d1372618aa937239c94 (accessed on 12th September 2025).

35. Park, H. W., Park, S., & Chong, M. (2020). Conversations and medical news frames on Twitter: Infodemiological study on COVID-19 in South Korea. *Journal of Medical Internet Research, 22*, e18897–e18897. https://doi.org/10.2196/18897 (accessed on 12th September 2025).

36. Patle, N., Raj, A. B., Joseph, C., & Sharma, N. (2021). Review of fibreless optical communication technology: History, evolution, and emerging trends. Journal of Optical Communications, 45, 679–702. https://doi.org/10.1515/joc-2021-0190 (accessed on 12th September 2025).

37. Qian, X., Xie, M., Ren, M., & Gao, H. (2024). Analysis and research on abnormal of shore-ship VoIP call forwarding. 2024 IEEE 4th International Conference on Data Science and Computer Application (ICDSCA). https://www.semanticscholar.org/paper/a491a4450567c96e67baa5856db8a2ba0e9a734c (accessed on 12th September 2025).

38. Raja, C., Ramachandran, M., Ramu, K., & Sivaji, C. (2024). Radio resource management satellite communication networks MCDM method. *Computer Science, Engineering and Technology*. https://www.semanticscholar.org/paper/705e03af4b8353feae9d716f6c7af71faff9c19b (accessed on 12th September 2025).

39. Saini, T. K., Kushwaha, N. K., & Roy, A. K. (2025). Media access techniques for efficient channel utilization in SATCOM interactive networks. 2025 First International Conference on Advances in Computer Science, Electrical, Electronics, and Communication Technologies (CE2CT). https://www.semanticscholar.org/paper/88a504249e6ceafc1b160a51017c63fab623378a (accessed on 12th September 2025).

40. Sinha, S. R., Sullivan, L. R., Sabau, D., San-Juan, D., Dombrowski, K., Halford, J. J., Hani, A. J., et al. (2016). American Clinical Neurophysiology Society guideline 1: Minimum technical requirements for performing clinical electroencephalography. Journal of Clinical Neurophysiology, 33, 303–307. https://doi.org/10.1097/wnp.0000000000000308 (accessed on 12th September 2025).

41. Smith, J. A., & Judd, J. (2020). COVID-19: Vulnerability and the power of privilege in a pandemic. Health Promotion Journal of Australia, 31, 158–160. https://doi.org/10.1002/hpja.333 (accessed on 12th September 2025).

42. St. Denis, R. W. (2025). Optical data transmission technology for fixed and drag-on STS payloads umbilicals. Volume 1: Executive Summary. https://

core.ac.uk/download/pdf/42861483.pdf (accessed on 12th September 2025).

43. Svynchuk, O., Barabash, O., Nikodem, J., Kochan, R., & Laptiev, O. (2021). Image compression using fractal functions. Fractal and Fractional, 5, 31. https://www.semanticscholar.org/paper/c60a511b90f587b8d3b0b09344543b5daf306e7b (accessed on 12th September 2025).
44. Tataria, H., Shafi, M., Molisch, A. F., Döhler, M., Sjöland, H., & Tufvesson, F. (2021). 6G wireless systems: Vision, requirements, challenges, insights, and opportunities. Proceedings of the IEEE, 109, 1166–1199. https://doi.org/10.1109/jproc.2021.3061701 (accessed on 12th September 2025).
45. Vetrivel, S. C., Sabareeshwari, V., & Sowmiya, K. C. (2024). Artificial intelligence in communications. Advances in Computational Intelligence and Robotics book series. https://doi.org/10.4018/979-8-3693-3775-2.ch008 (accessed on 12th September 2025).
46. Wu, H., Wang, J., & Cai, P. (2017). FPGA based research on safety information transmission mechanism of rail transit. https://www.semanticscholar.org/paper/77eec0643ad3cda0aebeb8a1c2c8fb7a2c317902 (accessed on 12th September 2025).
47. Zakaria, Y. A., Hamad, E., Elhamid, A. S. A., & El-Khatib, K. (2021). Developed channel propagation models and path loss measurements for wireless communication systems using regression analysis techniques. Bulletin of the National Research Centre, 45. https://www.semanticscholar.org/paper/6e1447fae264c461668910c522d62e6320c87f1c (accessed on 12th September 2025).
48. Zhou, H. (2016). Research on design and implementation of ZigBee gateway system based on Internet of Things. https://www.semanticscholar.org/paper/fa18a742fc1a62b08028f37d8528e862c8133f92 (accessed on 12th September 2025).
49. Калач, А. В., Смоленцева, Т. Е., & Акатьев, Я. А. (2024). The use of social media as the main communication tool with the user to support decision-making in the organizational system of higher education. Моделирование, Оптимизация и Информационные Технологии. https://www.semanticscholar.org/paper/2dea9f7e03fbee3c9a1ba65ff09d069a965b1cf6 (accessed on 12th September 2025).

CHAPTER 9

SPREAD SPECTRUM AND MULTIPLE ACCESS

Contents

INTRODUCTION

The evolution of communication systems has witnessed revolutionary technology that has revolutionized the manner in which we communicate and share information. The technique of spread spectrum and multiple access schemes is one of the major developments in the field, offering secure and efficient communication in more than one channel. Spread spectrum techniques use bandwidth in an innovative way, minimizing interference and enhancing security for the received signal. This introduction will address the origin principles of these technologies, taking into consideration their significance in modern communication systems and capabilities in facilitating applications from mobile phones to IoT. With increased network complexities in communication networks, understanding the principles of spread spectrum and multiple access becomes increasingly crucial for researchers and practitioners to effectively deploy and utilize in various environments. Additional analysis indicates that the application of spread spectrum methods is central in achieving dependable communication despite increasing demands and challenges. Such techniques as Frequency Hopping Spread Spectrum (FHSS) and Direct Sequence Spread Spectrum (DSSS) are vital in overcoming problems such as multipath fading and jamming of a signal. In addition, there are several access methods including Time Division Multiple Access (TDMA) and Code Division Multiple Access (CDMA) that support multiple users utilizing the same channel with little interference.

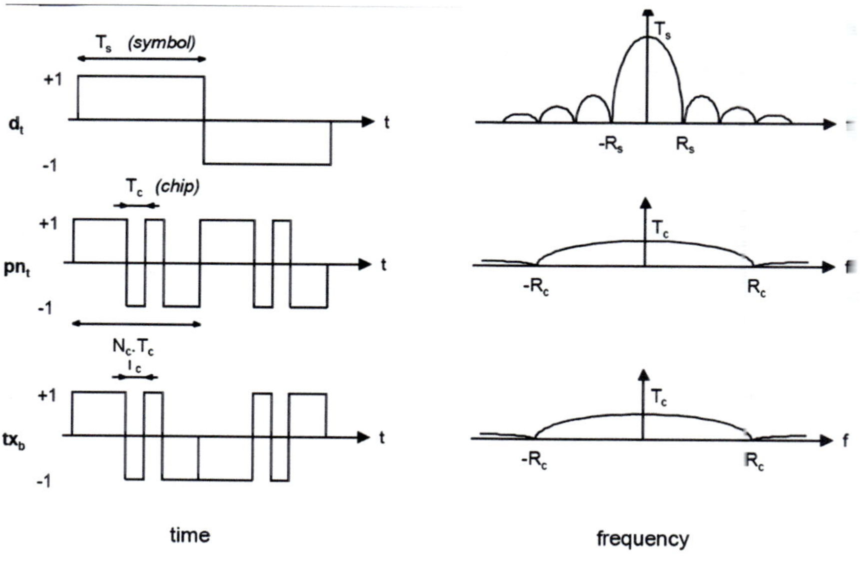

Figure 9.1. *Spread spectrum and multiple access.*

Source: https://image.slideserve.com/627738/spread-spectrum-multiple-access-8-l.jpg.

According to existing literature, combining these approaches is key to high-density applications anticipated with the future deployment of 5G technology, which not only seeks to optimize bandwidth usage but also enhances user experience. The high demand for seamless connectivity warrants ongoing research efforts to create such technologies and extend existing limitations in communications systems. Spread spectrum and multiple access techniques are significant beyond mere efficiency; they spur a new era of communications by facilitating smart and networked systems. Technologies such as the IoT take advantage of these advancements in ways that enable smart devices to interface and process information without human intervention, according to early research. The conjunction of these modes of communication with new technologies plays a key role in identifying future applications and promoting their application. Researchers must get past many problems, including optimizing integration and flexibility among communication protocols, and greater energy efficiency, in order to push the potential of these systems. By embracing the challenges and opportunities presented by spread spectrum and multiple access, the communications world can achieve a level of connectivity and capability never seen before, propelling society towards a digitally-fostered future (Figure 9.1).

Definition of Spread Spectrum

The spread spectrum is a fundamental breakthrough in communication systems, particularly in signal integrity enhancement and interference reduction. Spread spectrum techniques merely entail spreading the signal across a wider frequency band than is absolutely necessary for transmitting it. Through this method, the signal becomes more resistant to jamming and eavesdropping and can be accessed by several users on the same frequency channel without significant degradation in quality. This is achieved through modulation methods such as Frequency Hopping Spread Spectrum (FHSS) and Direct Sequence Spread Spectrum (DSSS), which respectively alter the carrier frequency or spread the signal by multiplying the same with a pseudorandom noise sequence. The ultimate objective of spread spectrum is not only to maximize security and reliability but also to optimize the use of available bandwidth, making communication channels more dynamic and efficient resources. Spread spectrum technology is also significant in multiple access communication systems, which are required to provide access of a number of users to one single communication medium simultaneously. Techniques such as Code Division Multiple Access (CDMA) use concepts of spread spectrum by allocating special codes to each user and allowing them to transmit information over the same band of frequencies without compromising the signal integrity. Co-channel interference is reduced, thus the quality of the communication channel is enhanced. In this case, the integration of spread spectrum and multi-access structures provides an avenue to maximize system capacity and efficiency, which is a significant advancement in telecommunications. The benefits of such systems have widely been utilized in numerous applications, from mobile telephony networks to satellite communications, and signify a paradigm shift in the way communication technologies evolve. Besides enhancing the efficiency and reliability of communication, spread spectrum technology is also credited with addressing security and privacy concerns in data transmission. Spread spectrum signals' vulnerability to being intercepted is because they are based on broadband; it requires tremendous resources for an unauthorized party to intercept and decode the signal. This protection feature is particularly vital for applications with sensitive information, including financial transactions and confidential messaging. In addition, the combination with advanced coding techniques enhances its noise and interference resistance even more. Therefore the evolution of spread spectrum technologies contributes not just to higher transmission quality but also to a safe and secure mode of communication systems. The ongoing advancements in this field continue to drive innovation in numerous fields, solidifying its pivotal role in modern-day communication systems, as well as its relevance to the broader implications of data release as conducted in studies such as Cinelli et al. (2020); Giannozzi et al. (201[illegible]) (Figure 9.2).

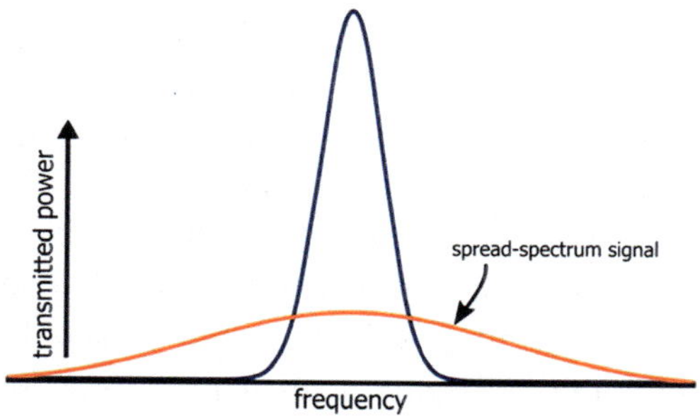

Figure 9.2. *Spread spectrum.*

Source: https://www.allaboutcircuits.com/uploads/articles/TB_SSRF_2.JPG.

Importance of Multiple Access in Communication

The need for multi-access techniques in communications systems is underscored by the rapidly growing number of mobile and internet subscribers around the world. As the number of unique mobile users approaches five billion, and total mobile connections approach over eight billion, efficient delivery of communications resources is more of an imperative than ever. Multi-access techniques permit numerous users to use the same frequency band in an efficient manner, optimizing utilization of available bandwidth and increasing the overall system capacity. Such advancements are particularly important in a scenario where demands for quality of service or experience (QoS/QoE) are becoming more and more demanding. As was exemplified through recent research, optimization of spectral efficiency by leveraging state-of-the-art multiple access technology allows communication networks to carry higher data rates and more users simultaneously, hence addressing the evolving demands of modern connectivity (Reddy, 2024). With the demands of the growing digital universe, it becomes crucial that several access strategies be investigated to maximize the performance of communication. Application of methods such as Overloaded Spread Spectrum Orthogonal Frequency Division Multiple Access (OFDMA) has been shown to be extremely efficient in maximizing spectral efficiency, with studies suggesting that an overload of 24% will lead to up to a 33% increase in efficiency. By the optimization of performance of mobile communication systems, such approaches not only alleviate congestion caused by greater demands from users but also offer mechanisms whereby increased-quality service may be delivered between networks. Developments in multi-access methods thus represent a very basic threshold in telecommunications necessary to ensure the quality and development of world communication systems (Reddy, 2024). In addition, the evolving nature of wireless communication calls for continuous investigation into various access methodologies to suit the variety of applications that emerge with technological developments. There are several areas that gain from such advancements, ranging from mobile computing to the Internet of Things (IoT). Exponential growth of multimedia content over the internet has a significant impact on user experience, raising the bar for efficient handling of multiple forms of data at the expense of various competing objectives such as delay, throughput, and power use. Providing easy connectivity through various methods of access opens up a more interconnected world, addressing ever-increasing levels of user experience, multiplicity of applications, and network efficiency. Therefore, the pursuit of better multiple access solutions remains core to the evolution and viability of communication systems in our age of digital transformation (Afzal et al., 2019).

9.1. FUNDAMENTALS OF SPREAD SPECTRUM

The spread spectrum technology is founded on the principle of placing a wide-bandwidth signal within a narrower channel, which actually maximizes communication system performance. Through a technique of spreading a signal across a wider frequency band than the lowest required, spread spectrum systems inherently provide resistance to most forms of interference as well as the general security of the transmission. This modulation scheme allows for simultaneous transmission of several signals through a single channel with negligible cross-talk, and thus it is a highly efficient technique for multiple access systems. Direct-sequence spread spectrum (DSSS), specifically, uses a pseudo-random noise code as the channel modulation of the data signal, thus creating a signature per channel. Such features not only utilize the maximum bandwidth most efficiently but also facilitate efficient communication under unfavorable environmental conditions. In addition to noise tolerance, adoption of spread spectrum techniques eliminates multipath fading effects to a great extent, which is a common wireless communication problem. Multipath fading occurs when signals arrive at the receiver by more than one path, and the interference is destructive in nature. The spread spectrum technology, through its diversity, enables the receiver to capture beneficial signals even if it is subjected to such interference. Contemporary systems with these technologies also have increased capacity and spectral efficiency. Spread spectrum protocols also often come with other technologies like Code Division Multiple Access (CDMA) so that several users utilize the infrastructure simultaneously with minimal interference. The coupling of wideband signaling and user-specific coding presents an effective means of processing communications in cluttered environments, demonstrating the adaptability of spread spectrum application. The incorporation of quantum principles in spread spectrum technology further widens its applications and research and development frontiers. Decomposition of creation operators provides a rigorous mathematical framework for investigating quantum direct-sequence spread spectrum multiple access communication systems, enlightening us into quantum signal dynamics. Employing coherent states as quantum signals to communicate, as highlighted in the literature, researchers show that inter-user interference manifests as only an additive component in the output coherent state, ensuring quantum transmission purity. Furthermore, in systems that use Fock states—such as single-photon states—entanglement potential provides complex applications in secure communications. Such new techniques, as described in contemporary literature, will solve major problems in the areas of quantum communications and signal processing and demonstrate the revolutionary evolutionary leap in the underlying spread spectrum technology knowledge (Dastgheib et al., 2024, 2025).

9.1.1. Historical Development of Spread Spectrum Techniques

Historical development of spread spectrum techniques is a result of 20th-century advancements in communication technology. Spread spectrum techniques were initially conceived in World War II and proposed to enhance military communication system security levels from jamming and interception.

Notable pioneering work by scientists George Antheil and Hedy Lamarr, who invented frequency-hopping spread spectrum (FHSS), set the stage for more sophisticated applications. Their study focused on the idea of changing frequencies rapidly to transmit information securely, something that was vital during war. That early technology paved the way for further research and experimentation, ultimately finding its way from military use into consumer goods. Spread spectrum techniques began to catch up in less unusual systems by the 1970s, providing the potential for greater bandwidth and improved immunity to signal interference. The 1980s witnessed a phenomenal milestone in the application of spread spectrum technology, most significantly the advent of code-division multiple access (CDMA). In this technique, unique spreading codes were applied to facilitate the use of the same frequency band by multiple users simultaneously. The arrival of CDMA revolutionized the face of mobile communication by providing greater capacity and call quality. One of the major highlights of this era was the standardization of CDMA as a telecommunication system, especially in the second-generation (2G) mobile phone networks.

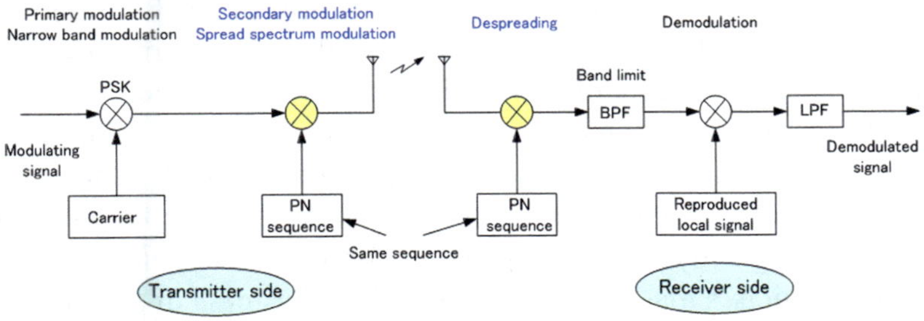

Figure 9.3. *Historical development of spread spectrum techniques.*

Source: https://www.edn.com/wp-content/uploads/contenteetimes-images-00wirelessnetworking-septoct-spread-spectrum-modulation-demodulation-psk.png?is-pending-load=1.

The practical benefits of spread spectrum technology became more prominent with the phenomenal increase in wireless communication demand, as it resulted in increased research and development investment. By highlighting the various aspects of these methods, telecommunications institutions adopted them, valuing their value in managing multiple users and enhancing overall system performance, which has long-term implications in the creation of communication infrastructures. With our advances into the 21st century, the role played by spread spectrum methods in communications systems has become progressively more vital as a result of the emergence of mobile technology as well as the necessity for rising data rates. The transition into 4G and 5G networks also leverages spread spectrum technologies to overcome the requirements introduced by population growth patterns and data traffic. It has transitioned into ultra-wideband signals and next-generation modulation techniques leveraging the presence of wide spectra for better performance. These innovations are needed with the increasing need for wireless applications and services, particularly with the anticipated rollout of 6G technologies, which will need access to previously unseen ranges of spectrum, as noted in recent wireless system research (Tataria et al., 2021). The course of historical development of spread spectrum techniques reveals continuous improvement aimed at accommodating the need for novel paradigms of communication, demonstrating their inherent character in shaping the future of connection (Figure 9.3).

9.1.2. Key Principles of Spread Spectrum

The main concepts of spread spectrum technology form the foundation of modern communication systems, particularly in terms of reliability and security improvement. Spread spectrum techniques tend to be characterized by the capacity to spread the transmitted signal over a wide frequency bandwidth as compared to narrowband techniques. The spreading is achieved through various modulation schemes such as frequency hopping and direct sequence spread spectrum. By dividing the information signal over a greater frequency range, these techniques greatly reduce the potential for interference and jamming, and the communication is made more resilient. The robustness of spread spectrum is essential where there is a requirement for the integrity of signals, such as with military communications or densely populated urban areas where there are numerous devices competing for the limited frequency resources. Another significant aspect of spread spectrum theories relies on multiple access methods which accommodate numerous users accessing a single communication channel simultaneously. This plays a significant role in managing the limited spectrum available for wireless communication. Some of the most popular multiple access schemes based on spread spectrum are Code Division Multiple Access (CDMA) and Time Division Multiple Access (TDMA). CDMA, for instance, uses personal spreading codes assigned to each user, which facilitate concurrent transmission without significant interference. This aspect enhances the capacity and efficiency of the systems, enabling many users and quality of communication. Other people just perceive each users data as noise, practically offering privacy and security. Consequently, the application of spread spectrum principles not only optimizes the use of available bandwidth but also enhances the security of information provided. In recent times, the development in radio frequency technology and the emergence of new radio frequency bands also renewed interest in spread spectrum techniques in light of the anticipated launch of 6G wireless networks. Since the systems are evolving to operate over frequencies in the THz spectrum, e.g., above 100 GHz, the application of spread spectrum principles becomes increasingly relevant. The high-frequency bands possess unique propagation features that necessitate unconventional approaches and necessitate novel solutions to enhance signal quality and performance. Concepts like reconfigurable intelligent surfaces are now starting to come into being, which have the ability to manage the physical characteristics of radio waves to neutralize the impact of natural propagation effects. This development in technology necessitates a reconsideration of traditional communication models, highlighting the continued applicability and adaptability of spread spectrum ideas in emerging wireless communication paradigms (Theodore et al., 2019; Ertuğrul et al., 2019).

9.1.3. Spread Spectrum Techniques

Spread spectrum techniques are critical in modern communication systems primarily due to their ability to make signals more immune to interference and eavesdropping. Direct Sequence Spread Spectrum (DSSS) is perhaps the best-known technique where the data signal is spread by modulating the data signal with a pseudorandom sequence of noise so that the output signal has a wider bandwidth. This spread makes the signal intrinsically more resistant to jamming and allows many users to share a common frequency channel simultaneously, hence multiple access. The strength of DSSS continues to resonate in the 5G networks, where it is the foundation for the non-orthogonal multiple access (NOMA) scheme that allows efficient sharing of resources by users within the same blocks, increasing system capacity and user experience as documented in the literature on NOMA innovation (Ding et al., 2017). One of the most popular techniques is Frequency Hopping Spread Spectrum (FHSS), that avoids interference by way of the frequency shift at a high rate according to a predetermined sequence. This technique disperses the signal over a vast frequency band over time, minimizing the impact of narrowband interference. In jammer-heavy environments, FHSS enjoys a significant advantage in the sense that it becomes difficult for adversaries to predict the transmission frequency. Its application within other wireless protocols, such as Bluetooth, demonstrates its value and effectiveness. Its resilience and flexibility remain significant in today's evolving landscape of communication systems, particularly in structurally dynamic systems such as those intended for application in mobile edge computing applications (Mao et al., 2017). Through the advantageous frequency changes made at strategic points in time, FHSS not only offers security but also optimizes frequency utilization in dense spectrum areas. A third approach, Time Hopping Spread Spectrum (THSS), employs time as the means of spreading the signal. In THSS, the timing of transmission of signal bits is modified according to a pseudorandom sequence, allowing the signal to cover vast bandwidths while eliminating opportunities for interception. The approach has a powerful application in secure communication since it provides an added layer of privacy and security against misuse. All of these spread spectrum methods have unique functions in supporting the broader goal of improving communication efficiency and security. As telecommunications evolve further, especially with technologies like mobile edge computing and advanced resource management frameworks, the use of spread spectrum methods is proportionally more necessary in bolstering the rising requirements for high-bandwidth, low-latency, and reliable connectivity (Mao et al., 2017).

9.2. TYPES OF SPREAD SPECTRUM TECHNIQUES

The finding of spread spectrum techniques reveals two dominant techniques: Direct Sequence Spread Spectrum (DSSS) and Frequency Hopping Spread Spectrum (FHSS). DSSS spreads in the eponymous manner by multiplying the data signal by a pseudorandom noise sequence, thus dispersing the signal across a larger bandwidth. This technique provides improved immunity against interference and multipath fading and hence is appropriate for applications where security and reliability are needed. FHSS, on the other hand, achieves spreading by rapidly hopping the carrier frequency between a predetermined set of channels, allowing the signal to "hop" across the frequency spectrum. This method offers resistance to narrowband interference because its high-frequency rates make it difficult for potential interceptors to capture the full signal, thus guaranteeing maximum communication stealth and security. Concurrently, these techniques have important functions in optimizing bandwidth usage and safeguarding the integrity of the signal in complex communication systems (Benvenuto et al., 2021). Spread spectrum techniques have real applications in most wireless communication areas, significantly influencing their performance and efficiency. More and more, as telecommunication progresses, managing bandwidth takes on greater significance. Particularly, Direct Sequence Code Division Multiple Access (DS-CDMA) employs DSSS methods such that various users can share the same spectrum with minimal interference such that capacity is increased. Similarly, hybrid techniques such as Multi-Carrier Spread Spectrum (MC-SS) blend the strengths of multi-carrier modulation and spread spectrum methods. These developments not only have improved spectral efficiency but also resilience and flexibility to accommodate advanced communication demands. The development of such cutting-edge systems brings to the forefront a deeper imperative: the optimal use of the frequency spectrum, a limited resource that remains essential to wireless applications today (Fazel et al., 2008). With ongoing advancements in spread spectrum and multiple access communication systems, new methodologies keep redefining operational models. The union of multi-carrier modulation with spread spectrum technologies has brought forth cutting-edge solutions like Orthogonal Frequency-Division Multiplexing (OFDM) applied in combination with CDMA methods. Next-generation technologies like these are designed to meet the pressing demands for increased data rates and reduced latency in fixed and mobile wireless communications. More importantly, the use of these hybrid schemes in infrastructure like broadband multimedia services positions them at the leading edge of telecommunication today. The convergence of such advanced methods across numerous global standards reflects an overall trend toward wireless system optimization to facilitate continued progress in spectral efficiency and connectivity as targeted by future communications generations (Fazel et al., 2008).

9.2.1. Frequency Hopping Spread Spectrum (FHSS)

FHSS is a significant communication technology innovation that addresses the growing need for security and interference in wireless systems. Through its capability to rapidly switch frequencies during transmission, FHSS lowers the risk of detection and interception and thus the robustness of the communication link. The technique divides the bandwidth into a number of channels and uses a pseudorandom sequence to hop among these frequencies such that the signal cannot be traced back to potential eavesdroppers. As the need for secure communication in sectors like the military and commercial applications continues to rise, FHSS's resistance to jamming and improvement of signal integrity grows more crucial and emerges as a pervasive characteristic in modern communication systems (Charitha et al., 2025). In addition to security benefits, FHSS provides effective multiple access characteristics, allowing several users to share the same frequency spectrum with minimal interference. This is particularly useful in high-density situation areas, e.g., city areas or events with high-density communication channels that may be overwhelmed. Each user is assigned a unique frequency-hopping sequence, which can facilitate communication simultaneously with guaranteeing data integrity. The adaptive potential of FHSS is crucial for achieving optimal performance, especially in dynamic Signal-to-Noise Ratio (SNR) environments and hence a general-purpose solution for all types of communication situations (Charitha et al., 2025). With evolving wireless systems, utilizing FHSS in combination with other modern techniques can be maximized in terms of system performance as a whole, with greater utilization of resources and more users. Moreover, the marriage of FHSS with emerging technologies, for instance, Quantum Random Number Generators (QRNG), provides new ways to enhance security and reliability of communication systems. Utilization of QRNG can produce randomly generated frequency-hopping sequences, which evidently enhances resistance against jamming and interception attacks. This new technology can especially be of significant use for communication in applications like drone communications, where reliable and secure data transmission is imperative. Through the application of secure communication protocols and robust error handling, the dependability of FHSS systems can be significantly enhanced, thereby enabling crucial operations within any sector, including military and emergency response. In most cases, the continuous development of FHSS, particularly when combined with cutting-edge technologies, makes it a fundamental building block in the development of secure and effective communication systems (Figure 9.4).

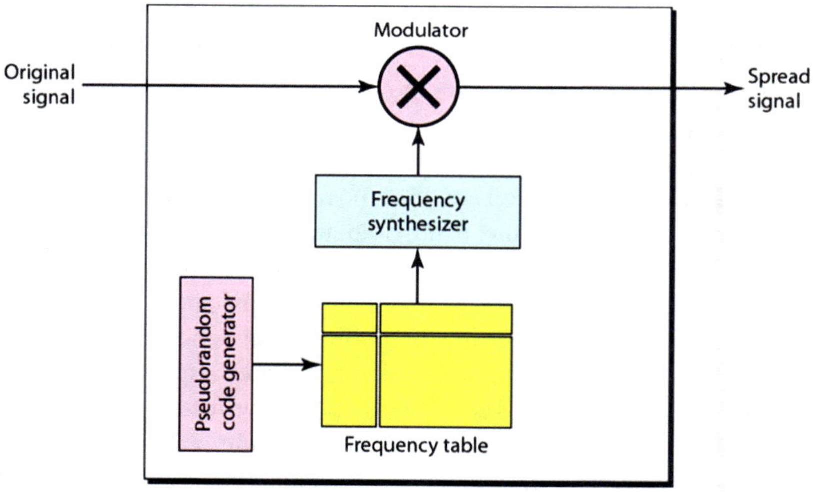

Figure 9.4. *Frequency hopping spread spectrum (FHSS).*

Source: https://image1.slideserve.com/2086210/frequency-hopping-spread-spectrum-system-l.jpg.

9.2.2. Direct Sequence Spread Spectrum (DSSS)

DSSS is one of the most important technologies in spread spectrum communication, characterized by its ability to minimize interference and enhance security in wireless systems. Basically, DSSS accomplishes this by spreading the original signal over a bandwidth larger than the minimum required, through the use of a unique spreading code. This technique makes the signal power distribute in a broader bandwidth, making it immune to narrowband interference and intercept. In real life, DSSS finds large-scale application in numerous applications, e.g., WLANs and GPS, and is found to be successfully used in real environments where communication has to be rugged. The inherent advantages of DSSS, including improved resistance to jamming and multipath fading, highlight the pivotal role that is played by DSSS in driving contemporary communication systems in the spread spectrum and multiple access approach context. It is on this basis that the DSSS technology rests, whereby the pseudo-random noise (PN) sequence is used for spreading and extraction of signals. The pseudo-random code, or the spreading code, is employed at a frequency greater than that of data to be sent, and the output produced by it is a combination that appears random in nature.

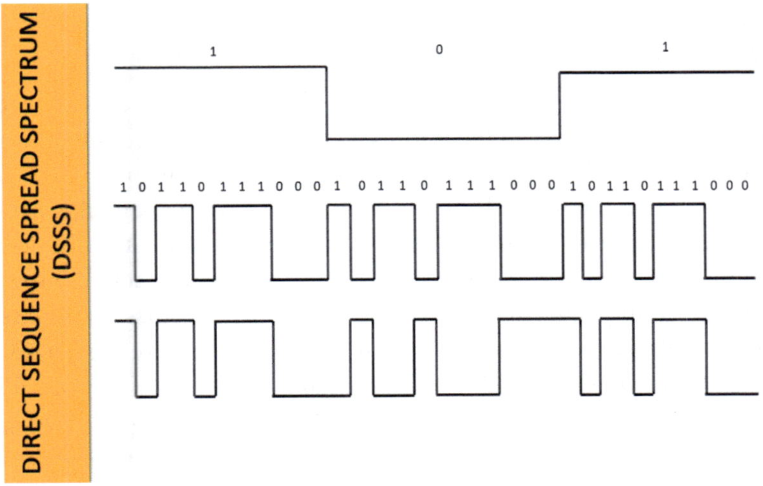

Figure 9.5. *Direct sequence spread spectrum (DSSS).*

Source: https://i.ytimg.com/vi/MhKeypgepOc/maxresdefault.jpg.

This aspect not only grants privacy to the communication but also allows multiple users to operate within the same band without significant interference, an extremely valuable feature in multi-access communication systems. For efficient processing and synchronization, the receiver must match the received PN sequence so that the desired signal could be filtered from the noise. The fact that DSSS can be used in combination with several modulation schemes only increases its practicality, offering a perfect solution for a broad variety of applications from military communication to smart city infrastructures, where secure and stable data transmission is a priority (Kanellopoulos et al., 2023).

Since urbanization and communication sophistication are constantly changing, the integration of DSSS in smart city systems and Internet of Things (IoT) strategies has been increasingly important. Smart cities use DSSS for efficient communication for large sensor networks and devices, making resource management simpler and enhancing the quality of urban life. For instance, deployment of DSSS in intelligent transport and smart grid applications guarantees improved data integrity and reduced error rates, which are essential for effective operational management in urban areas. Moreover, scalability provided by DSSS supports scaling of IoT applications for communication in various industries such as healthcare, environmental monitoring, and energy management (Eshrat et al., 2023). The combination of DSSS and smart city concepts is the best example of the technology's ability to ensure innovation within cities, showing its relevance to modern communication systems (Figure 9.5).

9.2.3. Time Hopping Spread Spectrum (THSS)

THSS is a pioneering work in the overall scenario of spread spectrum techniques, and particularly in the improvement of communication security and effectiveness. This technique allows frequency modulation so that interference is minimized to the maximum level, as well as improving channel capacity. By conducting a process of time-hopping, in which data bits are transmitted through a pre-determined sequence of time slots, THSS performs effectively to counter the downgrades in single-frequency transmission systems. This flexibility makes it particularly useful for application in situations where the conditions of the signals are fluctuating and there is the likelihood of jamming. The ability to make adjustments in the timing of transmission based on pre-patterned patterns significantly increases the strength of the communication system and makes THSS a critical commodity for application in military and commercial applications. This spread spectrum is an advancement due to ongoing response to growing demands for secure communications within today's crowded radio frequency environments. The significance of THSS goes beyond its technical operation; it also crosses over the concept of multiple access in communication. Specifically, THSS provides for more than one user to share a common transmission medium without compromising severe interference. This is accomplished by allotting specifically time slots to various users, promoting efficient use of the spectrum while exercising stringent communication. As the radio frequency spectrum has become increasingly crowded, there has been a greater demand than ever for innovative multiple access methods. THSS solves these problems by allowing for simultaneous transmission over the same bandwidth, thus overcoming the restrictions that come with traditional methods like Code Division Multiple

Access (CDMA) and Frequency Hopped Spread Spectrum (FHSS). Besides, dynamic transmission slot adjustment is a means of enhancing system-wide throughput and is a desirable option for usage ranging from cellular networks to Internet of Things (IoT) networks. In the face of ubiquitous security and reliability concerns that are present in today's communication, THSS is an appealing remedy. Its robustness against disruptions particularly appeals to individuals who want secure transmission under hostile settings. By integrating time-hopping sequences with random signal features, systems can offer a higher jamming resilience level, as evidently depicted in research indicating improved Bit Error Rate (BER) performance. These solutions represent a fusion of proven spread spectrum techniques with sophisticated methods in signal security, thus enhancing the development of dependable communication systems. With the advancement of technology, research into enhancements in THSS could yield worthwhile dividends, particularly in regards to spectral efficiency and dynamism within shifting operating environments. Continued research into these advancements remains crucial in a bid to ensure that communication technologies remain able to keep pace with evolving issues against a backdrop of ever-changing digital surroundings.

9.3. ADVANTAGES OF SPREAD SPECTRUM

Advantages of spread spectrum technology are numerous, especially in terms of enhancing the reliability and robustness of communication systems. One of its primary advantages is its phenomenal ability to minimize interference, and which is necessary in high-electromagnetic noise environments. By spreading the signal over a wider bandwidth, spread spectrum techniques such as frequency hopping significantly reduce the probability of interference interrupting the transmission to enable clearer communication, even in adverse conditions. This feature is particularly useful in military operations, wherein stealth of communications and anti-jam capabilities are critical. The use of pseudo-random frequency changes allows the system to provide operational integrity, and resolve possible interference without affecting performance or clarity (Wei et al., 2024). Consequently, spread spectrum is a vital technology used in a wide range of applications, and its application ranges from military use to commercial use in fields such as telecommunications. Another important advantage of spread spectrum is that it is very insensitive to interception as well as eavesdropping. The inherent intricacy of the method of modulation makes it challenging for illegal receivers to decipher the transmitted signals. This aspect is most essential in secure communication, where privacy is important. For instance, Direct Sequence Spread Spectrum (DSSS) and similar methods can effectively conceal the signal such that only intended receivers can access the information being transmitted. This advantage is most desirable in light of the escalating concern over data security in light of the escalating pattern of digital communication. Secondly, using spread spectrum techniques in advanced systems, such as those for the SS-GFDM-based integrated Satellite-Terrestrial communications, enhances end-to-end security and quality of service at low signal-to-noise ratio (SNR) (Yang et al., 2019).

The combination of strong security and high performance further enforces spread spectrum as one of the most important technologies for today's communication infrastructures. Besides, the efficiency and flexibility of spread spectrum methods render them easily applicable to various multiple access methods, allowing various users to use the same band of frequency without substantial interference. Such a capability is crucial in crowded networks, where bandwidth is usually a limitation. Spread spectrum methods, particularly Code Division Multiple Access (CDMA), enable efficient utilization of the available bandwidth through concurrent transmission by various users. A user is assigned a particular code utilized in hiding his or her signal, and the communications may be transmitted simultaneously without interference with one another. This is more relevant in modern systems where spectrum-efficient use is most critical, particularly in the aftermath of 5G networks and beyond. The complementary nature of spread spectrum with ALOHA-type protocols even makes its efficiency justifiable and places it in the limelight for future wireless communication systems (Yang et al., 2019). With the industry moving towards consolidated and fault-tolerant systems, the advantages of spread spectrum technology are bound to go a long way in shaping the direction of communications in the future (Figure 9.6).

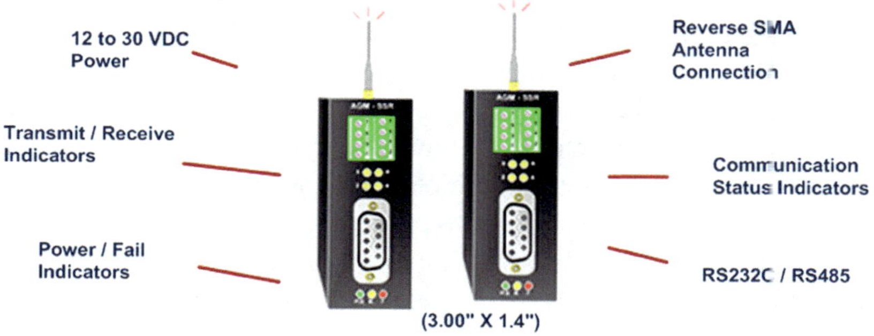

Figure 9.6. *Advantages of spread spectrum technology.*

Source: https://agmelectronics.com/Products/SSR/Images/GraphicSSRNew1Lge.jpg.

9.3.1. Resistance to Interference

Interference resistance is a critical parameter in the development of spread spectrum and multiple access communication systems. Such systems enhance signal integrity, particularly in electromagnetic noise-ridden environments and potential blocking from other communications systems. Spread spectrum techniques distribute signals across a wide frequency range, reducing interference impacts, enabling safe data transfer even in poor conditions. For example, employing direct sequence spread spectrum (DSSS) enables the signal being sent to be covered by a pseudo-random noise sequence, hiding the signal and making it less susceptible to interference. This technique not only enhances the immunity of the communication system but also enables the facility of numerous users to operate under a single frequency band without experiencing significant performance degradation, thereby overcoming a fundamental problem in wireless networks.

As time progresses and improvements in communication systems are made, the need for greater immunity against interference has grown, particularly with the emergence of 5G and the imminent 6G systems. With the increasing demand for high-bandwidth applications, the mobile traffic growth implies that interference reduction is more crucial than ever. Techniques such as millimeter-wave frequencies, as part of these future networks' crucial elements, are more prone to atmospheric absorption and scattering. Hence, researchers underline the importance of multiple access techniques with the ability to dynamically adapt to the dynamic interference conditions. The integration of high-level algorithms and machine learning methods into network management can be critical in optimizing resource utilization and user experience via the mitigation of the effects of interference on communication dependability and quality, as highlighted by Zhang et al. (2019). With this evolving context, accurate understanding and actualization of resistance to interference mechanisms are critical to the future of communication systems. The exploration of diverse frequency bands for 6G, i.e., frequencies up to the terahertz regime, underscores the intrinsic complexity imposed by higher frequencies such as increased sensitivity to interference and obstacles. Counterbalancing these issues necessitates a holistic system design approach including advanced modulation and coding techniques, novel multiple access methods, and real-time signal processing capabilities as outlined in Tataria et al. (2021). These advancements must be deeply researched to formulate viable solutions that can operate optimally in complex environments to enable communication systems to sustain a high level of performance amid growing demands and barriers to interference (Figure 9.7).

Figure 9.7. *Resistance to Interference technologies communication systems.*

Source: https://karlobag.eu/images/upload/y4kyk.jpg.

9.3.2. Enhanced Security Features

In communication systems, other features of security are most critical, particularly in adopting spread spectrum and multiple access techniques. Such systems, as they are intended to transmit information over several channels, are inherently privileged by security features adopted to protect the data integrity and confidentiality. Spread spectrum technology, which spreads transmitted signals over a wide frequency band, avoids threats of interception and jamming as the signal is no longer recognizable to unauthorized users as noise. Secondly, the application of advanced algorithms in multiple access systems ensures that only authorized users are allowed to use particular channels, thereby providing the least possible opportunities for unauthorized access or leakage of data. Secondly, the increased application of artificial intelligence in communications has encouraged the development of sophisticated security tools. For instance, with technologies like ChatGPT and other generative AI technologies emerging across industries, they offer opportunities and challenges for data protection and security. Sophisticated security features offer protection for secure use of such technologies through stringent measures against conceivable cyber threats and vulnerabilities. As the organizations are increasingly reliant on such AI driven models of communication and data management, addressing ethical aspects of privacy highlights the need for robust security mechanisms that can evolve and adapt in the fast-changing technological landscape (Yogesh et al., 2023). Finally, digital twins could be a prospective topic of improving security features in spread spectrum and multiple access communication systems. By creating virtual representations of physical infrastructures, digital twins enable real-time tracking and forecasting analysis, enhancing security management decision-making capacity. With this new method, it becomes possible to identify anomalies and potential security threats, providing businesses with an active means of securing their communication infrastructures. The modularity of digital twinning also enables the integration of multi-disciplinary security solutions in a manner that closes the loop in tackling intricate vulnerabilities that are possibly undetectable through traditional means (Rasheed et al., 2020). Where these technologies intersect, the potential of merging enhanced security attributes with adaptive communications patterns has profound implications for the secure information transfer of the future (Figure 9.8).

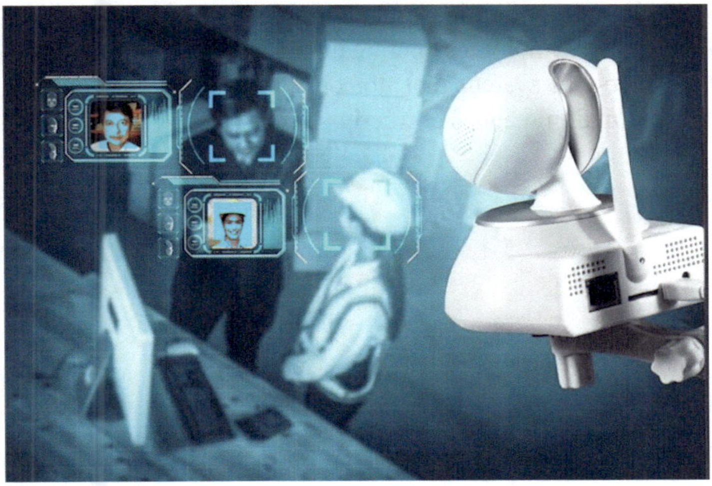

Figure 9.8. *Enhanced security features access techniques.*

Source: https://www.infolks.info/blog/wp-content/uploads/2024/04/AI-IN-SECURITY-05-1-1024x683.jpg.

9.3.3. Improved Signal Quality and Reliability

The evolution in the communication system has been aimed mostly at improving signal quality and reliability, especially with techniques such as Orthogonal Frequency Division Multiple Access (OFDMA). OFDMA employs multiple orthogonal subcarriers, which essentially remove interference that arises due to overlapping signals. This reduces carrier and symbol interference, hence making data transmission more efficient and guaranteeing clearer communication. Also, other innovations such as dynamic subcarrier allocation and adaptive modulation further make this process more accurate. The ability of OFDMA to adapt quickly in response to varying channel conditions facilitates a robust mechanism for maintaining quality across varying environments, an improvement over traditional methods like Frequency Division Multiple Access

(FDMA) or Time Division Multiple Access (TDMA). All these instances combine to create a secure system of communication, making OFDMA shine as it seeks to address the issue of signal degradation prevalent in wireless networks (Abdulwahab et al., 2024). Another key contribution to improving signal quality is the use of advanced error correction techniques. Error correction is a core element of wireless communications since it detects and corrects corrupted data resulting from interference or noise. Techniques such as Turbo and Low-Density Parity-Check (LDPC) codes greatly improve the trustworthiness of information being communicated, facilitating more reliable connections in areas prone to interference. These technologies operate in concert with methods like OFDMA, creating an end-to-end system that not only suppresses interference but also ensures the information reaches correctly.

Figure 9.9. *Improved signal quality and reliability.*

Source: https://fastercapital.co/i/Attenuation--Signal-Attenuation--The-Power-of-Negative-Feedback--Techniques-for-Improving-Attenuation-and-Signal-Quality-webp.

The resulting synergy between multi-access and error correction techniques translates to overall enhanced performance and mitigates dangers of information loss, strengthening the dependability of communications systems across a wide variety of applications (Abdulwahab et al., 2024). In the context of drone communications, improving signal quality and integrity can be achieved with the assistance of advanced technologies such as Quantum Random Number Generators (QRNG). Applying QRNG technology to Frequency Hopping Spread Spectrum (FHSS) systems allows the generation of truly random frequency sequences, which are extremely effective against interception and jamming attacks. This new combination not only guarantees enhanced spectral efficiency but also enhances the security and reliability of data communication among more than one UAV. With simulation studies and analyses, scholars have proved that QRNG-based systems have high levels of unpredictability as well as homogeneity in signal distribution. These are fundamental in providing consistent communication even in stressful situations. Therefore, the combination of QRNG with the traditional methods offers a realistic step towards the solution of the strongly interdisciplinary problems of wireless communication systems, thus making it an essential step forward towards improved signal quality and reliability (Figure 9.9).

9.4. MULTIPLE ACCESS TECHNIQUES

There are several access methods that are an essential concept in modern day communication systems and allow the bandwidth to be utilized efficiently for concurrent transmission by different users. CDMA is one such method, particularly when applied in spread spectrum systems. CDMA allows different users to transmit data on a shared frequency channel by encoding their information with different spreading codes. This method not only diminishes interference among users but also enhances the systems resistance to signal fading effects and multipath propagation. Furthermore, including quantum mechanics in CDMA, as outlined in quantum spread spectrum communication systems, will have the ability to enhance signal strength. With the intricate nature of quantum states, researchers have established a mathematical system that supplements traditional approaches, thereby augmenting communication capacity in next-generation technologies (Dastgheib et al., 2024). Development in mobile communication technology necessitates advancements in a range of many-to-one access techniques to address growing user needs and higher data rates. Overloaded Spread Spectrum Orthogonal Frequency Division Multiple Access (OFDMA) is a promising approach that achieves maximum spectral efficiency. Through this technique, bandwidth is divided into subcarriers that can be allocated to users on a dynamic basis according to their existing demand, optimizing capacity as well as throughput. In accordance with a recent analysis, utilization of a 24% overload in this context can lead to a 33% boost in spectral efficiency, establishing the method's ability to support additional users and address varying demands (Reddy, 2024). This adaptive nature of OFDMA facilitates efficient resource allocation, synchronizing multiple access techniques with the strategic goals of next-generation communication systems with priority to performance as well as user experience. The dynamic nature of multiple access techniques sheds light on the essential interrelationship between technological innovation and user requirements in communication systems. As demand for higher data rates and connectivity persists, integration of spread spectrum methods with new multiple access methods appears to be a must in order to ensure that the network is optimized. Techniques such as CDMA and

OFDMA are exemplars of this integration since they address interference and resource management issues of high density scenarios. Besides, as research investigates quantum-inspired access methods, the future of multiple access keeps advancing, presenting windows to unparalleled network capabilities and capacities. Such a comprehensive approach continues to be vital for the ongoing advancement of communication systems, showcasing the potential for innovative solutions to frightening communication challenges in the midst of a rapidly changing digital age.

9.4.1. Definition and Importance of Multiple Access

Multi-access technology is a fundamental component in communication systems in which multiple users can have access to a common channel for transmitting information simultaneously. This method is required in managing limited bandwidth efficiently, which becomes increasingly significant with the continually escalating data transmission demand under different applications such as mobile telecommunication, satellite transmission, and internet-based services. Through provision of multiple users to share bandwidth at a given time with minimized interference and maximum throughput, a number of access techniques enhance the utility and performance of communication networks. The importance of such technologies is enhanced in contemporary environments where the prevalence of congestion can have compounded impacts on performance; hence, various access methods are important in providing an effective communication network that is capable of addressing varied user requirements efficiently. There have been several methods devised under the umbrella of multiple access, each with different operational requirements and environments. Some of the most noted are Time Division Multiple Access (TDMA), Frequency Division Multiple Access (FDMA), and Code Division Multiple Access (CDMA). Each system allocates resources in a strategic manner in a fashion that reduces interference and optimizes performance. For instance, TDMA divides the time into individual slots and gives a single access to a user's own slot of time slot. The converse is when FDMA gives users particular frequency bands, while CDMA utilizes particular spreading codes to allow multiple signals to share the same spectrum but not be interfering. Such methods exhibit the versatility and usage of multiple access schemes in various demands in communication systems, thereby their extensive usage in contemporary networks. Additionally, the deployment of different access methods plays an important role in ensuring efficient management of wireless network resources, particularly as the landscape of IoT devices and service keeps evolving. With more devices accessing the internet, end-to-end seamless connectivity is a need that demands solid multiple access architectures. By ensuring efficient management of the communication among various devices, multiple access technologies allow data sharing without threatening the integrity and quality of the network. Further, algorithmic breakthroughs and processing capabilities continually enhance the effectiveness of these techniques through complementarity with spread spectrum technologies. In addition to minimizing the chance of interference, this synergy also increases overall robustness for communications systems, further underscoring multiple access relevance in light of constantly evolving technological needs and user expectations (Reimer et al., 2020; Eyring et al., 2016) (Figure 9.10).

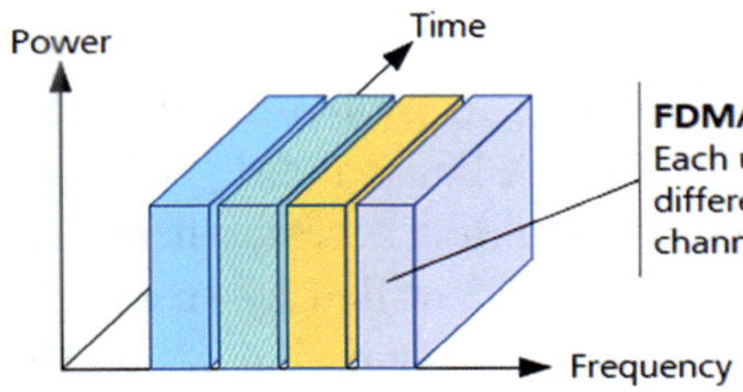

Figure 9.10. *Definition and importance of multiple access in communication systems.*

Source: https://www.researchgate.net/profile/Devasis-Pradhan-2/publication/349348812/figure/fig1/AS:991889025150976@1613496180993/Fig-31-Frequency-Division-Multiple-Access-FDMA.ppm.

9.4.2. Types of Multiple Access Techniques

The study of multiple access methods illustrates the complex framework that dictates modern communication systems, where multiple users need simultaneous access to a shared medium. TDMA is one of the leading techniques that allocates fixed time slots to users to transmit data, thereby effectively managing bandwidth allocation and interference. By organizing communication through discrete time slots, TDMA supports determinable performance, which is critical for applications with real-time data transmission. In addition, this method enhances spectral efficiency significantly, particularly beneficial in multiuser systems, such that even at maximum use, communication is seamless and reliable. With the advancement of mobile computing, especially with regard to trends such as the Internet of Things, the efficiency of TDMA and its application in other domains of technology such as mobile edge computing are merely representative of its role in reaching the potential of next-generation communication systems (Mao et al., 2017). FDMA, however, utilizes different frequencies for each user so that individual channels of transmission can take place at the same time without interference. A unique frequency band is assigned to each user, essentially slicing the spectrum into separate pieces. This method is advantageous where latency is critical, as parallel communication guarantees that there will be little delay. FDMAs immunity from frequency-selective fading makes it an excellent option in instances where unbroken signal quality is needed, i.e., radio broadcasting and satellite communication. Further, that it dynamically allocates the frequency bands makes it better suited to handle systems with fluctuating traffic patterns, an adaptive bandwidth management approach. In the Internet of Things, where numerous devices will need to go online and communicate, FDMA's capacity to enable them to transfer data without interruption is critical in order to facilitate real-time applications and enhance operational effectiveness (Al A-Fuqaha et al., 2015). CDMA is another sophisticated method of supporting multiple access, wherein all the users share a common frequency channel but are identified by specific spreading codes. This technique supports simultaneous transmission without reserved time or frequency slots and therefore reduces interference and increases capacity. The superiority of CDMA lies in its ability to support an enormous number of users on a limited bandwidth through spread spectrum technology to encode data, thus securing it and making it more reliable. CDMA has matured significantly with its application being greatly expanded in mobile communication, particularly in 3G and higher, where data rates are high and spectrum efficiency is very important. As the demand for data-intensive applications continues to rise, particularly with the introduction of 5G technologies, CDMA's principles are at the center of formulating creative solutions for scalable and robust communication systems (Mao et al., 2017).

9.4.3. Comparison of Multiple Access Methods

In communication systems, multiple means of access are essential to allow efficient transfer of data between numerous users under a single frequency band. Within such ranges, Code Division Multiple Access (CDMA) and Frequency Division Multiple Access (FDMA) stand out, each having its advantages and limitations. CDMA, particularly in its Direct Sequence Spread Spectrum (DSSS) form, employs unique spreading codes to facilitate coexistence transmission, boosting capacity and security, especially in interference environment scenarios. In contrast, FDMA distributes available bandwidth into distinct frequency bands, where each user is assigned one band. This method is straightforward and effective for managing users' coexistence but leads to wastage of bandwidth during off-peak loads, introducing a highly crucial trade off that impacts overall system efficiency. Performance of various multiple access methods can be critical examined through a number of parameters including Bit Error Rate (BER) and Signal to Noise Ratio (SNR). For CDMA in specific to DSSS, the most important factors are the ability to maintain communications quality in the presence of narrowband interference and jamming. For example, effective coding methods such as employing Barker codes can optimize system performance by enhancing anti-jam capability, thus leading to low BER. All evidence indicates how various signal powers and optimized spreading factors significantly influence power spectral density (PSD) and how their adjustments can enable jamming-free transmission in such vulnerable environments. Such analysis transcends theory, presuming its practical value in today's security-focused communication environments, where immunity to interference is crucial (Manna et al., 2024).

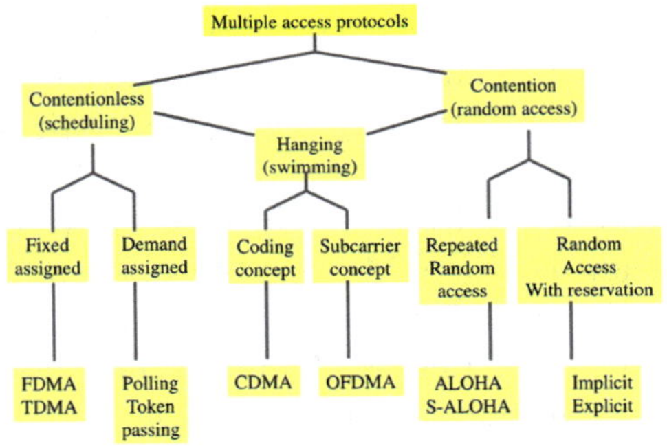

Figure 9.11. *Comparison of multiple access methods.*

Source: https://image2.slideserve.com/4811410/classification-of-multiple-access-protocol-l.jpg.

Moreover, evaluating the likelihood of mistakes in asynchronous CDMA systems is a difficult task due to the need for precise modeling under varying conditions. Techniques such as Standard Gaussian Approximation (SGA) and Simplified Improved Gaussian Approximation (SIGA) have been developed to facilitate this evaluation. Using simulation algorithms, scientists have been capable of comparing these methods and approximating their accuracy when predicting communication performance in cases involving multiple accesses. Such computational emulation not only provides an insight into the dynamics of various access methods but also aids in designers' optimization of system parameters so that they can be more efficient to use in real-world contexts. Ultimately, it is important to comprehend these relative advantages and technical intricacies in order to design communication technology, render systems resilient and efficient on various operation lands (Yip, 2025) (Figure 9.11).

9.5. TYPES OF MULTIPLE ACCESS TECHNIQUES

In communications systems, different methods of multiple access are of paramount significance in enabling numerous disparate users to share the same communication medium in an effective manner. Among the most prevalent forms of multiple access schemes are Time Division Multiple Access (TDMA), Frequency Division Multiple Access (FDMA), and Code Division Multiple Access (CDMA). TDMA operates on the principle of allotting a specific time slot to a specific user within a given frequency channel, thus preventing signal overlap and ensuring clear communication. FDMA, in contrast, divides the available frequency spectrum into separate independent channels for each user. This process allows simultaneous communication but can lead to inefficiency in spectrum usage. By contrast, CDMA takes a more sophisticated approach by assigning unique codes to each user, under which multiple signals may be put on the same channel simultaneously without colliding with one another. This underlying distinction unites the differing capabilities of these techniques. Improvements in technology in mobile communication have necessarily resulted in a shift towards more sophisticated multiple access techniques. Among these, Orthogonal Frequency Division Multiple Access (OFDMA) has received significant attention due to its inherent advantage over conventional methods. OFDMA allows users to send in parallel over orthogonal subcarriers, which leads to increased spectral efficiency and stronger immunity to interference. The technique can allocate bandwidth dynamically based on user demand, hence achieving maximum system performance. More specifically, recent research indicates that Overloaded Spread Spectrum OFDMA can enhance spectral efficiency by approximately 33% for an overload of 24%, with a considerable improvement in throughput support for mobile communication networks. Such advances indicate the necessity of continuously optimizing multiple access techniques to serve the increasing demands of users and provide support to the diverse applications that are being realized in the networked environment of today (Reddy, 2024). Quantum communication has introduced revolutionary possibilities to a number of access techniques, most significantly in the form of Quantum Spread Spectrum Code Division Multiple Access (QSS-CDMA). This novel method employs quantum mechanical effects to spread the spectrum of the signals carried, significantly enhancing the capacity and security of communications systems. By manipulating the quantum states of light, QSS CDMA inherits not only the strong interference control characteristic of conventional spread spectrum technology but also new advantages contributed by quantum physics. The mathematical theory explaining the quantum spread spectrum techniques is based on elaborate theoretical models that manifest the feasibility of their operations. With increasing research in this field, QSS-

CDMA may pave the way for upcoming multi-user quantum networks, which is a step toward rapid progress in various methods of multiple access (Dastgheib et al., 2024). Hence, exploration into both traditional and unconventional methodologies is the need of the hour for ongoing development in effective communication systems (Figure 9.12).

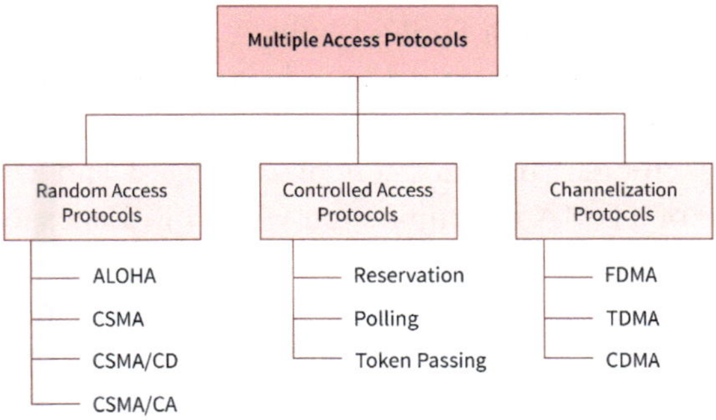

Figure 9.12. *Types of multiple access techniques.*

Source: https://external-content.duckduckgo.com/iu/?u=https%3A%2F%2Fscaler.com%2Ftopics%2Fimages%2Fmultiple-access-protocols.webp&f=1&nofb=1&ipt=94271e637fb7bfb37ecbad3551bc10161b696dfffe139106c5cef9e3c7aa0c4b.

9.5.1. Time Division Multiple Access (TDMA)

Within the modern communication networks, Time Division Multiple Access (TDMA) has become one of the major multiple access techniques best suited for scenarios where maximum use of resources is essential. TDMA achieves communication among multiple users by allocating individual time slots to each user to avoid interference and utilize the channel in its maximum capacity. This systematic process comes in very useful in usage such as cellular telephony and satellite communications, where bandwidth is a costly commodity. Users transmit within their assigned time slots, which not only enhances the quality of communication but also allows for efficient management of the network. Such an efficiency is essential in high-traffic environments, delivering every user guaranteed service without stealing bandwidth, which is essential for both quality of service and user satisfaction. The application of TDMA with cutting-edge technologies like Direct Sequence Spread Spectrum (DSSS) can greatly enhance the efficiency of communication systems. Even though DSSS offers advantages such as anti-jamming and low intercept probability, it faces problems in handling multiple accesses with the addition of more users. Thus, combining TDMA with DSSS unites a hybrid system with the advantage of both systems. Both methods complement each other in order to facilitate smoother communication and data transmission while maintaining the integrity of the signal. As can be seen from various studies, such as the adaptation of IEEE 802.11b standards, use of the hybrid system results in an improved Bit Error Rate (BER) performance, which concludes in more robust and dependable communications under various conditions, such as adaptive white Gaussian noise, and channels of Rayleigh fading (Sajid et al., 2019; Priya et al., 2022). In addition, the uses of TDMA extend to industries where remote control and monitoring are key. An example is in unmanned aerial vehicle communication where secure and efficient transmission of command and control data is a necessity. The use of TDMA in conjunction with DSSS is especially useful in unmanned aerial systems where multiple aircraft must be used in coordination without interference. By allocating specific time slots for transmission by individual UAVs, TDMA ensures that critical communications are collision-free, enhancing operational efficiency. This synchronizing not only maintains the scalable nature of UAV operations, but also complies with stringent demands of defense and commercial applications. As the communication environment continues to evolve, hybridization of TDMA with other access methodologies requires, such as DSSS, will increasingly be key in enhancing the functionality and dependability of modern-day communication systems (Priya et al., 2022; Sajid et al., 2019) (Figure 9.13).

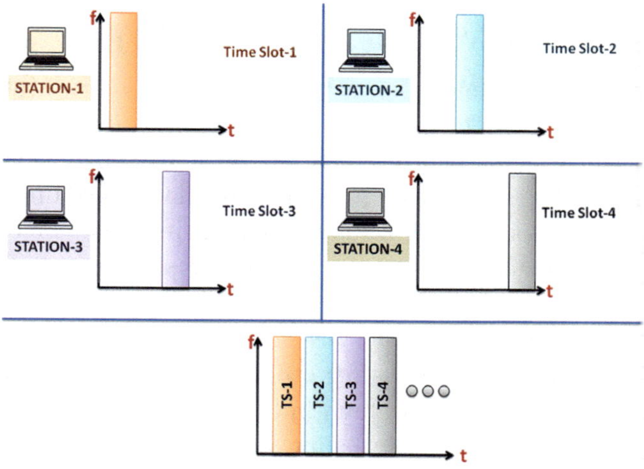

Figure 9.13. *Time division multiple access (TDMA).*

Source: https://www.sanfoundry.com/wp-content/uploads/2024/05/time-division-multiple-access-protocol.png.

9.5.2. Frequency Division Multiple Access (FDMA)

FDMA seems to be a crucial technology in the context of communication systems, particularly in cases involving individual assignment of a channel among multiple users. Dividing available bandwidth into separate frequency bands, FDMA assigns separate frequencies to different users for transmitting simultaneously without interference. This organized assignment provides higher capacity realization while avoiding signal overlap, a problem often encountered in dense communication environments. The reliance on continuous transmission in allocated bands promotes straightforward introduction of FDMA in traditional analog as well as digital systems. Advantages are balanced with compromises, however; linear scalability and wasteful spectrum usage can result under conditions of different user demands, making adaptive platforms with the ability to optimize performance in a variety of scenarios necessary. Even though FDMA is positive, it also has its grandest challenges, especially with regard to spectrum efficiency and dynamic resource allocation. Fixed frequency channels define traditional FDMA systems, and these have the potential to lead to underutilization, especially in unpredictable traffic environments. In case user demand is unpredictable, then assigned frequencies become idle or congested, impacting performance and overall throughput. Consequently, the addition of advanced techniques, such as dynamic frequency assignment or hybrid access technologies, can be more flexible. By establishing a culture in which user assignments may be changed based on real-time communication requirements, these technologies can reduce some of the inherent scalability challenges that are typical of standard FDMA systems. Therefore, the evolution of FDMA in current communication protocols is critical in a bid to achieve a balance of efficiency and user demand, proportional to the general aims of network optimization. The introduction of technologies such as Orthogonal Frequency Division Multiple Access (OFDMA) has put competition on the FDMA agenda through demanding more efficient interference management and spectrum utilization. While FDMA provides a straightforward solution, OFDMA derives more flexibility by employing orthogonal subcarriers to support numerous users sharing the same band of frequencies without creating severe interference. This resource's adaptive adjustment based on current traffic demand is a resolution to some limitations proposed by traditional FDMA framework. Especially, studies on chaos-based schemes being conducted also create new possibilities for improving FDMAs, particularly for secure communications where jamming and interference by hostile forces may be terribly dangerous (Charitha et al., 2025). Ultimately, FDMA's future needs to embrace these trends if it is to remain useful as part of a robust, multi-layered communication scheme in our increasingly interconnected world (Abdulwahab et al., 2024) (Figure 9.14).

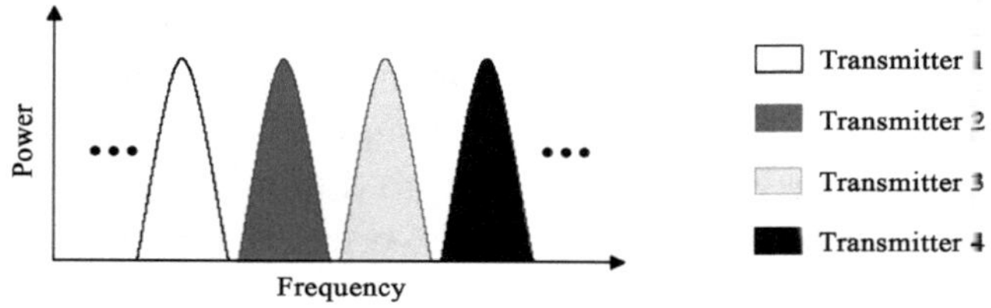

Figure 9.14. *Frequency division multiple access (FDMA).*

Source: https://image.slideserve.com/627738/frequency-division-multiple-access-fdma-2-l.jpg.

9.5.3. Code Division Multiple Access (CDMA)

CDMA is a significant wireless communication advancement that enables multiple users to share the same frequency band simultaneously. Unlike traditional time-division or frequency-division multiple access methods, CDMA employs spread spectrum technology that scatters the signal over a large frequency range. This enhances the quality of the communication link against jamming and interference and, simultaneously, allows for higher user capacities in a given spectrum. In CDMA, every user has been assigned a unique spreading code to modulate their signals. This unique coding makes it possible for several transmissions to occur on a single bandwidth because each transmission can be separated at the receiving end based solely on its unique code, forming a secure communication channel. The power of CDMA lies not only in multiple-user support but also in its inherent security characteristics. With the signal spread over a larger bandwidth and interleaved with unique codes, interception and decoding by any unauthorized user are very hard. Moreover, the use of pseudonoise codes for each user introduces yet another layer of encryption, making eavesdropping highly unlikely. The design principles of the CDMA utilize mathematical concepts such as orthogonality wherein the cross-correlation of each of the users' codes is minimized. The outcome is that the system can handle co-channel interference very effectively, resulting in better overall as well as user performance. Such an advantage turns out to be critical in today's cellular communication with ever-mounting demands for bandwidth and connectivity. In recent years, quantum communication has introduced new paradigms that could further enhance CDMAs performance. Quantum Spread Spectrum Code Division Multiple Access (QSS-CDMA) is an area being explored that employs quantum mechanical techniques to manage photon wavepackets to ensure signal integrity and capacity. This new paradigm is ready to introduce the best aspects of classical CDMA—robustness against interference—along with the quantum advantage of secure information transmission and efficient multi-user access. Through analysis of the mathematically established environment of QSS-CDMA, researchers have already started proving its potential to be viable in future quantum communication networks, setting the stage for highly efficient and secure communication systems that may transform the industry (Dastgheib et al., 2024, 2025) (Figure 9.15).

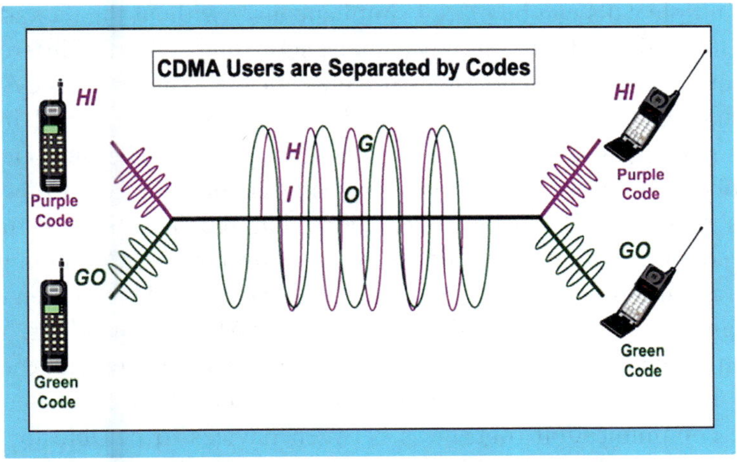

Figure 9.15. *Code division multiple access (CDMA).*

Source: https://chrisbell.com/images/SNHU/CDMA-code-division-multiple-access-protocol.jpg.

9.6. APPLICATIONS OF SPREAD SPECTRUM AND MULTIPLE ACCESS

The application of spread spectrum techniques in modern communication systems has had a profound effect on the reception and transmission of data. Spreading data with wideband interference-free signals, Direct Sequence Spread Spectrum (DSSS) and Frequency Hopping Spread Spectrum (FHSS) techniques are capable of providing better security and reliability of communications. The techniques allow signals to spread over a wide bandwidth, hence resisting jamming and interception more than before. Further, in frequency-space-congested regions, DSSS and FHSS provide many users with access to the same frequency space and minimize the chances of signal collision. For this reason, these technologies are critical in military communications, secure communication systems, and consumer devices like Wi-Fi and Bluetooth, revolutionizing the wireless communications scene. There are various methods of access required for optimal use of communication resources by many users and devices to optimize network efficiency. Two of the most recognized multiple access techniques are Time Division Multiple Access (TDMA) and Code Division Multiple Access (CDMA). TDMA allows users to access a shared frequency channel by allocating particular time slots, thereby preventing interference among users. This method is widely utilized in cellular networks so that multiple calls can be placed simultaneously without the degradation of quality. Unlike this, CDMA employs unique spreading codes for different users so that multiple simultaneous transmissions on the same frequency channel are facilitated. This provides efficient bandwidth utilization and increased capacity, making it very suitable for cellular systems. The combination of these different multiple access technologies with spread spectrum techniques enhances security and performance, addressing the increasing demand for efficient communication within dense urban environments. Combining spread spectrum and multiple access technologies has also spurred innovation in future technologies, particularly in IoT applications and MEC applications. With the expansion of IoT, enabling billions of devices to communicate with each other, efficient utilization of the spectrum is critical. Spread spectrum techniques reinforce communication robustness between devices to be able to transmit information securely and without interference. MEC, nonetheless, brings computational capability closer to the user's location, creating low latency and more efficient energy use. Both complement each other as it is fundamental in aiding latency-critical applications like self-driving cars and real-time analytics. Recent research has focused on the integration of the disciplines of wireless communications and mobile computing, with the focus on the need for innovative designs in mobility, resource management, and privacy. Ultimately, coordination among various access and spread spectrum techniques will be used to counter the issues brought about by today's communications demands, as identified in recent academic journals (Mao et al., 2017; Al A-Fuqaha et al., 2015).

9.6.1. Use in Wireless Communication Systems

Spread spectrum techniques in wireless communication systems play a crucial role in capacity improvement and reducing interference. Techniques based on them, including Frequency Hopping Spread Spectrum (FHSS) and Direct Sequence Spread Spectrum (DSSS), enable the transmission of a signal across a broad frequency range, thereby restricting the effect of multipath fading and jamming. Spread spectrum systems can achieve increased resistance to intentional and unintentional interference by dispersing the signal energy across multiple bands. This characteristic is particularly valuable in systems where many users need to share a common frequency spectrum, allowing good communication in high-density systems. Additionally, these techniques provide inherent security advantages since the wideband signal is impossible to intercept easily. Hence, application of spread spectrum technology in wireless communications ensures better performance and reliability in overcoming common problems like interference and security threats. With advancement in wireless communications in the future, there is an increasing demand for effective multiple access techniques. Multiple access technologies like Code Division Multiple Access (CDMA) and Orthogonal Frequency Division Multiple Access (OFDMA) depict how multiple signals can be shared in the same frequency without much interference. CDMA utilizes unique spreading codes assigned to each subscriber with the ability to carry data with multiple users sending at the same time using different interpretations of their signals within the same bandwidth. OFDMA, in contrast, utilizes the available spectrum to maximum channel capacity by dividing the allocated spectrum into orthogonal subcarriers that can be dynamically allocated based on demand and channel conditions. Such a dynamic approach to resource allocation significantly reduces the possibility of interference and optimizes utilization of spectrums. Such multiple access techniques' efficiency and dependability are paramount with data traffic in wireless networks growing, making them indispensable parts of using today's communication systems. Recent research in wireless networks, Cognitive Non-Orthogonal Multiple Access (C-NOMA), offers a promising new technique in using bandwidth to its best potential. By incorporating advanced algorithms and cooperative relay strategies, C-NOMA addresses issues of spectrum sharing among heterogeneous users with varying channel conditions. The implementation of SADEA further optimizes antenna parameters, resulting in reduced Outage Probability (OP) and Bit Error Rate (BER). Optimizations such as these are crucial to maintaining communication quality, particularly for systems that employ high-order modulation schemes such as 64 QAM or 128 QAM. The special merit of C-NOMA lies in its ability to enhance spectral efficiency while ensuring reliable signal reception through Successive Interference Cancellation (SIC) and Channel State Information (CSI) mechanisms. The innovation represented by C-NOMA holds significant future prospect for 5G wireless communication networks and is thus a cornerstone in the development of ongoing wireless communications systems, helping to manage increasing user demands and complex operation demands (Ntonya et al., 2025; Abdulwahab et al., 2024).

9.6.2. Applications in Military Communications

The convergence of spread spectrum and multiple access technologies has significantly enhanced military communications to enable secure and reliable operations under combat. Through the application of spread spectrum technologies, military communications are resistant to jamming and interference, which are top priorities in hostile environments. These techniques employ wideband signals that occupy a broad bandwidth of frequencies such that it is difficult for adversaries to tap or intercept communications. Furthermore, signal propagation assures that information is encrypted, insulating sensitive military operations from interception. The adaptability of different access schemes, including Code Division Multiple Access (CDMA), allows the military to send messages on the same frequency band at the same time, achieving maximum spectral efficiency while maintaining confidentiality and security, ultimately enhancing operational effectiveness.

Technological advancements are again paving the way for progress in military communications, particularly with the transition towards the use of higher-frequency bands, such as those indicated for 6G and later. Examining frequencies between 100 GHz to 3 THz has plenty of scope for utilization in the military sector in terms of increased data transmission speeds, enhanced sensing systems, and enhanced targeting systems. This transition to higher frequency bands not only accommodates additional devices and services but is also optimizing utilized spectrum resources needed for modern military operations requiring real-time communications to make decisions. With advancing military tactics through technology, the success of spread spectrum techniques will be supplemented, providing secure and reliable communication links regardless of harsh operating environments (Ian et al., 2020). Furthermore, the convergence of artificial intelligence and pervasive network automation within military communications is introducing a new era of operational capability.

The technologies can optimize network performance, automate frequency management, and provide dynamic spectrum access that will allow military units to adapt to evolving battlefield environments. More advanced communication infrastructures will allow for advanced applications such as autonomous drones and unmanned vehicles that require uncompromised connectivity in order to offer effective surveillance and engagement. With the military sector further embracing wireless communication technologies, spread spectrum and multiple access technologies will not only increase readiness levels in operations but also endow them with more strategic benefits in combat operations with sophisticated situations (Theodore et al., 2019).

9.6.3. Role in Modern Networking Technologies

In the context of advanced networking technology, spread spectrum methods have transformed the communication system with enhanced bandwidth efficiency and signal immunity. The key advancement discovered in mobile networks is Overloaded Spread Spectrum Orthogonal Frequency Division Multiple Access (OFDMA). This approach attains huge enhancements in spectral efficiency, accommodating more users and improved data rates, which are necessary as consumer needs continue to escalate. The effectiveness of such a method is seen through research showing that through a 24% overload, there is a 33% increase in spectral efficiency, marking a step towards the optimization of cellular communication systems. With such superior access methods employed, the existing networks are well positioned to accommodate the dynamic character of user connections and data exchanges, largely determining the future of communication technologies (Reddy, 2024). With the advancement in frequency division techniques, Code Division Multiple Access (CDMA) plays an important role in today's networks by enabling multiple users to access simultaneously within a common frequency band. CDMA systems utilize spreading codes that reduce interference and increase user capacity with sophisticated correlation features. However, achieving the right balance between autocorrelation and cross-correlation features is still a major issue in such systems. Improvements in this area can lead to optimal performance, especially with growing demand for higher data rates. In the recent past, improvements in code generation, particularly in the case of Golay sequences, have led to innovations that reduce cross-correlation values and thus enhance communication quality. This innovation not only encourages better signal integrity but also plays an important role in guaranteeing performance in multi-user systems. Hence, CDMAs evolution is a unifying theme in enhancing the efficiency of wireless networks in the present era (João et al., 2024). Moreover, bypassing limitations with peak-to-average power ratio (PAPR) is still crucial to making modern communication systems effective. Circuits like the Orthogonal Perfect Discrete Fourier Transform Golay (OPDG) power transmission circuit are proof that signal processing technologies can mitigate high PAPR while minimizing energy use. By optimizing PAPR, the solutions presented above maximize the energy efficiency as well as the reliability of operation of communication systems and are thus vital for wireless and mobile networks based on high performance. Recent experiments have validated the usability of such optimizations in practice, hence validating their applicability in contemporary networking. Collectively, such advancements in spread spectrum and multiple access techniques represent a significant shift toward more efficient and resilient patterns of communication infrastructures, essential to supporting the expanding demands of users in an increasingly integrated world (João et al., 2024) (Figure 9.16).

Figure 9.16. *Role in modern networking technologies.*

Source: https://www.smartdatacollective.com/wp-content/uploads/2020/09/ne-working-technologies.jpg.

SUMMARY

In conclusion, the study of multi-access and spread spectrum technology proves their fundamental role in enhancing resistance and security in communication systems. The integration of these technologies allows for more efficient use of bandwidth while at the same time resisting jamming and interference, particularly in multichannel environments. Through methods such as Frequency Hopping Spread Spectrum (FHSS), as in our discussions, systems are able to introduce more unpredictability in their communication order, which results in better data integrity and confidentiality. The innovations proposed in systems that employ Quantum Random Number Generators (QRNG), as discussed in Curtò et al. (2024), show how the integration of advanced technology can prove highly beneficial towards secure communication, particularly in situations like the functioning of drones, where reliability plays a pivotal role. Moreover, the implication of using multiple access schemes in conjunction with spread spectrum techniques has broad implications for wireless communication systems of today's era. As highlighted by Chen et al. (2024), the necessity of common ground measurement and control stations working together highlights the importance of such techniques not only for improving performance but also for enabling the unobstructed operation of complex systems. Such a multifaceted approach allows various subsystems to make use of common facilities without interposing, which is essential where synchronized communication is necessary, e.g., satellite tracking and control. Hence, the combined application of these technologies builds strong infrastructure that increasingly becomes essential to our interconnected world. Ultimately, the continuous innovation of spread spectrum and multiple access methods is a revolution in communications systems. Continued research and development play a vital role to address the complexities introduced by interference and attacks on security in adaptive operational environments. As suggested by findings presented here, particularly those relating to QRNG technology and the secure channel ramifications of same, potential for future development is vast. It not only enhances the performance of operational systems but also provides the foundation for next-generation communication infrastructures with the capability of supporting the requirement of various industries, thereby substantiating the role of spread spectrum and multiple access in wireless communication in the future.

The history of spread spectrum and multiple access communication systems uncovers significant innovations in wireless technology critical to ongoing applications. With the development of mobile communications from generation to generation, it highlights growing needs for high-capacity networks that cater to the diverse needs of users. With the introduction of fifth-generation (5G) networks, there is an obvious direction towards sixth-generation (6G) networks with enhanced performance using more frequency between 100 GHz and 1 THz. The developments indicate not only the technological leaps needed for the provision of increasing bandwidth requirements but also the need for new forms of modulation and codes to effectively handle a broader spectrum. Knowledge of these advancements is at the core, as they are the basis for upcoming communication systems to achieve unparalleled levels of service and connectivity (Tataria et al., 2021). The process of spread spectrum techniques and multiple access methods combined is at the core of today's communication systems' functionality. Combining these technologies produces higher data rates while restricting possibilities of interference, thus ultimately offering improved overall network reliability. Techniques such as Code Division Multiple Access (CDMA) and Frequency Hopping Spread Spectrum (FHSS) are ideal examples of how spectrum efficiency can be maximized so that numerous users can share the same frequency band without any significant reduction in quality of service. Furthermore, as systems become more sophisticated to more complex architectures, the application of advanced antenna arrays, real-time signal processing, and adaptive modulation techniques becomes increasingly vital. These innovations ensure that network structures developing can change in relation to environmental conditions, and therefore remain at optimal operation when environmental conditions vary (Reimer et al., 2020). Lastly, the continued innovation of spread spectrum and multiple access communication systems is irretrievably linked with social demands as well as technological innovators. When new networks come into existence, it is essential that one examines the strengths as well as the loopholes of these systems. A path to developing 6G networks that are operational requires overcoming technological hurdles while simultaneously retrofitting legacy infrastructures to support new applications and services. This includes overcoming latency levels and rearchitecting core networks to accommodate time-sensitive applications more effectively. As scientists go on to uncover next future innovations in various layers of the world of communications, the coming together of sound methodology will be pivotal in maximizing system performance and achieving goals set in next-generation network development (Tataria et al., 2021).

REFERENCES

1. Abdulwahab, N., & Baban, A. (2024). Utilizing orthogonal frequency division multiple access technology to mitigate carrier and symbol interference in wireless communication networks. Muthanna Journal of Engineering and Technology. https://www.semanticscholar.org/paper/4bb016a2171874decd9d4a2c0e84ea2d1cab6cce (accessed on 12th September 2025).
2. Afzal, M. H., Rehmani, Y. B., Zikria, Y., & Ni, Q. (2019). Data-driven intelligence in wireless networks: Issues, challenges, and solution. Transactions on Emerging Telecommunications Technologies, 30. https://www.semanticscholar.org/paper/99c29d99b39e2761527702d720b3659469e0e992 (accessed on 12th September 2025).
3. Akyildiz, I. F., Kak, A. C., & Nie, S. (2020). 6G and beyond: The future of wireless communications systems. IEEE Access, 8, 133995–134030. https://doi.org/10.1109/access.2020.3010896 (accessed on 12th September 2025).
4. Alahi, M. E. E., Sukkuea, A., Tina, F. W., Nag, A., Kurdthongmee, W., Suwannarat, K., & Mukhopadhyay, S. C. (2023). Integration of IoT-enabled technologies and artificial intelligence (AI) for smart city scenario: Recent advancements and future trends. Sensors, 23, 5206. https://doi.org/10.3390/s23115206 (accessed on 12th September 2025).
5. Al-Fuqaha, A., Guizani, M., Mohammadi, M., Aledhari, M., & Ayyash, M. (2015). Internet of Things: A survey on enabling technologies, protocols, and applications. IEEE Communications Surveys & Tutorials, 17, 2347–2376. https://doi.org/10.1109/comst.2015.2444095 (accessed on 12th September 2025).
6. Andrews, J. G., Buzzi, S., Choi, W., Hanly, S. V., Lozano, A., Soong, A. C. K., & Zhang, J. C. (2014). What will 5G be? IEEE Journal on Selected Areas in Communications, 32, 1065–1082. https://doi.org/10.1109/jsac.2014.2328098 (accessed on 12th September 2025).
7. Başar, E., Di Renzo, M., de Rosny, J., Debbah, M., Alouini, M. S., & Zhang, R. (2019). Wireless communications through reconfigurable intelligent surfaces. IEEE Access, 7, 116753–116773. https://doi.org/10.1109/access.2019.2935192 (accessed on 12th September 2025).
8. Benvenuto, G., Cherubini, S., & Tomasin, S. (2021). Spread-Spectrum Systems. https://doi.org/10.1002/9781119567998.ch10 (accessed on 12th September 2025).
9. Charitha, M., & Hosur, S. (2025). Enhanced BER optimization and jamming resilience in chaos communication systems using MIMO-OFDM

and adaptive spreading factors. Engineering, Technology & Applied Science Research. https://www.semanticscholar.org/paper/37b0e9e72c6c6366f0dbe83a41e09d07408f615 (accessed on 12th September 2025).

10. Chen, T., Guo, S., Yang, T., Zhao, K., & Dong, F. (2024). An inter-correlation method for measuring multi-access interference theory and simulation analysis. 2024 International Conference on Sensing, Measurement & Data Analytics in the Era of Artificial Intelligence (ICSMD). https://www.semanticscholar.org/paper/dedc899f9bb8a18bbc5edf41fe1208090e3aa434 (accessed on 12th September 2025).

11. Cinelli, M., Quattrociocchi, W., Galeazzi, A., Valensise, C. M., Brugnoli, E., Schmidt, A. L., Zola, P., et al. (2020). The COVID-19 social media infodemic. Scientific Reports, 10, Article 16598. https://doi.org/10.1038/s41598-020-73510-5 (accessed on 12th September 2025).

12. Curtò, I., Zarza, I., Cano, J. C., & Calafate, C. (2024). Enhancing communication security in drones using QRNG in frequency hopping spread spectrum. Future Internet, 16, 412. https://www.semanticscholar.org/paper/9f76c6c61a70731f94511ca817629ce955698d89 (accessed on 12th September 2025).

13. Dastgheib, M., Salehi, J., & Rezai, M. (2024). Quantum spread-spectrum CDMA communication systems. 2024 IEEE Middle East Conference on Communications and Networking (MECOM). https://www.semanticscholar.org/paper/abd9813346b5bad8904fef557abfc386ccfd94a1 (accessed on 12th September 2025).

14. Dastgheib, M., Salehi, J., & Rezai, M. (2025). Quantum direct-sequence spread-spectrum CDMA communication systems: Mathematical foundations. IEEE Transactions on Quantum Engineering, 6, 1–40. https://www.semanticscholar.org/paper/dbae3cb4cc6c4ca195d5f348c00b6601526d0471 (accessed on 12th September 2025).

15. Ding, Z., Lei, X., Karagiannidis, G. K., Schober, R., Yuan, J., & Bhargava, V. K. (2017). A survey on non-orthogonal multiple access for 5G networks: Research challenges and future trends. IEEE Journal on Selected Areas in Communications, 35, 2181–2195. https://doi.org/10.1109/jsac.2017.2725519 (accessed on 12th September 2025).

16. Dwivedi, Y. K., Kshetri, N., Hughes, L., Slade, E., Jeyaraj, A., Kar, A. K., Baabdullah, A. M., et al. (2023). Opinion paper: “So what if ChatGPT wrote it?” Multidisciplinary perspectives on opportunities, challenges and

implications of generative conversational AI for research, practice and policy. International Journal of Information Management, 71, 102642. https://doi.org/10.1016/j.ijinfomgt.2023.102642 (accessed on 12th September 2025).

17. Eyring, V., Bony, S., Meehl, G. A., Senior, C. A., Stevens, B., Stouffer, R. J., & Taylor, K. E. (2016). Overview of the Coupled Model Intercomparison Project Phase 6 (CMIP6) experimental design and organization. Geoscientific Model Development, 9, 1937–1958. https://doi.org/10.5194/gmd-9-1937-2016 (accessed on 12th September 2025).
18. Fouad, H. S., Elsayed, H., & Girgis, S. K. (2019). Direct sequence and frequency hopping signals classification based on co-occurrence matrix and clustering techniques. Journal of Computer Science. https://www.semanticscholar.org/paper/3aab402e8be09fc4e02cad6459f319ba34bc8d9a (accessed on 12th September 2025).
19. Giannozzi, P., Andreussi, O., Brumme, T., Bunău, O., Buongiorno Nardelli, M., Calandra, M., Car, R., et al. (2017). Advanced capabilities for materials modelling with Quantum ESPRESSO. Journal of Physics: Condensed Matter, 29, 465901. https://doi.org/10.1088/1361-648x/aa8f79 (accessed on 12th September 2025).
20. Kaiser, S. (2008). Multi-Carrier and Spread Spectrum Systems. https://doi.org/10.1002/9780470714249 (accessed on 12th September 2025).
21. Kanellopoulos, D., Sharma, V. K., Panagiotakopoulos, T., & Kameas, A. (2023). Networking architectures and protocols for IoT applications in smart cities: Recent developments and perspectives. Electronics, 12, 2490–2490. https://doi.org/10.3390/electronics12112490 (accessed on 12th September 2025).
22. Manna, T., Das, A., Bhattacharjee, B., & Mukherjee, S. (2024). Robust communication strategy for overcoming narrowband jamming in low-frequency CDMA-DSSS system using MATLAB. Fact Universitatis, Series: Electronics and Energetics. https://www.semanticscholar.org/paper/1cb515c57145bd95fbea55421f8082c839a00d9e (accessed on 12th September 2025).
23. Mao, Y., You, C., Zhang, J., Huang, K., & Letaief, K. B. (2017). A survey on mobile edge computing: The communication perspective. IEEE Communications Surveys & Tutorials, 19, 2322–2358. https://doi.org/10.1109/comst.2017.2745201 (accessed on 12th September 2025).
24. Ntonya, Y. J., Musyoki, S., & Oduol, V. (2025). Cognitive non-orthogonal multiple access with cooperative relaying and antenna optimization using SADEA for efficient spectrum sharing. Engineering, Technology

& Applied Science Research. https://www.semanticscholar.org/paper/9eb15a8ee443c6cb842f3b66c7ffe6b2a561cf4f (accessed on 12th September 2025).

25. Pereira, J. S., & Ferreira, H. (2024). Perfect sequences based on Golay codes for communication systems. Digital Technologies Research and Applications. https://www.semanticscholar.org/paper/8c496006d4f7c39582f8af98acf9c7975d6969ab (accessed on 12th September 2025).

26. Priya, G. S., Sai, G., Gowthami, G., Singh, I., Tayal, S., & Khan, M. J. (2022). Analysis of bit error rate for multi-user TDMA-based communication system. 2022 First International Conference on Electrical, Electronics, Information and Communication Technologies (ICEEICT). https://www.semanticscholar.org/paper/25dd131725f9f29fe378c21ee74200f08cb27331 (accessed on 12th September 2025).

27. Rappaport, T. S., Xing, Y., Kanhere, O., Ju, S., Madanayake, A., Mandal, S., Alkhateeb, A., et al. (2019). Wireless communications and applications above 100 GHz: Opportunities and challenges for 6G and beyond. IEEE Access, 7, 78729–78757. https://doi.org/10.1109/access.2019.2921522 (accessed on 12th September 2025).

28. Rasheed, A., San, O., & Kvamsdal, T. (2020). Digital twin: Values, challenges and enablers from a modeling perspective. IEEE Access, 8, 21980–22012. https://doi.org/10.1109/access.2020.2970143 (accessed on 12th September 2025).

29. Reddy, Y. A. K. (2024). Overloaded spread spectrum OFDMA in outdoor environment in the more interleaving scenario. International Journal of Advanced Research in Science, Communication and Technology. https://www.semanticscholar.org/paper/398430051c05b5f77f0da882a0a7e6261264ea0e (accessed on 12th September 2025).

30. Reimer, P., Austin, W. E. N., Bard, É., Bayliss, A., Blackwell, P. G., Bronk Ramsey, C., Butzin, M., et al. (2020). The IntCal20 Northern Hemisphere radiocarbon age calibration curve (0–55 cal kBP). Radiocarbon, 62, 725–757. https://doi.org/10.1017/rdc.2020.41 (accessed on 12th September 2025).

31. Sajid, A., Habib, A., Ali, S., & Ejaz, S. (2019). Development of multi-user TDMA-based DSSS system. 2019 2nd International Conference on Communication, Computing and Digital Systems (C-CODE). https://www.semanticscholar.org/paper/767456afd8e4631057579e5a0bb6c870[illegible]0e36862 (accessed on 12th September 2025).

32. Tataria, H., Shafi, M., Molisch, A. F., Döhler, M., Sjöland, H., & Tufvesson, F. (2021). 6G wireless systems: Vision, requirements, challenges, insights,

and opportunities. Proceedings of the IEEE, 109, 1166–1199. https://doi.org/10.1109/jproc.2021.3061701 (accessed on 12th September 2025).

33. Yang, Y., Zhu, L., Mao, X., Tan, Q., & He, Z. (2019). The spread spectrum GFDM schemes for integrated satellite-terrestrial communication system. China Communications, 16, 165–175. https://www.semanticscholar.org/paper/7c6fa637de510ee16e0eea5e00fc914a45318fea (accessed on 12th September 2025).
34. Yip, L. (2025). Comparison of error probability analyses for asynchronous DS-CDMA satellite communication systems. 2025 IEEE Aerospace Conference. https://www.semanticscholar.org/paper/613a0931753865230d2c13981ed7b0b85f38412b (accessed on 12th September 2025).
35. Zhang, C., Patras, P., & Haddadi, H. (2019). Deep learning in mobile and wireless networking: A survey. IEEE Communications Surveys & Tutorials, 21, 2224–2287. https://doi.org/10.1109/comst.2019.2904897 (accessed on 12th September 2025).
36. Zhao, W., Jia, M., Liang, T., Wu, H., Tan, Q., & Si, J. (2024). Key technologies of ultra-wideband frequency-hopping anti-interference system based on microwave photons. Proceedings of SPIE, 13502, 135020P–1–135020P–8. https://www.semanticscholar.org/paper/4617150dfc01e1fd7e0d779f53909228ea907f6d (accessed on 12th September 2025).\

CHAPTER 10

ADVANCED COMMUNICATION SYSTEMS

Contents

INTRODUCTION

The foundation of advanced communication systems engineering lies in the intricate design and functioning of communication systems, which establish a method of information exchange across different mediums. The subject encompasses theoretical and practical aspects in data transmission, reception, and processing. Communication technology history has been characterized by revolutionary breakthroughs that have significantly impacted the interaction between society and economic development. The European Unions ACTS program is a classic example of this advancement through its funding of research activities that enhance communication infrastructures. As outlined in Richardson et al. (1997), ACTS aims at the development of complex advanced communication systems and services with regard to the complexity of technological evolution and regulatory innovation. This emphasis on usage demonstrates the programs commitment to strengthening economic growth and social cohesion across Europe. Grounded in these basic principles, deployment of advanced communication systems demands extensive understanding of various technical components and their interplay. In the field of communication systems, a major focus on interactive digital multimedia services, high-speed networking, and mobility reflects evolving customer requirements. Challenges in integrating these technologies into complete systems are salient and can be overcome by innovative research.

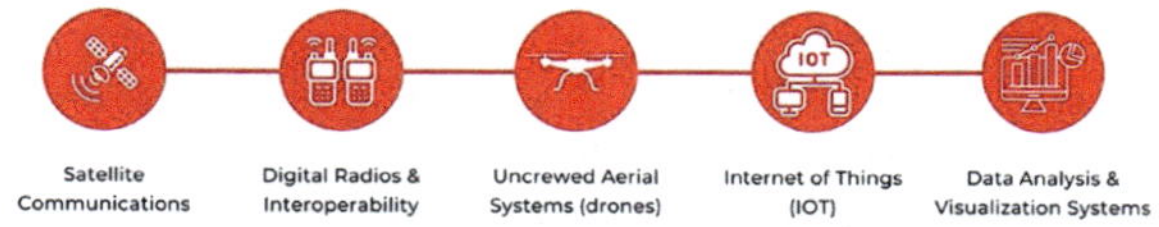

Figure 10.1. *Advanced communication systems.*

Source: https://www.basecampconnect.com/wp-content/uploads/2024/01/blog-scr-jan-2024-advanced-communication-systems-1024x614.jpg.

For instance, the MISAT project provides the potential applications of microsystems technology (MST) in building minute and optimal satellite platforms for space and terrestrial missions. As illustrated in Gill et al. (2007), such advancements pave the way for innovative advances in distributed systems and sensor technologies that further enhance the capabilities of communication infrastructures. Finally, the path to the realization of sophisticated communication systems is characterized by concerted action that combines expertise from numerous areas and areas of study. These interdisciplinarily are seen from the participation of various organizations in initiatives like ACTS and MISAT, fostering an attitude of innovation and sharing of knowledge. Public and private sector collaboration facilitates scientists to address the intricacies of issues surrounding the enhancement of communication technology. The constant engagement in real-world trials and demonstrations is an important feedback loop that steers further improvements and developments. Thus, as communication systems engineering advances, not only does it contribute towards making technology more sophisticated, but it also actively helps bridge gaps and strengthen connectivity in society, harmony that our increasingly digital life demands (Figure 10.1).

Definition and Importance of Communication Systems

Communication systems define a wide range of mechanisms and technologies employed for information transfer between parties. These systems can range from simple mechanisms of communications like spoken exchanges to complex setups like digital networks and satellite communications. In future systems engineering for communication, it is of utmost significance to build an insight into the system complexity, particularly how systems function in complex environments. Notably, communication systems not only enhance connectivity but also contribute to the overall resilience of network infrastructures. The intricacy inherent in such systems tend to arise to offer means to generate robustness, but it does so while also producing new fragility, demonstrating the subtle trade-offs present in system design (Alderson et al., 2010). Communication systems cannot be more vital as they become particularly crucial in industries where efficiency and dependability take the lead, like aerospace. Effective communication is at the core of structuring operations, managing resources, and accomplishment of organizational goals in harmony with all stake groups. For instance, BAE SYSTEMS recognizes the criticality of structured plans for formal communication in achieving seamless information flows in the midst of change management activities in its engineering projects. The realization precipitates an understanding of production stages, which becomes a must when accomplishing enhancements of processes in the aerospace sector. Therefore, communication systems create a foundation for collaboration, making it easier for organizations to leverage technology for enhanced performance at work (Barker et al., 2005). During periods of rapid technological shifts, communication system development reflects broader social change and demands. With the proliferation of digital communication aids, the form by which information receives transmission widens and becomes increasingly diversified. This evolution necessitates continuous redefinition of past communication practices to make them updated and efficient. Furthermore, with companies wanting to implement "world-class" changes in procedures, the role of communication systems becomes even more critical. The addition of good communication channels not only makes processes smoother but also helps in adapting new ideas and techniques, thereby fostering an innovative culture. Hence, understanding the definition and importance of communication systems is important for advancing engineering practice in various industries (Alderson et al., 2010).

Historical Development of Communication Technologies

The evolution of communication technologies has greatly transformed the ways in which information gets shared, and the impact has been felt across all domains of human interaction. From the beginning of spoken language to the advent of script, each development was a human connective leap. The 15th-century discovery of the printing press was one such milestone, breaking open access to information and facilitating mass literacy. This growth offered the foundation for more advanced systems of communication, ultimately culminating in the emergence of telegraphy during the 19th century and the means to transmit messages across large distances virtually simultaneously. This era demonstrated the importance of communication as a force of social and economic change, ultimately laying the groundwork for communication technologies in the contemporary world by establishing fundamental methods for the dissemination of information. In the 20th century, radio and television brought a new era of mass communication, enabling live broadcasting to the masses. Besides entertainment, these technologies enabled important roles in education, propaganda, and public debate during important moments of history. With the evolution of communication technologies came increased complexity and sophistication in the infrastructure necessary to enable them. The development of satellite technology in the second half of the century further enhanced worldwide connectivity and enabled instantaneous communication between continents.

The continuous development of the systems culminated in the digital revolution, which was best symbolized by the introduction of computers and the Internet. This revolution not only expanded the avenues of communication but also provided new trends of interaction that transformed social culture and business practices at their essence. Communication technology has also developed to match challenge and ongoing innovation requirements. Recent developments in the building of communication systems, such as the embedding of the smart grid technology, illustrate how engineering and communication converge today. The project at UC Berkeley to build a Building-to-Grid testbed is also a sign of how an effective communication infrastructure is needed to facilitate energy management systems, illustrating how interdisciplinarity operates to resolve complex societal needs. Furthermore, the evaluation of metering technology highlights implementation barriers that continue to dominate the industry, as seen from prior studies based on identifying barriers to higher adoption (Levy et al., 2002). Such historical experiences justify enhancing communication technologies not only to make connectivity possible but also to enable sustainability and efficiency in operations within modern communication systems (Peffer et al., 2010).

Overview of Modern Communication Systems

Modern communication systems form the center of activities of contemporary society through their ability to provide information on different platforms with high efficiency and speed. The history of ICT is interwoven with the evolution of Information and Communication Technologies (ICT), which have re-engineered historical modes of communication into networks. Not only are these networks helping to enable real-time communication, they also enable a myriad of applications, from social media to cloud computing. One of the most important aspects of contemporary systems is their scalability, making it possible for different segments—ranging from urban to rural regions—to integrate effortlessly. As an integral part of smart grid innovations, such systems maximize energy distribution and management, serving to highlight the crucial role ICT plays in contemporary infrastructure (Borghetti et al., 2013). The incorporation of communications systems into learning frameworks has also seen tremendous growth, with digital technology being the impetus behind their spread. Learning management systems (LMS) are a testimony to this, offering platforms that cater to the diverse needs of learners and teachers alike. With capabilities to provide new means of learning and teaching, they assist in bridging the educational gap, particularly in rural areas where access is limited. Through the establishment of advanced resource centers interconnected by the internet, distant communities can be provided with modern educational content and pedagogy.

This not only improves inclusive education but also facilitates a culture of collaborative learning that enhances overall quality in education, hence highlighting the pivotal position of ICT in transforming education spaces (Hussain et al., 2013). Besides their use in education, modern communication networks have transformed the energy management sector through smart grids. The new networks involve the application of digital communication technology on the electric grid, making electricity distribution monitorable and controllable in real time. This is a far cry from traditional electric networks that were quite inefficient and unreliable. The incorporation of smart grids facilitates the use of sustainable energy since it encourages the integration of alternative sources of power, addressing existing problems such as rising electricity demand and energy-saving policies. The introduction of smart grids, therefore, illustrates the wide-ranging impact of the contemporary communication system on non-telecommunications industries, boosting the function of such systems in fostering both societal necessities and environmental sustainability (Borghetti et al., 2013) (Figure 10.2).

Figure 10.2. *Overview of modern communication systems.*

Source: https://i.ytimg.com/vi/Jl7SBkioCbQ/maxresdefault.jpg.

10.1. FUNDAMENTAL PRINCIPLES OF COMMUNICATION

The underlying philosophy of communication is the basis of successful communication on different media, particularly in the context of communication systems engineering. Communication entails the exchange, reception, and perception of information, which is crucial in the creation and installation of systems to facilitate hassle-free transmission of data. Efficient communication essentially relies on clarity, brevity, and cohesiveness. These experts have to understand how to construct messages that can be interpreted by different audiences, and employ appropriate language and jargon. This is evident in efforts by interdisciplinary groups, such as those by the Park City Math Institute, to develop guidelines that center on clear communication in data science education (Agarwal et al., 2017). By establishing these basic principles, engineers can craft communications systems that not only deliver information but also facilitate understanding among the users. A second equally important aspect of communication principles involves the flexibility of the strategy to messaging in order to accommodate various contexts and technologies. As communications systems evolve, engineers must be quick to modify their methods to accommodate different forms and modes of conveying information. The discipline of computational science and engineering emphasizes this need, where the practitioner has to straddle the theoretical and experimental aspects of the profession. Good communication is needed not just for the purposes of reporting the results but also in order to facilitate interdisciplinary collaboration as technology transforms the current paradigms (Biros et al., 2016). Appreciation of the context sensitivity of communication enables engineers to use technological tools optimally, spreading the reach and influence of the message. This versatility, in turn, stimulates ongoing innovation and fosters a climate of knowledge sharing. Lastly, the intrinsic cyclical character of communication again emphasizes its value in advanced communication systems engineering. Feedback loops are critical to the simplification of messages and the systems sending them. Through anticipatory soliciting and incorporation of user feedback, engineers can improve system functionality as well as user experience. This cyclical process is very much a reflection of the principles of effective communication since it generates a continuous loop of exchanging, receiving criticism, and adjusting based on the criticism. The changing relationships that are initiated as a consequence of this communication process ultimately lead to more effective and better systems. Accentuating the need for feedback and changeable messaging, the communication principles demonstrate the ever-evolving nature of technology and its influence on engineering practices in this field (Figure 10.3).

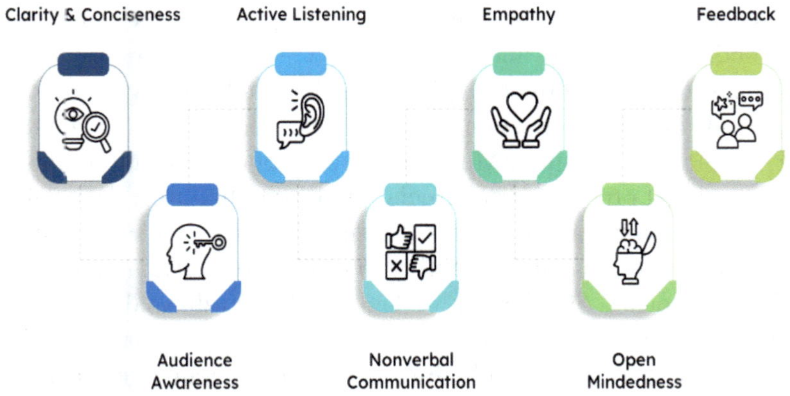

Figure 10.3. *Fundamental principles of communication serve.*

Source: https://kapable.club/blog/wp-content/uploads/2024/08/Core-Principles-Of-Effective-Communication.webp.

10.1.1. Signal Processing Techniques

In Advanced Communication Systems Engineering, signal processing techniques are key to enhancing transmission of data to be efficient and reliable. The techniques not only assist in recovering useful information from complex signals but also resist noise and interference common in wireless systems. One of the notable applications is in Cloud Radio Access Networks (C-RAN), where fronthaul transmission link optimization innovation is essential. In such a case, signal processing methodologies such as compression and quantization are employed to increase spectral efficiency (SE) and energy efficiency (EE). Network operator application of these methods can assist in countering the impact of fronthaul constraints on user quality of service, thus achieving a better-reliable and stable communication experience. This incessant evolution leads one to focus on the necessity of integrating superior signal processing techniques into the future communication paradigms. The advancements in signal processing are further revealed by their applications in space communication systems, and a prime example in this regard is the Deep Space Network (DSN) system. As high-quality data transmission requirements escalate, the reliance on intricate signal processing methods has become the sole feasible alternative. Technologies such as automated control and monitoring systems, along with specialized processing of telemetry, provide integration-free communication between spacecraft and ground stations. The DSN system design approach targets architecture integration from front-end antennas to end users, providing interoperability through open-systems standards. Such design models employ advanced signal processing in order to enhance data integrity and security, further emphasizing the need for ongoing technology development in communication systems, according to telecommunication strategy studies in the future. Moreover, new resource allocation optimization methods in signal processing are crucial in order to optimize the performance of advanced communication systems. These techniques enable dynamic control over network resources and adaptability to meet shifting user requirements and channel conditions. With the use of large-scale coordinated processing and clustering, communications networks can significantly enhance capacity and user experience, particularly in high-density metropolitan areas. The focus on software-defined networking and network function virtualization nowadays highlights how these signal processing techniques deal with rapidly changing technology environments. Implementation of such strategies not only addresses the issue of low bandwidth but also promotes the way towards smart, cost-efficient, and scalable communication systems that are capable of serving future demands, as evident in the studies of fronthaul-limited C-RANs and their models of operation (Lau et al., 2015; Atkinson et al., 1992).

10.1.2. Modulation and Demodulation Methods

Modulation techniques play a pivotal role in advanced communication systems for effective information transfer over various media. Modulation is tasked with transforming baseband signals into an compatible format for transmission, generally enhancing channel capacity and reducing the effect of interference. Techniques such as Amplitude Modulation (AM), Frequency Modulation (FM), and Phase Modulation (PM) are utilized for varying purposes depending on context, impacting signal integrity and spectrum efficiency. Advanced methods, including Orthogonal Frequency Division Multiplexing (OFDM), employ multiple carrier waves to offer enhanced resistance against multipath disturbance and improved spectral efficiency. Such a modulating technique adaptability is crucial in digital communication where reliability and high data rates of utmost concern, highlighting the persistence of modulation technique innovation critical to contemporary engineering practice. Demodulation, or extraction of the original information from the modulated carrier wave, is as crucial as modulation in the communication system. It employs a variety of techniques which are counterparts to their modulation counterparts to ensure the integrity of the received signal is maintained.

For instance, coherent demodulation is employed where phase information is critical, while non-coherent techniques could be employed for applications with less critical phase demands. The latest developments, such as the development of time- and frequency-domain demodulation techniques in models like triple-layer hybrid optical orthogonal frequency division multiplexing (THO-OFDM), also enhance signal recovery performance and efficiency even more. The new methods not only improve bit error rates but also significantly reduce computational complexity, confirming the interdependence between demodulation and modulation in achieving better communication performance (Ghassemlooy et al., 2020). Moreover, the use of new algorithms and signal processing has tremendously enhanced the performance of modulation and demodulation processes. Recent research has utilized innovative strategies to detect occasional faults in communication systems using injection of a constant frequency sinusoidal signal that modulates the occasional faults, enabling appropriate monitoring.

The function is particularly relevant in channel characteristics and identifying precursors of imminent failure (Ahmad et al., 2015). The ability to distinguish among various fault types and channel effects using advanced demodulation techniques not only increases the reliability of communication systems but also results in greater understanding of signal integrity in real-time systems. With every step in the field of communication systems engineering, modulation and demodulation interaction continues to be the backbone of research and development, looking forward to overcoming future challenges and enhancing system reliability (Figure 10.4).

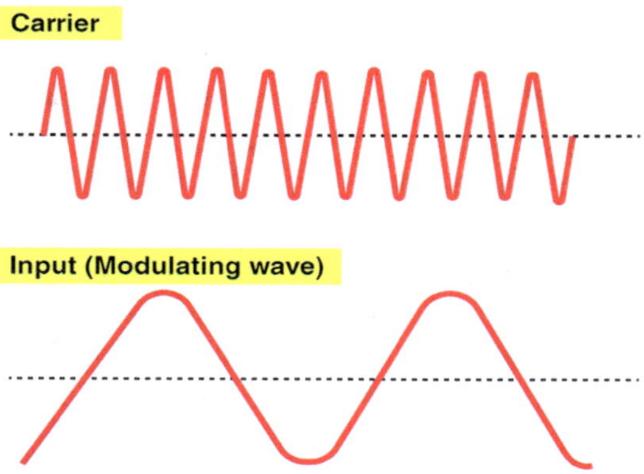

Figure 10.4. *Modulation and demodulation methods.*

Source: https://cdn1.byjus.com/wp-content/uploads/2020/02/Modulation-And-Demodulation-1-1.png.

10.1.3. Noise and Interference Management

Noise and interference management is one of the key subjects of research in advanced communication systems engineering, particularly due to increased demand for efficient wireless communication. Natural and man-made noise in modern networks can significantly influence signal integrity, leading to communication inefficiency. In an effort to counter this challenge, engineers apply various methods designed to enhance the integrity of data transmission. Techniques like error correction coding and advanced modulation techniques play a crucial role in reducing the ill effects of noise. Furthermore, cloud radio access network (C-RAN) advancements are a good example of how advanced architectures can yield greater spectral and energy efficiency along with an ability to resolve noise problems in wireless communication. By maximizing resource usage and employing coordinated processing techniques, C-RANs facilitate better noise interference control together with the provision of quality services to users, ensuring more uniform communication experience (Lau et al., 2015). Besides conventional noise control methods, the integration of wide-area monitoring systems into smart grids illustrates the influence of noise and interference on such key systems as well. Such systems rely heavily upon synchronized monitoring facilitated by satellite-based global positioning systems (GPS) for accurate timing references. Such systems are increasingly at risk from radio frequency interference capable of compromising their operational integrity. The threat of cyber-attacks from outside makes it critical to have detailed analysis with priority placed on GPS-dependent monitoring system vulnerabilities. By establishing these vulnerabilities, the engineers are able to design strong safeguards against interference so that communication and synchronization will be reliable despite the possibility of external forces. This research not only illuminates the challenges of working against noise but also prompts innovations to put together self-healing grids that would be able to maintain functionality even in the presence of interference (Ahmed et al., 2014). In the end, interference and noise management convergence with revolutionary technologies drives communications system evolution. Software-defined networking and network function virtualization integration provides new opportunities to enhance interference management capabilities. With dynamic resources and adaptive algorithms, systems can cope with changes in noise levels, providing personalized solutions to maintain signal integrity. Constant advances in techniques such as compression and quantization also lead to the realization of fronthaul-constrained C-RANs with improved spectral efficiency. As communication systems evolve, a visionary solution for noise and interference management will increasingly be important, highlighting the importance of wide-ranging research and innovative design. Lastly, these strategies will shape the future of communication engineering, leading to more resilient networks that will be able to operate in extreme environments (Lau et al., 2015).

10.2. TYPES OF COMMUNICATION SYSTEMS

Communication systems form the core of the art of advanced communication systems engineering in that they facilitate the transmission of information across various networks and platforms. Analog and digital systems are among the principal classes of communication systems. Analog communication systems transmit information in the form of continuous signals, commonly defined as voltage or frequency change. However, with the advent of technology, digital communication systems have taken center stage since they can convert information into binary code. Digital systems are of higher fidelity and less susceptible to noise, thereby being better for those applications that require more data integrity. Additionally, they facilitate ease of integration with modern computing technologies, resulting in sophisticated communication protocols required in telecommunications and data networking applications. The second important classification of communications systems is between wired and wireless methods of communication. Wired communications systems employ physical media such as coaxial cable, fiber optic, or twisted pair line to transport information, offering security and stable connections. Wireless communication systems convey information through radio waves by utilizing satellites, transmitters, and receivers to interconnect people without employing physical wires. While wired systems are typically faster and less prone to interference, they are geographically limited and expensive to install. Wireless systems provide convenience and portability and enable long-range coverage in uses such as mobile communications and IoT networks. The ongoing advancement of wireless technologies, with the integration of high-performance computing techniques, further enables their usage in modern communication systems to efficiently handle robust and effective network design (Alias et al., 2017). Further, the utilization of communication systems in niche applications brings forth new categories such as machine-to-machine (M2M) and smart grid communications.

These systems are identified with their focus on enhancing efficiency, reliability, and sustainability in various sectors. For instance, intelligent grid communication systems utilize two-way communication to facilitate efficient distribution of energy through the application of data analytics and advanced monitoring technology to regulate energy usage and facilitate renewable energy integration (Ghazi et al., 2018). M2M communication enables devices to communicate with each other automatically, delivering real-time information to enhance operational efficiencies in industries like manufacturing and healthcare. The sophistication of such communication systems underscores the importance of aggressive security mechanisms in addition to performance optimization, requiring engineers to design highly evolved frameworks that respond to both functionality and insensitivity to possible weaknesses in the underlying network design (Figure 10.5).

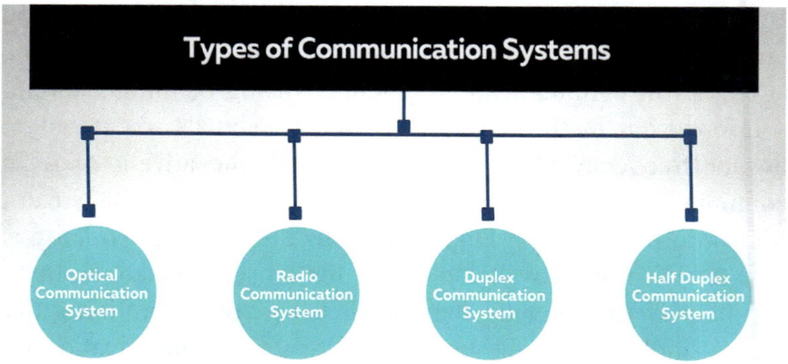

Figure 10.5. *Types of communication systems.*

Source: https://www.chtips.com/wp-content/uploads/2022/01/Types-of-Communication-Systems-1024x536.webp.

10.2.1. Analog vs. Digital Communication

Development of communication technology has brought in the paradigm shift from analog to digital methods with respective advantages and disadvantages. Analog communication, transmitting information in continuous signals, has been the backbone of telecommunications. It is at a disadvantage due to susceptibility to noise and degradation, which reduces quality over long transmission paths. Digital communication packages information into binary forms with superior immunity to interference and the ability to correct errors. This change has not only improved picture and sound quality but also enabled combining various programs, such as video streaming and high-fidelity audio. Application of digital methods is particularly evident in advanced communication systems design, where digital signal processing is employed for performance and reliability improvement. The strength of the field of digital communications is its ability to utilize high-tech techniques that minimize the drawbacks usually associated with analog systems. For instance, the construction of optical communication systems has revolutionized data transmission through the use of fiber optics, essentially eliminating distortions encountered in signal transmission. As shown in the literature, coherent detection systems with digital signal processing algorithms can have the ability to significantly improve spectral efficiency of transmission, thus enabling high-speed 100 Gbps or even higher data transfer. The revolutionary technology does not just revolutionize bandwidth usage but also augments the capacity for multiplexing, making it possible for new generations of communication standards. Therefore, although analog systems lay the groundwork, the sophisticated abilities of digital systems make them central to modern communication infrastructure and make them indispensable in the field of engineering. Another central element in the analog/digital communication debate has to do with the role played by hybrid systems that borrow from elements of both schemes. More recent advances, indeed in the context of 5G technology, have favored emphasizing the capabilities of hybrid analog-digital precoding techniques. These systems leverage both the analog signal processing continuity and digital modulation efficiency to achieve maximum performance. Studies have established that the modeling of hybrid systems demonstrates significant radio-frequency loss and component limitations effects on their performance. Notably, Butler networks have been shown to offer improved performance compared to fully connected analog beamforming networks, especially under scenarios with numerous RF chains. This implies that hybrid structures might offer real-world benefits in the realm of increasing demands on high data rate and reliable connectivity, eventually reiterating the relevance of integrating both analog and digital communication methodologies into engineering designs.

10.2.2. Wired Communication Systems

In Advanced Communication Systems Engineering, wire communication systems remain the mainstay of both individual and organizational connectivity. Physical wires such as fiber optics and copper wire are employed in these systems to transfer data with minimal interference and maximum reliability. Unlike their wireless counterparts, wired systems guarantee dependable performance, which is critical in environments requiring stable connectivity, including telecommunications and data centers. The integrity of the signals passing through wired mediums gives them more robust data integrity so that they are most appropriate for uses involving high bandwidth and low latency. The fiber optic technology has significantly increased capacity and speed, thereby revolutionizing the scenario of wired communication systems by enabling larger data transmission and greater transmission distances. Despite the advancements in wireless communication technologies, the significance of wired systems cannot be overlooked. They are the backbone of most critical infrastructures, offering secure channels of communication in urban centers as well as rural regions. Wired systems are extremely widespread in applications where security and protection of data is a consideration, as they are less vulnerable to eavesdropping or signal weakening. Additionally, certain industries such as manufacturing and healthcare depend on wired communication to exchange data as well as monitor systems in real-time. Traditional power grids, as they are converted to Smart Grids, also bring out the importance of stable wired communication infrastructures. As has been highlighted in recent research, Smart Grids employ a number of devices to provide uninterrupted connections in order to function successfully, thereby solidifying the always-enduring position of wired infrastructures in the improvement of communication systems (Copeland et al., 2019). The development of communication systems still goes on, with wired technology still contributing in an ever-wireless world. Presently, challenges such as the cost of cabling and installation, along with periodic maintenance, act as stumbling blocks to mass consumption. Nonetheless, research and development keep pushing past the drawbacks by enhancing fiber optic capability while exploring the convergence of wired and wireless technologies. Of special note is the realization of Wireless Intelligence Network (WIN) technology that unveiled directions to innovation in wireline systems, achieving flexible solutions that could bridge the performance shortfalls between wireline and wireless communication systems (Adesuyi et al., 2014). As demands for quicker, more secure communication continue to rise, wireline communication systems will indeed play a critical role in shaping the future of advanced communication systems engineering (Figure 10.6).

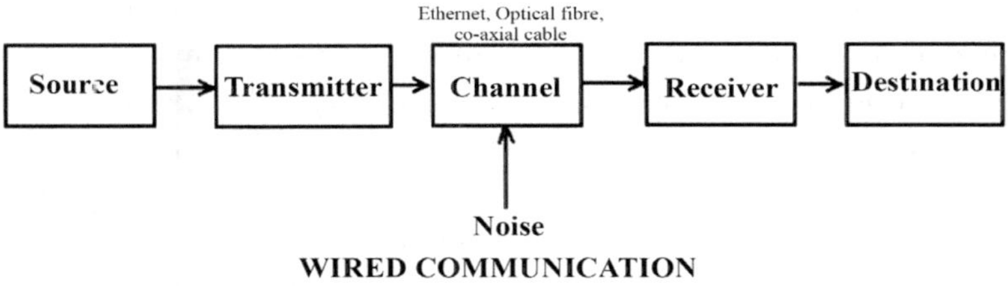

Figure 10.6. *Wired communication systems.*

Source: https://electricalvoice.com/wp-content/uploads/2020/06/wired-communication-block-diagram.jpg.

10.2.3. Wireless Communication Systems

In the last few decades, wireless communication systems development has transformed the landscape of information and communication technology. The core of this transformation lies in the pervasiveness of mobile computing devices that have made convenient connectivity possible across a range of applications. Research over the last half-decade indicates significant progress in the deployment of wireless systems, particularly sensor network-based systems for environmental monitoring and intelligent transportation systems. These technologies underscore the flexibility of wireless communication technologies in addressing complex social needs, ranging from biomedical applications to intelligent city planning. Increased use of frequency bands, particularly within the ISM band, is indicative of an evolutionary trend towards optimizing communication efficiency and ultimately enhancing the operation of modern systems (Elyas, 2025). With the continued innovation within the industry, it is fertile ground for further study and development of wireless communication practices. The application of self-positioning buoys represents another significant leap in the abilities of wireless communication systems, showing their application in coastal areas. Traditional buoys, though possible, have high costs, including maintenance and safety risks with respect to drifting. The application of GPS-located, autonomous moving ships is a progressive method to circumvent the challenges presented by such restrictions, with real-time observation and communication. They apply state-of-the-art networking technologies, including Delay Tolerant Networking (DTN), that enhance the reliability of wireless communications in dynamic and challenging situations. By enabling autonomous systems for data transmission through satellite and direct ship communication, these technologies significantly contribute to maritime domain awareness and national security (Ballou et al., 2006). This approach not only solves previous operational handicaps but also speaks of a future when autonomous systems can potentially enhance our capacity to manage and monitor marine ecosystems efficiently. With advancing wireless communications systems, its ramification in curriculum planning in engineering studies is large. The rapid advances in mobile computing and autonomous systems call for rigorous rethinking of previous and current courses of studies in telecommunication and informatics engineering. With the incorporation of the latest innovations and technology applications into the education curriculum, schools are able to equip learners with the abilities necessary to excel in a dynamic industry. Furthermore, a study of cross-disciplinary approaches could help offer a better understanding of how wireless communications technologies interact with other fields, varying from environmental monitoring to transport management. Tackling real-world issues, such as effective buoy maintenance and environmental issues, will improve the learning process and prepare students to provide solutions to problems of contemporary wireless communication systems (Elyas, 2025; Ballou et al., 2006).

10.3. ADVANCED TECHNOLOGIES IN COMMUNICATION

The technology for communication has been strongly influenced by better research and development activities. More specifically, the European Union's Advanced Communications and Technology Services (ACTS) program is now a single determining factor in advancing communication facilities and services across Europe. The program centers on the development of new systems of communication that not only encourage economic growth but also social integration. By shifting the focus from underlying theory frameworks to the installation of high-end systems itself, ACTS encourages collaborative efforts among different organizations. Backed by a total of approximately 670 MECU supporting nearly 160 projects, the undertaking speaks volumes about the size and commitment needed to push communication technological developments. This all-encompassing approach addresses the demands of modern communications in the form of a structured research methodology that facilitates uses of advanced communications technology in everyday life (Richardson et al., 1997). The most significant building block of advanced communication system engineering is the integration of wireless technologies, which have spread their applications to various sectors of society at a very rapid pace. As recent studies indicate, the ubiquitous nature of mobile computing has enabled the implementation of never-before-seen interconnectivity between heterogeneous objects and the worldwide information and communications technology (ICT) network. For instance, new types of applications such as wireless environment monitoring, intelligent parking systems, and smart transport systems are the optimal means of portraying how mobile computing is implemented to create smarter cities.

Figure 10.7. *Advanced technologies in communication.*

Source: https://europeyou.eu/wp-content/uploads/2019/05/ICT-1024x737.jpg.

Besides, sophisticated sensor network deployment guarantees higher reliability and efficiency in the systems, demonstrating the convergence of autonomous technology and smart design. Mobile computing research continues advancing with higher frequency bands to employ for communications and leveraging strong operating systems, which together redefine the manner in which individuals interact with technology (Elyas, 2025). Apart from improving communication capabilities, advanced communications also emphasize the need for security and quality assurance. The implementation of sound quality, safety, and security measures is crucial in communication system engineering. With the growing sophistication of these systems, the vulnerability and cyber threats are also heightened. Optimal security mechanisms must be integrated throughout the communication infrastructure in order to ensure data integrity and prevent illegal access. ACTS places very much emphasis on such factors, as its activities not only focus on enhancing the technical status of communication systems but also on addressing the ethical issues and regulatory policies involved in their deployment. By prioritizing security in the case of modern communication systems, researchers and engineers can design secure infrastructure, which facilitates trustworthiness and dependability in digital communication, eventually leading to greater user acceptance and utilization of new-generation technologies (Richardson et al., 1997) (Figure 10.7).

10.3.1. Fiber Optic Communication

With the discovery of fiber optic communication, telecommunications have undergone a radical makeover, providing unprecedented advantages over traditional copper-based communication. At the heart of it is the ability of optical fibers to transmit information at high speeds over long distances without significant deterioration in the signal integrity. This is because of the low attenuation rates of glass fibers and their immunity to electromagnetic interference, giving easier and more reliable communications channels. As the photonic technology further improves, the capacity of fiber optic systems to accommodate greater capacity is sure to increase, making fiber optics cope with the fast expansion of interactive video, voice communications, and data services that are not only economical but also green. The potential for fiber optics to satisfy the increasing traffic demand is underscored by projections of colossal growth of global internet users, which demands innovative means of transmitting data (Abdollahi et al., 2010). In sophisticated communications systems engineering, communications using fiber optics is a significant infrastructure component that is in line with global trends of digital interconnectedness. While organizations and end-users increasingly shift toward bandwidth-intensive applications, traditional access networks are seriously strained in delivering that. In such a context, fiber optics offer solutions that not only have the potential to bridge these gaps but also leverage photonic signal processing capability to yield better network performance and greater data rates. These innovations are vital in accordance with the evolving paradigms of network infrastructure, whereby wireless and wired networks converge to optimize provision of services. The feasibility of integrating fiber optic technology with wireless communication networks indicates an emphasis towards a robust and flexible form of communication, thus setting the stage for a greener and slimmer approach to global connectivity (Abdollahi et al., 2010). Moreover, the use of fiber optic systems in various applications suggests their convenience and adaptability in various industries, including telecommunications, medicine, and industrial automation. A good example is the use of fiber optics in medical diagnostic devices for high-quality imaging and real-time data transfer, which are essential for modern healthcare technology. Moreover, the work that is being carried out on control system applications for various functions, like developing DC-AC converters for windmills, depicts the potential of fiber optics to facilitate innovations enhancing energy efficiency and stability in operations. Utilizing advanced control methods, this convergence is an excellent example of the interdisciplinarity of fiber optic communication, offering bases for environmental-friendly technological innovation across industries. Thus, fiber optics not only enhances the effectiveness and speed of communication systems but also allows for broader environmental and energy objectives in the light of contemporary communication system engineering.

10.3.2. Satellite Communication Systems

Satellite communications systems are a very essential aspect of modern communication engineering, as they provide the means for sending and receiving data at great distances without the constraints imposed by terrestrial networks. The flexibility offered by satellites accommodates diverse applications, including tele-education, tele-medicine, and broadcasting, that are most vital in remote or underserved regions. For instance, Japan's Engineering Test Satellite-5 (ETS-5) helps play a key part in bringing a telecommunications network across the Asia-Pacific, illustrating how satellite technology can bridge communication gaps. In enabling real-time engagement through TV conferencing and other services, satellite communication enhances educational access and healthcare delivery in manners conventional systems cannot in many cases, thereby underscoring potential to change with these technologies (Iida et al., 2025). The development of advanced satellite systems continues to advance with further focus on efficiency and functionality. Projects like the ETS-6 demonstrate the evolution in inter-satellite communication and mobile application, expanding the horizons of satellite application from conventional broadcasting to encompass dynamic communication demands. These developments provide a window for testing millimeter wave personal communication, and that suggests continued work in new technology. This holds true to the current trend of embracing miniaturization techniques in satellite architecture, which result in smaller and more efficient platforms. Satellite system engineering must, therefore, focus on versatility and scalability so that they are able to accommodate a growing number of applications and address the growing demand for dependable communication networks (Iida et al., 2025). Micro System Technology (MST) projects, such as the MISAT program, underscore the importance of developing high-tech satellite platforms suited for space use as well as terrestrial applications. Through a focus on distributed systems and a use of miniaturization, MST projects make satellite operations more effective while reducing energy consumption and overall physical presence. The introduction of deep space nano-satellites, on the basis of cutting-edge technologies like re-configurable radio front ends and plug-and-play, is a huge jump in terms of satellite communications capabilities. The advantages offered by these advancements are not only to enable rapid deployment but also continuous technology demonstration on orbit. As research keeps developing in this field, these developments will lead to increasingly sophisticated functionalities for satellite communication systems, ultimately opening research and applications vistas even wider in scientific discovery and commercial endeavors.

10.3.3. 5G and Future Wireless Technologies

The advent of 5G technology is a key achievement in wireless communication that is characterized by record speed and connectivity. Unlike its predecessors, 5G can accommodate the increasingly high demands for services and data that define the modern environment. The technology employs a new range of frequencies such that it can offer lower latency as well as higher throughput. Particularly, its ability to connect an incredibly large number of devices simultaneously plays a critical role in the development of IoT, where intelligent devices communicate seamlessly. 5G has extensive implications that transcend single communication; it has the ability to revolutionize sectors such as healthcare, transport, and smart cities and hence enhance the quality of life. As companies move towards these new technologies, the infrastructure and procedures needed will be key in actualizing the entire potential of the capabilities offered by 5G. As the foundation of coming wireless networks, the use of machine learning (ML) in 6G networks is an innovation that is at the forefront. ML is able to enhance many communication processes, transforming how networks respond to real-time information and user demands. The future opportunity in zero-touch optimization lies in the fact that networks can be configured and optimized automatically by themselves, thus removing human intervention and error. With these advancements, not only does the network become more efficient but also enables the provision of sophisticated services like autonomous vehicles and dynamic resource allocation. The relationship between ML and wireless communication also gives adaptive algorithms for better signal processing and security features. Such an alignment is a prime key to tackling the complex challenges of our ever more interconnected world, in which the demand for stable and real-time connectivity has never been higher (Abbas et al., 2020). Furthermore, the role of smart grids in the contexts of 5G and future wireless technologies cannot be overemphasized. Traditional power grids are faced with widespread problems due to their unidirectional information flow and increasing energy demand. Conversely, smart grids utilize IoT devices to allow energy to travel in both directions and receive real-time monitoring. This shift is needed because it maximizes the efficiency and reliability of energy distribution. Through connectivity provided by advanced wireless technologies like 5G, smart grids can seamlessly modify energy demand and supply variations quickly. The use of IoT devices such as sensors and smart meters also provides the consumer with the authority over real-time data on their energy consumption, which results in sustainability (Copeland et al., 2019). Therefore, interaction between 5G technology and smart grids is at the core of identifying a future where the energy system is both efficient and smart (Figure 10.8).

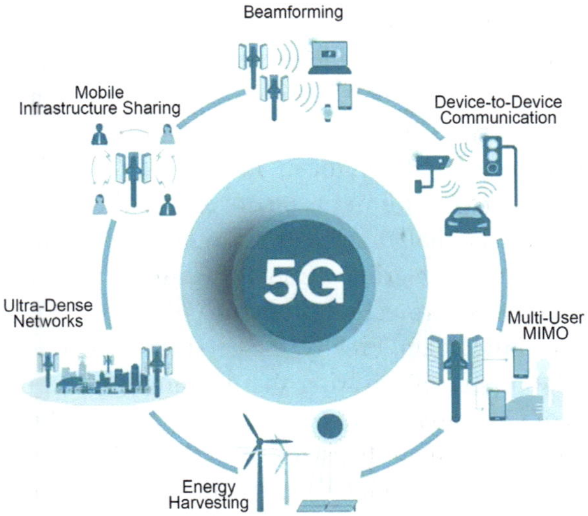

Figure 10.8. *5G and future wireless technologies.*

Source: https://res.cloudinary.com/engineering-com/image/upload/w_640,h_649,c_limit,q_auto,f_auto/image001_o5ojin.jpg.

10.4. COMMUNICATION PROTOCOLS AND STANDARDS

It is necessary to develop communication standards and protocols to facilitate the better performance and reliability of advanced communication systems. As technology continues to advance, interoperability across various types of systems is more imperative. Communication protocols are the underlying guidelines through which data can be exchanged, allowing devices and systems made by various manufacturers to communicate without any issues. Such is the case with the shift towards smart grids, where hundreds of devices from generation to transmission to distribution must interact efficiently so that maximum efficiency and reliability may be achieved. The use of standardized communication protocols does not only foreclose potential mistakes but also necessitates a well-organized structure that facilitates real-time communication and energy management among different components of the grid, making overall energy system efficiency in a technologically progressing setting greater. Moreover, as the complexity of communication systems increases, problems of security and privacy also emerge in proportion to the complexity. Smart grids, which rely heavily on digital communications, need more security since digital communications are mostly vulnerable to cyber vulnerability. The implications of these vulnerabilities are profound; compromised systems can potentially lead to devastating economic loss and energy supply stability danger, as indicated in the research that smart grid communications vulnerabilities have the potential to lead to disastrous outcomes for consumers and utilities (Qian et al., 2012). Addressing these security challenges requires a profound understanding of the cyber threats to which smart grids are open, resulting in formulating secure frameworks that enhance protection from unauthorized access and assured integrity of the transmitted data. Finally, the need for concerted standardization is imperative in the aspect of ensuring that communication protocols evolve with the demands of innovative technologies. The interoperability idea in smart grids focuses on the requirement for standardized standards that are capable of crossing technology and geographical boundaries. Research indicates that the application of consistent communication standards significantly increases the integration of various elements in the energy network along with security and privacy concerns (Chin et al., 2011). As countries and regions consolidate in the push to create a shared standard, advancements towards cutting-edge energy management systems become apparent, eventually leading the way towards stronger and more effective smart grids. This forward-thinking process not only addresses current technology challenges but also lays the groundwork for tomorrow's breakthroughs in communication systems engineering.

10.4.1. Overview of Communication Protocols

The evolution of communication protocols has dramatically changed the dynamics of data exchange between various platforms and technologies. Essentially, communication protocols are rules of standardization governing data transfer to ensure interoperability between various devices and systems. The pace has been driven by the increasing requirement for effective, secure, and reliable means of communication in an age of digitalization. Protocol components such as the Transmission Control Protocol (TCP) and Internet Protocol (IP) are an ideal representation of the networking infrastructural layers that provide ease of end-to-end connectivity. Moreover, whenever new technologies emerge, the adoption of cognitive radio technology poses new methodologies for improved spectrum utilization with dynamic spectrum access, thereby solving the problem of bandwidth scarcity. The success of such protocols, therefore, depends heavily on their ability to be tailored to evolving user needs and technology, requiring ongoing research and refinement in this area. Appreciation of the pivotal role played by security in communication protocols is key, especially as systems become increasingly networked. When intelligent systems like the smart grid, which utilizes robust two-way communication to enhance energy distribution, become operational, new threats emerge. As suggested, the implementation of information technology to electric systems not only optimizes efficiency but also creates novel and serious security risks. Traditional classes of cyber-attacks, normally handling confidentiality, integrity, and availability, normally overlook extremely critical factors such as accountability. A well-rounded security architecture that integrates various countermeasures is thus absolutely essential. This process endeavors to diminish threats triggered by potential attacks, ensuring that communication protocols are fortified against a variety of threats, thereby securing data exchanges among platforms. Furthermore, the consistency of various communications systems presents an inter-operability challenge since inter-protocol consistency is required. The arrival of advanced concepts, such as cognitive radio networks, brings into focus the need for advanced mechanisms in spectrum sharing, significantly impacting wireless communications efficacy and security. In such networks, devices utilize a shared control channel to talk to each other but are also under threat of security attacks such as licensed user emulation attacks. The emphasis on spectrum sharing security emphasizes the importance of safeguarding the common control channel, allowing for secure data exchange between cognitive radio nodes while minimizing the risk of potential misbehaviors. Consequently, communication protocol-related security concerns—particularly in complex and dynamic environments—are a critical area of research in advancing communication systems engineering, emphasizing the convergence of technological innovation and security concerns within the field (Figure 10.9).

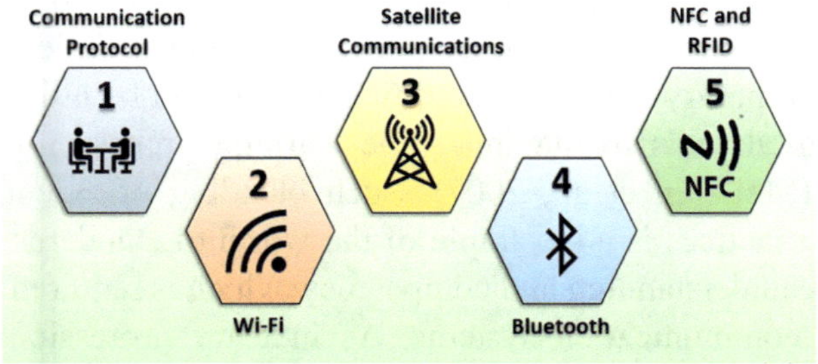

Figure 10.9. *Overview of communication protocols.*

Source: https://cdn.educba.com/academy/wp-content/uploads/2019/11/iot-communication-protocol-1.png.

10.4.2. Importance of Standards in Communication

Within the area of advanced communication systems engineering, setting standards is most important in allowing efficient data exchange between various platforms. Standards facilitate interoperability, by which various devices and systems can easily collaborate, and this enhances user experience as well as operational efficiency. The interoperable system allows for the incorporation of new technology that can operate within a system, thereby leading to more powerful communication channels. Further, conformity to established procedures reduces the risk of miscommunication, an occurrence that can cause project delays and costly errors. Thus, standards in communication are not technical details; it is the foundation of facilitating collaboration among stakeholders on the basis of shared knowledge of system capabilities and restrictions. The growth of computer communication technologies has brought the necessity of having standardized approaches to sustaining coherence and reliability into focus. For instance, in technical disciplines, such as telecommunications and aerospace, applying benchmark protocols guarantees a mutual understanding of complex information.

As has been observed in recent research on process improvement within aerospace organizations, the establishment of generic principles aids in information flow and enhances the quality of collaboration between engineers and project managers (Barker et al., 2005). By integrating practices with universal standards, organizations can be confident that all individuals are empowered with the basic skills of effective communication, hence enhancing innovation and system design enhancements. The convergence of engineering techniques with standard communications practice ultimately results in streamlined processes and increased productivity. Besides that, communication standards also serve to facilitate the transfer of knowledge and skill acquisition across numerous industries. As seen in studies about technical drawing and its use in contemporary education, traditional skills can be utilized in addition to existing digital tools to maximize the learning process for students and professionals (McLaren et al., 2007). Such blending of essential skills and contemporary practices is an example of the worth of standards in promoting comprehensive understanding and competency, which is required when dealing with intricate communication systems. As industry professionals adhere to standard communication modes, they not only facilitate exchange of ideas but also contribute to the broader diffusion of knowledge. The emphasis on standards under frameworks of communication is therefore a critical component in the facilitation of growth and development in engineered systems, emphasizing their core significance within the industry (Figure 10.10).

Figure 10.10. *Importance of standards in communication.*

Source: https://www.21kschool.com/wp-content/uploads/2025/02/Top-10-Importance-of-Communication-Skills.png.

10.4.3. Key Protocols in Modern Communication Systems

In the era of modern communication systems, bundling of major protocols is the pillar for enhanced information exchange and interoperability between different platforms. The protocols, from TCP/IP to HTTP and others, play a significant role in enabling reliable and efficient communication in the age of interconnectedness. The efficiency of such protocols is evidenced through a thorough comprehension of intricate systems engineering, which puts unique focus on the need of architectural strength in biological and technological systems. Such strength is extremely crucial in striking the balance of trade-offs in the design of networks, especially in confronting contemporary challenges in terms of data transmission and system resilience. As Alderson et al. (2010) suggests, the intricacy that results in such systems tends to produce new vulnerabilities, and thus there is the need for protocols that not only establish the means of communication but also system reliability and responsiveness. The use of smart grids is a prime example of the importance of communication protocols in contemporary applications. In smart grid systems, there is a need for real-time communication among different components of the grid for optimizing energy use and enhancing efficiency. This integration is founded upon advanced protocols with the objective of ensuring interoperability among generation, transmission, and distribution components of the energy infrastructure. Effective communication in this setting is difficult and promising, particularly if one looks at security and privacy issues. As Chin et al. (2011) suggest, the smart grid communication system optimization demands continuous observation of today's networking technologies that allow the development of standards for the future with the ability to adapt to evolving energy management needs while supporting the integration of heterogeneous communication systems. Finally, the evolution of communication protocols continues to inspire innovation in technology, particularly in fields like Internet of Things (IoT) and artificial intelligence. Interoperability procedures of communicating devices are at the forefront of IoT's breadth, where the capacity of different devices to exchange information comfortably relies on standardized procedures of exchanging data. This compatibility is not just important in enhancing the user experience but also in rendering these systems scalable. With the ongoing evolution of engineering approaches, emphasis on the creation of coherent architectures that leverage distributed networks will offset adverse effects of system complexity, making them resilient against potential disruptions. The dependency among such protocols and how they are applied within strongly structured technological systems underscores the ongoing necessity for research and innovation as noted in Alderson et al. (2010).

10.5. NETWORK DESIGN AND ARCHITECTURE

In the engineering of advanced communication systems, network architecture and design are important cornerstones upon which the seamless exchange of information between different platforms takes place. Good architecture not only aims for integrating the best in technologies but also interoperability and scalability. For instance, the architecture has to support computing, automation, and telecommunication advancements in the future, making it a highly scalable system. The architectural solution proposed for the Deep Space Network embodies this vision through an integrated approach that extends from the front-end environments at the antennas to end-user applications such as spacecraft teams and archival storage systems. This integration to strategic planning enables organizations to anticipate the rapid revolution in technology such that communication systems are robust, secure, and efficient in the long term (Atkinson et al., 1992). Even more daunting to the design process is the very complexity of designing networked systems, especially when the systems nest inside other sophisticated frameworks. Complexity is such a two-edged sword: it generates immunity to failures but can also introduce fragility, making systems all too susceptible in an unexpected way. Resilient but fragile trade-offs are the nexus of architecture discussions these days. By studying the interdependency between various components of network systems engineers are able to balance complexity such that networks are sophisticated but within manageable limits. This is a central issue in devising advanced communication systems, where integrative paradigms and predictive models could be used to counter the failures of high levels of complexity (Alderson et al., 2010). Integration of these ideas into an architecture ultimately leads to a communication system that is able to address challenges from new technologies and unpredictable operating requirements. Solid network design needs to incorporate sound security, efficient data processing, and welcoming interfaces to enhance overall functionality and user experience. The use of open-systems standards is also a crucial role in ensuring that diverse technology can work in a synergistic manner, and thus derive optimal performance and flexibility. As communications systems engineering continues to evolve, focus on robust architecture not only addresses current needs but also prepares the ground for future developments, facilitating easy evolution of communications practice with new technologies coming in Atkinson et al. (1992).

10.5.1. Types of Network Topologies

Network topologies provide the most important learning in understanding how data moves within an assortment of communication systems. Topology defines the shape of different elements within a network and influences its performance, scalability, and reliability. Star, bus, ring, mesh, and hybrid are the primary types of network topologies. Each has unique characteristics, advantages, and disadvantages. For example, the star topology focuses network functionality around a single hub, making it easier to administer and debug. The bus topology, however, relies upon a single central cable, which can be problematic in terms of scalability and fault tolerance. With evolving communication technologies, these network topologies remain relevant in determining the efficiency of distributed systems and connectivity networks, requiring continued research in the field of advanced communication systems engineering. Of the various topologies, the mesh topology is notable since it is marked with excessive redundancy and fault tolerance. In mesh topology, all the devices are interconnected in several ways, providing redundant routes for data transfer. This has the advantage of dependability as the loss of an individual node does not considerably impact the operation of the whole network. However, an expense and complexity of setting up an entirely meshed network might be too high, especially in larger infrastructures. The robustness of a mesh topology provides it with very high appeal in the event of demands for high availability and redundancy, such as critical infrastructures and data centers. By discerning benefits and constraints that come with mesh networks, this discussion highlights the delicate balance between functionality and complexity that characterizes modern networking systems, a position that remains valid in light of the axioms established in recent critiques of networked architecture (Alderson et al., 2010). In addition to the more conventional topologies, hybrid network configurations are increasingly salient as technological imperatives shift. Hybrid topologies borrow elements of different types and apply each of their strengths while reducing their weaknesses. In this way, organizations can form networks suitable to their own specific needs of operation. A hybrid network, for instance, could combine the scalability of a star topology with the redundancy of a mesh topology in order to achieve optimal performance and reliability. As communication systems become more complex, so does the need for architectures that can be versatile enough to meet varying requirements. This factor highlights the importance of consideration of not only intrinsic characteristics of each topology but also the long-term impact of architectural choice on network performance and efficiency (Alderson et al., 2005). By studying these diverse configurations in detail, we can more effectively address the problems of modern communications systems (Figure 10.11).

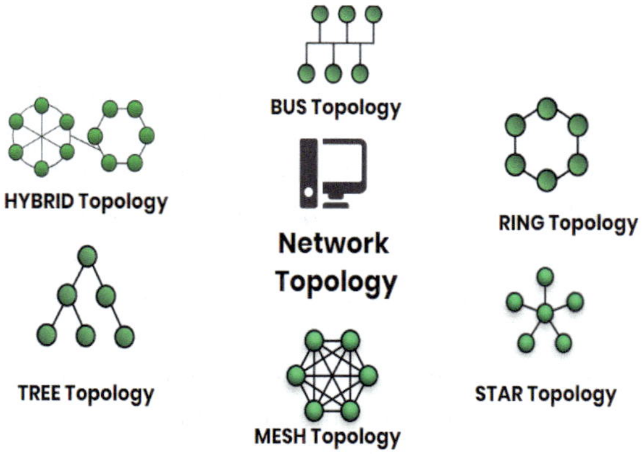

Figure 10.11. *Types of network topologies.*

Source: https://stl.tech/wp-content/uploads/2022/09/img1-2.png.

10.5.2. Role of Routers and Switches

Router and switch operation is the foundation of modern communications systems, playing crucial roles in data transmission and traffic management. Routers, with their name implying it, enable different networks to communicate and exchange information. Routers receive incoming data packets, determine the most effective path to transmit it, and send the data as such. This capacity to make routing choices in real time is critical to maintain the integrity and speed of data transport under layered network designs. In practice, routers avoid congestion and provide access in a reliable manner, impacting the functioning of the network overall. For this reason, their role is not transactional in nature but facilitative of smooth functioning of increasingly intricate systems that characterize high-level communication engineering. Unlike routers, switches operate at a different level of the networking stack and are more focused on internal traffic management within a local area network (LAN). Switches forward data packets to destination devices on the network, enhancing efficiency while reducing latency. Unlike routers operating on separate networks, switches create specific paths between end devices, facilitating faster communication. This variation is significant in low latency and high bandwidth settings where switch architecture can play a determinative role in performance and throughput. Here, switches routinely leverage virtual LANs (VLANs) in order to partition traffic, introducing additional optimization of network assets and enhanced security. The inclusion of switch technology within comprehensive network systems is an indication of the more complicated stratagems employed in advanced systems communication engineering. The merging of routers and switches into network infrastructure is also consistent with ongoing efforts to innovate and simplify communication paradigms, particularly through virtualization and programmability. Network resource virtualization, as shown in a number of studies, is a paradigm-shifting approach to managing networks where resources can be allocated on a demand basis dynamically. This evolution enables the concept of "slices," or abstract divisions of physical resources for enabling tailored user experiences with isolation and manageability. Furthermore, the evolution of programmable networks, as outlined in existing literature, offers fertile ground for potential to enhance flexibility and operational efficiency both in wired and wireless worlds. These developments confirm the central role of switches and routers because they adapt to address the issues of evolving technology environments to remain as essential components of advanced communication systems engineering (Cervelló et al., 2008; Ahad et al., 2014).

10.5.3. Network Security Considerations

In advanced communication systems engineering, it is essential to realize network security problems. The shift from closed telecommunications networks towards open market environments has introduced tremendous vulnerabilities, particularly on the privacy and data protection areas. The utilization of open networks allows sensitive information sharing between increasingly more entities, thereby facilitating higher chances for unauthorized entry and data exposure. Since privacy laws have been extended to encompass internet applications, system architects and engineers must consider not just the technological demands but the legal demands that govern data security. Maintaining the privacy of communications and protection of traffic data are not only ethical demands but legal demands that must be addressed in the design and development of complex communication systems (Hoven et al., 2005). Moreover, the intersection of telecommunications and computing networks has further dissipated the boundaries between different types of data handling. For instance, the development of cellular mobile technology and consequent processing of location information creates new privacy concerns. Each time one makes use of a mobile network, a trail of location data points is formed that traces a sophisticated pattern of an individual's activity and routines. This immediacy to access sensitive information demands robust security mechanisms and privacy management controls to protect against eavesdropping or misuse of data without authorization.

Figure 10.12. *Network security considerations.*

Source: https://blogger.googleusercontent.com/img/b/R29vZ2xl/AVvXsEj-c3C-kwgXFgQM-YjeShNFLyNcDaG6M8cQVw26sFwXG7vTDYcfAckmKrHN9kV-LETOMl335V_xtCxDRJe_KCDPVIZ5SRxQFT9jdi1TBJJp_m_qr2vdonjLbdUC-a3I3WTb5r95TrWp66ySdsjuUoNY-BONywEhJyE1j8MnaTjkW7nNjBrHvv-p-F7XHHecg/s1920/network%20security%20devices.webp.

Engineers must have greatest concern regarding such matters while rolling out mobile technologies and allied services. Determining the probable threats and establishing rules on the basis of evolving privacy norms is crucial while maintaining public trust and compliance in telecommunication (Hoven et al., 2005). Finally, the actual use of robust coding schemes in network systems only adds to the importance of safeguarding data exchanges. Applying Quasi-Cyclic Low-Density Parity-Check (QC-LDPC) codes can potentially be utilized to enhance error correction capabilities, thereby enhancing communication security even in hostile environments. These methods not only enable memory preservation in System on a Chip (SoC) designs but also neutralize the effect of threats due to untrustworthy or potentially harmful communication environments. Engineers are able to confirm that data integrity is ensured and that adaptive action is enabled against evolving security threats by analyzing the performance of modified systems using these advanced coding methods. This multi-dimensional approach to network protection ultimately serves to strengthen the overall architecture of the communication system, safeguarding it against both technical malfunctions and illegal access (Figure 10.12).

10.6. APPLICATIONS OF ADVANCED COMMUNICATION SYSTEMS

The application of sophisticated communication systems has profound implications across all sectors, constituting a revolutionary shift in the manner information is collected, shared, and utilized. Its widest application has been through mobile computing, which has witnessed extraordinary development in the past two decades. Innovative research has led to forays into many areas such as wireless environment monitoring, intelligent transport systems, and biomedical engineering. Specifically, sensor network integration enables data collection and analysis in real-time to assist in improved decision-making within these systems. The quest for ubiquitous connectivity is reflective of the broader trend towards looking to Information and Communication Technology (ICT) in facilitating interconnectivity, securing personal and societal functioning through effective communication and data management. These advancements point to the pivotal position played by advanced communication systems in contemporary society, influencing varied facets of life and operational processes (Elyas, 2025). Furthermore, the evolving landscape of advanced communication systems is primarily linked to the evolution of sound network infrastructures. The intricacy entailed by technology in contemporary times calls for advanced systems that are capable of dealing with multiple interrelated components productively. Network-oriented technology research has demanded greater understanding and mastery of complexities in the systems. The synergy of strength and architecture in networked systems becomes paramount as the systems must not just perform optimally but also be resilient against any occurrence of disruption. As the experts in the practice attempt to find universal laws that work in highly sophisticated systems, the ultimate challenge is always one of balance between complexity and stability of performance. This quest for improved methodologies underscores the desire for integrative and predictive models that can enable the unique demands of contemporary communication systems (Alderson et al., 2010). Secondly, new communication systems have revolutionized industries through facilitation of innovations like smart cities and Internet of Things (IoT) applications. These innovations have hastened the integration of automated systems into cities to ensure efficient management of resources and services. For instance, intelligent parking systems are employing real-time data to control available spaces, thereby boosting city convenience and reducing traffic congestion. Equally, intelligent transport systems employ sophisticated communication systems to improve traffic performance and safety features on the roads. The effect of such systems not only makes logistics a walkover and enhances the users' experience but also contributes to sustainability efforts by cutting down resource utilization. Besides, with research further enhancing the functionality of new communication technology, the advantage of implementing these systems across various domains becomes more evident, ushering in new frontiers and practical applications (Elyas, 2025).

10.6.1. Telecommunications Industry

The telecommunication industry is at the forefront of the current technological revolution, always revolutionizing the way individuals and businesses communicate. With an increase in the demand for more dependable and speedier connectivity, the industry has turned its attention to developing sophisticated wireless technologies. The move away from fixed wire to wireless systems is a prime example of this development, providing more mobility and user accessibility. Telecommunication networks of today not only enable voice communication as in the past but also enable an array of data services such as internet connection and streaming devices. This has formed a dynamic setting where companies in the industry are required to innovate vigorously to serve customers' needs and manage the sophistication that comes with wireless technology integration. The problems of the industry identify the necessity for the adoption of innovative methods like Wireless Intelligence Network (WIN) technology, with the emphasis placed on the necessity of flexibility in this fast-paced industry (Adesuyi et al., 2014). Moreover, globalization and deregulation have also drastically altered the telecommunications landscape, compelling local enterprises to compete on a global level. The traditional monopolistic models are breaking and giving way to a diversified marketplace full of various service providers that compete for clients' retention. This shift calls for advanced communication systems design that must handle the complexities introduced by a heightened competitive environment. With additional start-up companies vying to enter the market, sustainable business model (BM) development presents a significant challenge to them. This is particularly intense as cellular technology continues to transform with pressures on organizations to strike a balance between operational effectiveness and innovation. As such, an analytical evaluation of emerging business strategies is necessary for telecommunications firms looking to develop a competitive edge as they address the emerging needs of customers in this ever-dynamic global setting (Al-Debei et al., 2009). The impacts of technological progress in the telecommunication industry are even more pronounced in developing nations, often serving as the driving force for overall economic growth. Countries like Nigeria, for instance, demonstrate the transformative capability of telecommunication technology and which has gained ground by leaps and bounds in previously underserved regions. With improved telecommunication networks, a revolution is thus brought about, allowing nearby communities access to basic services and knowledge. This democratization of communications allows for more participation in the global digital economy, social networking, and entrepreneurship. More and more, however, as the market widens, providers must continue to deal with customer needs and strive toward innovative solutions that improve user experience. The ongoing evolution of mobile communication in the developing markets points to the necessity of adaptive strategies and robust engineering platforms in order to ensure sustainability and success within the telecommunication industry (Adesuyi et al., 2014) (Figure 10.13).

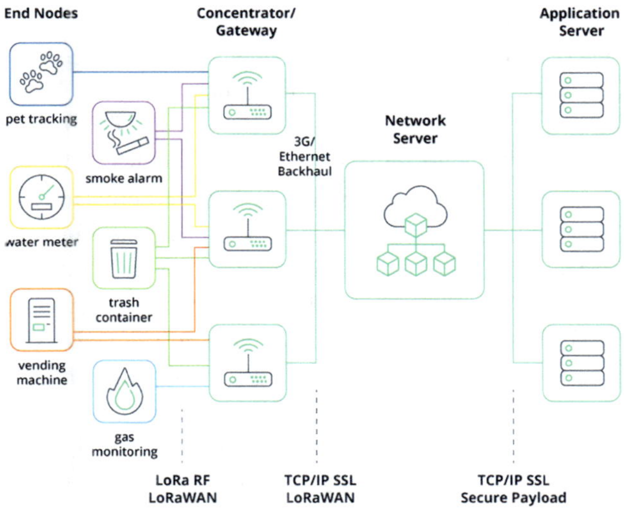

Figure 10.13. *Telecommunications industry of modern technological.*

Source: https://d17ocfn2f5o4rl.cloudfront.net/wp-content/uploads/2022/01/Byline-Telecom-Technology-Trends-in-2023-body-graphics_1-1.png.

10.6.2. Internet of Things (IoT)

The IoT represents a revolutionary communication systems engineering paradigm in which everyday objects are connected over the Internet to facilitate smooth exchange of information and interaction. Such connectivity enables the potential for intelligent solutions across a wide variety of fields, from industrial automation to medicine. By integrating many sensors and actuators, such objects are made cognitive, enabling them to sense and regulate real-time activity, thereby propelling operational efficiency and decision-making processes. As businesses continue to adopt IoT technologies, they unleash the potential for significant economic growth, as seen in Big Data convergence debates. However, the actual potential of the IoT is not only connectivity, as it opens up opportunities for new services and applications that cater to the intricate requirements of modern-day consumers, creating a fertile ground for research and development in the field of communication systems engineering (Perera et al., 2018). Despite its great potential, the widespread use of IoT technology poses huge challenges, namely data privacy and security. With many devices collecting and disseminating user data, location privacy and misuse of sensitive information problems emanate. Traditional models of security are not typically adequate, as they are not constructed to accommodate the unique requirements of IoT devices, which are low-cost and low-power. In this case, novel solutions must protect users' privacy while allowing them to control information disclosure based on contextual factors such as time or location. The development of purpose-designed middleware, such as the Internet of Things Management Platform (IoT-MP), is directed towards correcting such shortcomings through offering a more robust basis for managing heterogenous devices and their interactions. This paradigm shift necessitates re-examination of existing security measures and also demands further studies on adaptive privacy controls that are appropriate for the dynamic IoT environments (Cheung et al., 2016). Moreover, the IoT future is all about seamless evolution as communications technology advancements, such as 5G and edge computing, become immersed in IoT infrastructure. Bringing onboard high-speed networks and in-situ processing abilities makes IoT applications more responsive, enabling real-time data analysis and decision-making. This convergence of technologies not only improves user experiences but also facilitates the development of high-end systems that can efficiently handle high-end societal challenges, such as urban planning and resource management. As the IoT market matures, interaction among stakeholders from academia, industry, and government becomes essential in shaping standards and practices ensuring sustainable growth and innovation. It is important that engineers and researchers understand the intricacy of such interactions and their implications to communication systems when they enter the highly promising yet complex domain of the Internet of Things (Perera et al., 2018; Cheung et al., 2016) (Figure 10.14).

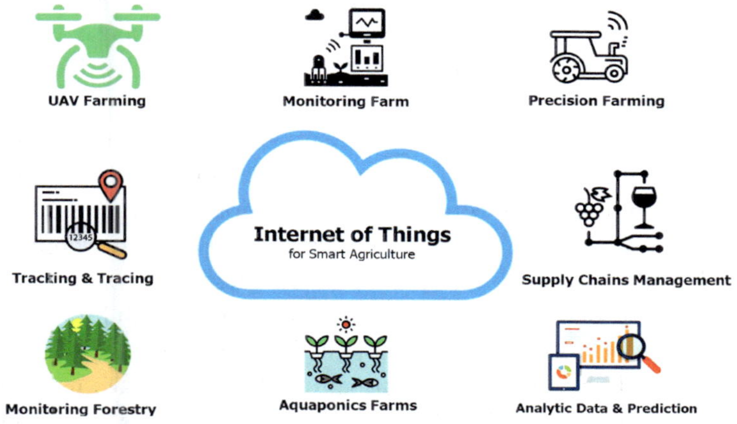

Figure 10.14. *Internet of things (IoT).*

Source: https://robots.net/wp-content/uploads/2023/09/what-does-the-internet-of-things-iot-enable-1694416957.jpg.

10.6.3. Emergency and Disaster Communication Systems

The integration of Emergency and Disaster Communication Systems into the framework of Advanced Communication Systems Engineering is significant to ensure effective response to crisis situations. Since disasters may occur suddenly and rapidly escalate, there needs to be a robust system of communication to facilitate timely and accurate transmission of information. This need is underscored by the Homeland Security Presidential Directive 20 (HSPD-20), which establishes provisions for continued operation through the implementation of Continuity of Operations (COOP) and Continuity of Government (COG) plans. The plans are essential for local, state, and tribal government agencies and private sector entities that own and control critical infrastructure. A lack of standards for how to incorporate these plans within the National Incident Management System (NIMS) indicates a huge operational readiness shortfall, which needs to be addressed through increased research and emergency operations staff training (Edwards et al., 2011). In addition, developments in communication technologies have resulted in heterogeneous distributed sensor networks (HDSN), which are the basis for emergency management. These networks enable various types of sensors to operate under a common platform, enhancing situational awareness and resource allocation in impacted areas. By using secure membership verification procedures in sensor networks organizations can ensure that only legitimate members participate in critical communications and operations. This security is essential in the prevention of unauthorized access, which would be harmful to response efforts. With the communities and agencies relying increasingly on such advanced sensor technologies, there is a need to grasp their operational complexities in order to effectively handle emergency response situations and strengthen resilience overall (Heo et al., 2007). Effective emergency communication systems rely heavily on regular training and operation coordination across governance levels. Preparation of targeted training programs, as reported in the research on COOP and COG, is called for in preparing emergency operations center (EOC) staff. Training programs focus on core competencies that align with FEMA's Incident Command System (ICS) and Emergency Support Functions (ESF). Equipping staff with detailed checklists and information makes agencies more prepared and responsive to emergencies. Including data and studies that hypothesize cooperation between levels of government ensures that communication systems not only exist to survive emergencies but also function in harmony while the emergency prevails, thereby ensuring protection of communities as well as securing order amidst adversity (Edwards et al., 2011).

SUMMARY

In the overview of the research in advanced communication systems engineering, it is critical to note the revolutionary influence of modern communication technology. Not only do these systems augment connectivity but also have a dramatic long-term influence on social, economic, and environmental aspects. With the observation of numerous methodologies involved in the discipline, such as digital signal processing and fiber optic technology, one can appreciate the constant innovation that results in improved efficiency and performance. Understanding the overlap of user requirements and engineering technologies is at the heart of understanding how modern society leverages these technologies to advance communication capacity. With improvements in communication networks, their dependability and coverage enable degrees of interaction and collaboration not experienced before, which accentuates the importance of ongoing research in this field, especially in safety and dependable environments, as seen through the research abstracts presented at the 2019 Engineering and Computer Science Research Conference (Adams et al., 2019). The other key feature highlighted in the next generation of communication systems is the role of some technical solutions such as discrete multitone (DMT) modulation. The technique holds tremendous potential for breaking new ground beyond existing limits with traditional optical fiber links. The suitability of DMT application in intensity-modulated, direct-detection systems has been reiterated based on theoretical analyses and experimental observations to outline the potential for higher transmission rates through step-index plastic optical fibers (SI-POF). These advances represent a shift in the ability to enhance the quality, diversity, and speed of communication. Furthermore, the findings of the research indicate the need to enlarge the boundaries of current technology and to explore new solutions as a means to address new problems of communication. Ongoing development is imperative as societies continue to rely on operational communication networks both in private life and public affairs. Lastly, the integration of knowledge from cutting-edge communication systems engineering reiterates calls for multidisciplinary efforts. The complexity of the discipline calls for input from various fields, including computer science, engineering, and social sciences. As smart technologies and Internet of Things (IoT) applications evolve, communication needs shift and it becomes essential for researchers and engineers to adopt adaptive approaches that encompass a broad range of perspectives. This is a methodology that helps strengthen systems that can handle the complexity of today's communication ecosystems. Lastly, the knowledge gathered from such an extended conversation ought to be applicable not only to academic studies but also to actual applications to better everyday communication experiences, as reflected by the innovative steps undertaken at forums such as the ECS Research Conference (Bluschke et al., 2009).

In considering the advancements achieved in communication systems engineering, it is worth noting what points highlight the revolutionary impact of new technology. The integration of metering technologies, as noted in various studies, plays a key role towards determining challenges to widespread implementation of advanced systems. Levy et al. (2002) says that such challenges need a precise knowledge of the accessible terrain, which directs the design and deployment of new communication infrastructure. The aggregate of information about such technologies allows engineers to anticipate problems and suggest solutions that are effective and efficient. It is not only a matter of the ability of such technologies to be created but also of the willingness of the stakeholders to adopt new ways and conform to change. One notable aspect of communication systems engineering is its application in certain fields such as mining. The increasing productivity demands of mining operations have driven the shift towards automation, possibly resulting in enhanced safety and efficiency. In particular, the use of remote-controlled and autonomous haulers operated by LiDAR sensors is a good example of how technological advances can revolutionize standard procedure. As described in current reviews, the utilization of LiDAR measurements to ensure accurate mine maps is promising for increasing operation efficacy. Borg et al. (2017) discusses the development of a system-of-systems (SoS) platform, which harmonizes various technological components to achieve maximum productivity while managing the political side involved in the involvement of stakeholders. This integration of disparate systems demonstrates a visionary approach that not only enhances security but also promotes innovation in engineering practices. Additionally, case study in the context of system-of-systems engineering emphasizes the importance of collaborative settings among stakeholders. Through the consideration of different business models and the breakdown of development work, the SoS debate emphasizes the intricate aspects of communication systems engineering. The comparison of scenarios developed for Bolidens mining operations highlights the significance of trade-off analysis involving control, cost, risk, and opportunities for innovation. Borg et al. (2017) effectively describes the manner in which these considerations interplay within the broader environment of stakeholders and require meticulous analysis to identify the most viable strategy. Thus, the active participation of all the stakeholders is needed, not only for technical enhancement but also for attaining agreement on the operational business processes in a realistic manner so that there can be sustainable advancement in communication systems.

REFERENCES

1. Abbas, R., Ahmad, I., Ali, S., Amuru, S., et al. (2020). 6G White Paper on Machine Learning in Wireless Communication Networks. http://arxiv.org/abs/2004.13875 (accessed on 12th September 2025).
2. Abdollahi, S. R., Al-Raweshidy, H., Darzi, A., Nilavalan, R., et al. (2010). Future Broadband Access Network Challenges. https://core.ac.uk/download/337968.pdf (accessed on 12th September 2025).
3. Adesuyi, R. S. (2014). The Future of Mobile Industry. https://core.ac.uk/download/19447812.pdf (accessed on 12th September 2025).
4. Agarwal, M., Averett, M., Baumer, B., Bray, A., et al. (2017). Curriculum Guidelines for Undergraduate Programs in Data Science. https://core.ac.uk/download/287356701.pdf (accessed on 12th September 2025).
5. Ahad, N., Ahmed, N., & Qadir, J. (2014). Building Programmable Wireless Networks: An Architectural Survey. http://arxiv.org/abs/1310.0251 (accessed on 12th September 2025).
6. Ahmad, S. W., Jennions, I. K., Perinpanayagam, S., Samie, M., et al. (2015). A Carrier Signal Approach for Intermittent Fault Detection and Health Monitoring for Electronics Interconnections System. https://core.ac.uk/download/42144211.pdf (accessed on 12th September 2025).
7. Ahmed, F., Zobaa, A., Vaccaro, A., Behrendt, K., Celano, T., Dickerson, B., Dion, A., Formato, G., et al. (2014). Vulnerability Analysis of Satellite-Based Synchronized Smart Grids Monitoring Systems. https://core.ac.uk/download/29204066.pdf (accessed on 12th September 2025).
8. Al-Debei, M. M., & Avison, D. (2009). Business Model Requirements and Challenges in the Mobile Telecommunication Sector. https://core.ac.uk/download/336341.pdf (accessed on 12th September 2025).
9. Alderson, D. L., & Doyle, J. C. (2010). Contrasting Views of Complexity and Their Implications for Network-Centric Infrastructures. https://core.ac.uk/download/4885216.pdf (accessed on 12th September 2025).
10. Alderson, D., Doyle, J. C., Li, L., & Willinger, W. (2005). Understanding Internet topology: Principles, models, and validation. https://core.ac.uk/download/4870423.pdf (accessed on 12th September 2025).
11. Alhakami, W., Mansour, A., & Safdar, G. A. (2014). Spectrum Sharing Security and Attacks in CRNs: A Review. https://core.ac.uk/download/29822044.pdf (accessed on 12th September 2025).
12. Alias, N., Malyshkin, V., & Sahnoun, R. (2017). High-Performance Computing and Communication Models for Solving the

Complex Interdisciplinary Problems on DPCS. https://core.ac.uk/download/154349994.pdf (accessed on 12th September 2025).

13. Amari, A., Ciblat, P., Dobre, O. A., Jaouën, M., et al. (2017). A Survey on Fiber Nonlinearity Compensation for 400 Gbps and Beyond Optical Communication Systems. http://arxiv.org/abs/1708.06313 (accessed on 12th September 2025).
14. Atkinson, D. J., Beswick, C. A., Cooper, L. P., Crowe, D., et al. (1992). Deep Space Network Information System Architecture Study. https://core.ac.uk/download/pdf/10438278.pdf (accessed on 12th September 2025).
15. Ballou, P., Curcio, J. A., Fall, K., Kitts, C., et al. (2006). Self-Positioning Smart Buoys, the 'Un-Buoy' Solution: Logistic Considerations Using Autonomous Surface Craft Technology and Improved Communications Infrastructure. https://core.ac.uk/download/72054531.pdf (accessed on 12th September 2025).
16. Barker, S. G., & Hendry, L. (2005). Process Improvement in BAe Systems and the Wider Aerospace Sector. https://core.ac.uk/download/155385.pdf (accessed on 12th September 2025).
17. Biros, G., Bungartz, H., Corones, J., Cramer, E., et al. (2016). Research and Education in Computational Science and Engineering. https://core.ac.uk/download/148025463.pdf (accessed on 12th September 2025).
18. Bluschke, A., Antonius, M. J., Bosco, G., Breyer, F., Gaudino, R., et al. (2009). Discrete Multitone Modulation for Maximizing Transmission Rate in Step-Index Plastic Optical Fibers. https://core.ac.uk/download/pdf/11406250.pdf (accessed on 12th September 2025).
19. Borg, M., Olsson, T., & Svensson, J. (2017). Piggybacking on an autonomous hauler: Business models enabling a system-of-systems approach to mapping an underground mine. http://arxiv.org/abs/1705.05087 (accessed on 12th September 2025).
20. Borghetti, G., Menga, G., Procaccianti, G., Kok, K., Ardito, L., Morisio, M., Mo, Y., et al. (2013). Smart grid technologies in Europe: An overview. https://core.ac.uk/download/pdf/11431923.pdf (accessed on 12th September 2025).
21. Cervelló Pastor, C., & Machado, R. (2008). Deliverable JRA1.1: Evaluation of current network control and management planes for multi-domain network infrastructure. https://core.ac.uk/download/18726491.pdf (accessed on 12th September 2025).
22. Cheung, H., Elkhodr, M., & Shahrestani, S. (2016). A middleware for the Internet of Things. http://arxiv.org/abs/1604.04823 (accessed on 12th September 2025).

23. Chin, W. H., Efthymiou, C., Fan, Z., Gormus, A., et al. (2011). Smart grid communications: Overview of research challenges, solutions, and standardization activities. http://arxiv.org/abs/1112.3516 (accessed on 12th September 2025).
24. Copeland, R., Crespi, N., Rehmani, M. H., Saleem, Y., et al. (2019). Internet of Things-aided smart grid: Technologies, architectures, applications, prototypes, and future research directions. http://arxiv.org/abs/1704.08977 (accessed on 12th September 2025).
25. Dobos, J., Farkas, P., Palenik, T., Rakus, M., et al. (2015). Analysis of minimal LDPC decoder system on a chip implementation. https://core.ac.uk/download/30312872.pdf (accessed on 12th September 2025).
26. Edwards, F. L., Goodrich, D. C., & Medigovich, W. M. (2011). Generic continuity of operations/continuity of government plan for state-level transportation agencies, research report 11-01. https://core.ac.uk/download/70413212.pdf (accessed on 12th September 2025).
27. Elyas, P. (2025). Ubiquitous wireless computing: Current research progress, challenging, and future directions. https://core.ac.uk/download/pdf/25495316.pdf (accessed on 12th September 2025).
28. Garcia-Rodriguez, A., Masouros, C., Rulikowski, P., Venkateswaran, V., et al. (2016). Hybrid analog-digital precoding revisited under realistic RF modeling. https://core.ac.uk/download/79509106.pdf (accessed on 12th September 2025).
29. Ghazi, H. E., Ghazi, H. E., Kaabouch, N., Mrabet, N., et al. (2018). Cyber-security in smart grid: Survey and challenges. http://arxiv.org/abs/1809.02609 (accessed on 12th September 2025).
30. Gill, E., Monna, G. L. E., & Scherpen, J., et al. (2007). MISAT: Designing a series of powerful small satellites based upon micro systems technology. https://core.ac.uk/download/pdf/12941826.pdf (accessed on 12th September 2025).
31. Heo, G., Hong, C. S., & Pathan, A. S. K. (2007). A secure lightweight approach of node membership verification in dense HDSN. http://arxiv.org/abs/0712.4177 (accessed on 12th September 2025).
32. Hoven van den, J., Penders, J., Thompson, P. B., Warren, S., & Altes, W. F. K. (2005). Privacy in (mobile) telecommunications services. https://core.ac.uk/download/8766994.pdf (accessed on 12th September 2025).
33. Hussain, S., Rahim, S., & Wang, Z. (2013). E-learning services for rural communities. http://arxiv.org/abs/1308.4820 (accessed on 12th September 2025).

34. Iida, T., Obara, N., & Wakana, H. (2025). ETS-5, ETS-6, and COMETS projects in Japan. https://core.ac.uk/download/pdf/42807110.pdf (accessed on 12th September 2025).
35. Lau, V., Peng, M., Poor, H. V., Wang, Y., et al. (2015). Fronthaul-constrained cloud radio access networks: Insights and challenges. http://arxiv.org/abs/1503.01187 (accessed on 12th September 2025).
36. Levy, R. (2002). Meter scoping study. https://core.ac.uk/download/288432784.pdf (accessed on 12th September 2025).
37. McLaren, S. V. (2007). Exploring perceptions and attitudes towards teaching and learning manual technical drawing in a digital age. https://core.ac.uk/download/9016897.pdf (accessed on 12th September 2025).
38. Peffer, T. (2010). UC Berkeley's Cory Hall: Evaluation of challenges and potential applications of building-to-grid implementation. https://core.ac.uk/download/288432727.pdf (accessed on 12th September 2025).
39. Perera, C., Chen, F., Suciu, G., Hashem, I. A. T., Xu, L. D., Belgaum, M. R., Sun, Y., et al. (2018). Challenges of Internet of Things and big data integration. http://arxiv.org/abs/1806.08953 (accessed on 12th September 2025).
40. Qian, Y., Sharif, H., Tipper, D., Yan, Y., et al. (2012). A survey on cyber security for smart grid communications. https://core.ac.uk/download/12211131.pdf (accessed on 12th September 2025).
41. Richardson, D. J. (1997). European Union Acts project MIDAS: Objectives and progress to date.
42. Yonis, B., & Mohammed, Y. (2011). Development of a DC-AC power conditioner for wind generator by using neural network. https://core.ac.uk/download/12007862.pdf (accessed on 12th September 2025).

INDEX

D

E

F

G

H

I

S

T

U

V

W

Z